Systems & Control: Foundations & Applications

Jeffery C. Allen

H^∞ Engineering and Amplifier Optimization

Springer Science+Business Media, LLC

Jeffery C. Allen
Space and Naval Warfare Systems Center
Naval Command Control and Ocean Surveillance
San Diego, CA 92152
U.S.A.

Library of Congress Cataloging-in-Publication Data
Allen, Jeffery, C.
H-[infinity] engineering and amplifier optimization / Jeffery C. Allen.
p. cm. – (Systems & control)
On t.p. "[infinity]" appears as the infinity symbol.
Includes bibliographical references and index.
ISBN 978-1-4612-6478-1 ISBN 978-0-8176-8182-1 (eBook)
DOI 10.1007/978-0-8176-8182-1
1. Broadband amplifiers–Mathematical models. 2. H [∞] control. 3. Hardy spaces. 4. Multidisciplinary design optimization. I. Title: Hardy spaces engineering and amplifier optimization. II. Title. III. Series.

TK7871.58.B74A43 2004
621.3815'35–dc22 2004047749
CIP

AMS Subject Classifications: 94C65, 47N70, 46N10, 47A48, 47A56, 47A20

ISBN 0-8176-3780-X Printed on acid-free paper.

9 8 7 6 5 4 3 2 1 SPIN 10954560

www.birkhasuer-science.com

To Martin and Esther for friendship and family,
to Elaine for insight and faith,
and to the memory of Kathryn.

Preface

Amplifying a weak, noisy, wideband signal is a canonical problem in electrical engineering. Figure 0.1 illustrates the classical solution where an amplifier is connected to the input and output matching circuits. The amplifier designer is given the amplifier and must design matching circuits that boost the input signal. The design problem is that the amplifier does increase the power of the weak input signal—but also amplifies the input noise, adds its own noise, and can be unstable for certain matching circuits.

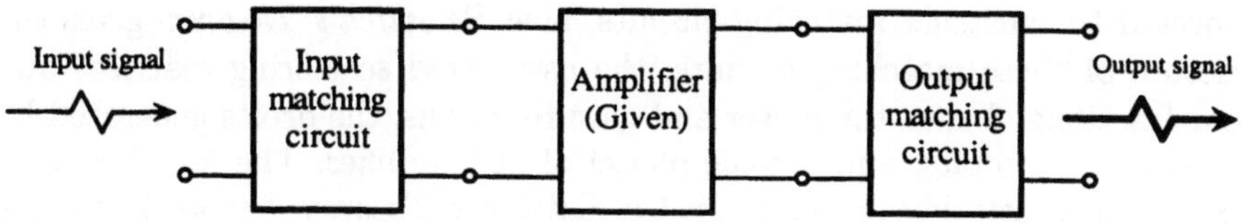

Fig. 0.1. The amplifier and matching circuits.

For example, anyone fiddling with a car radio learns that turning up the volume of a noisy signal simply produces a louder version of the same noisy signal. In addition, the amplifier must be stable. The characteristic "howl" or feedback of a public address system is an ear-splitting example of an unstable amplifier. The amplifier designer tinkers with the matching circuits to simultaneously trade gain for noise while maintaining stability. Thus, the Amplifier Matching Problem is really a multiobjective optimization problem [22].

The bulk of the amplifier literature is devoted to narrow-band solutions using the circuit of Figure 0.1 [106], [53], [140]. These solutions become sub-optimal as the bandwidth of the amplifier increases, stability margins become more stringent, and as multiple amplifiers are added to get greater gain.

This monograph shows how recent developments in H^∞ engineering equip the amplifier designer with new tools that

- compute the best possible performance available from any matching circuits,
- benchmark existing matching solutions, and
- generalize to multiple amplifier configurations.

This monograph is aimed at two groups: applied mathematicians with an interest in electrical engineering and electrical engineers with an interest in H^∞ theory. For mathematicians, there are few engineering subjects where an advanced topic like H^∞ theory has such an immediate connection to actual physical devices. For engineers, many H^∞ methods are still at the first level of development. The flexibility of the H^∞ methods coupled with the plethora of electrical devices offer splendid research opportunities for electrical engineers. Finally, both groups can attack the fundamental questions connecting circuits and H^∞ functions. We hope our readers realize a rich harvest from the research opportunities explicitly called out in this monograph.

Chapter 1 reviews the necessary circuit theory. The intended audience is the mathematician with a modicum of electrical engineering background. The scattering formalism is the framework for this discussion. In the scattering formalism, we review the lumped elements, how the lumped elements may be connected to construct matching circuits, how *Belevitch's Theorem* gives the structure of these matching circuits, the associated scattering matrices and linear-fractional forms, the power and gain functions, the orbits generated by the matching circuits, and a basic model of an amplifier. The foundation of these circuit-scattering techniques is the *Existence Theorem for the Scattering Matrix.* Roughly speaking, any passive circuit admits a scattering matrix that belongs to the unit ball of H^∞. We will see that the matching circuits correspond to the inner functions. In this sense, circuit design really corresponds to optimization over specified classes of H^∞ functions.

Chapter 2 makes explicit that the gain, noise, and stability functions of an amplifier are functions of the matching circuits. The Amplifier Matching Problem is to find input and output matching circuits that simultaneously trade off the competing objectives:

AMP-1 maximize the gain,
AMP-2 minimize the noise,
AMP-3 guarantee stability.

Because the matching circuits correspond to H^∞ functions, the Amplifier Matching Problem is really the multiobjective optimization of the gain, noise, and stability functions over specified classes of H^∞ functions.

Chapter 3 sets out the necessary tools of H^∞ engineering. The intended audience is the electrical engineer with some background in control theory. *Nehari's Theorem* is the fundamental result that essentially defines H^∞ engineering. J. W. Helton and his colleagues have made enormous progress adapting Nehari's Theorem to electrical engineering [65], [64], [9], [10], [11], [66], [15]; control theory [62], [63], [68], [69], [71], [74], [75]; and signal processing [85]. This chapter collects the technical results aiming at the interplay between H^∞ theory, which supports the computations for amplifier optimization, and the disk algebra, which corresponds to lumped matching circuits.

Chapter 4 presents the classes of matching circuits. At this point, the distinction between mathematicians and electrical engineers is starting to blur. The intellectual orientation is provided by the *Circuit-Scattering Correspondence* and the key question:

What elements in H^∞ correspond to a circuit and conversely?

The *State-Space Representation Theorem* characterizes this correspondence for the lumped circuits. There are nontrivial open questions when a correspondence between distributed circuits and H^∞ is attempted. These research questions are made explicit with a warning on some of the pitfalls.

Chapter 5 gathers the preceding material into a high-level presentation of the Amplifier Matching Problem. The amplifier designer optimizes over selected classes of matching circuits. Each collection of matching circuits pushes the input and output loads around to generate the input and output reflectances. These collections of reflectances are called the *orbits* of the input and the output loads generated by the particular class of matching circuits. Thus, the amplifier designer can either optimize over the matching circuits directly or optimize over the corresponding orbits. Likewise, the gain, noise, and stability functions are viewed as either functions on the matching circuits or functions on the associated orbits. Both the matching circuits and the orbits are H^∞ functions. *Darlington's Theorem* specifies when these orbits are dense in the unit ball of the real disk algebra, which is a strict subset of H^∞. Nehari's Theorem solves the optimization problems on H^∞. The challenge is to link the computable Nehari bound to the optimization problem on the orbits.

Chapter 6 presents the multiobjective optimization of several amplifiers. These examples acquaint us with the amplifier data and the numerical multiobjective optimizers. As such, these examples are the "raw material" for the subsequent analysis. The Amplifier Matching Problem then generalizes to an H^∞ multiobjective optimization problem that leads to the *Stable Amplifier Conjecture* and the H^∞ Multidisk Method.

Chapter 7 presents the H^∞ Multidisk Method, which is a direct generalization of the single-frequency disk method found in every amplifier text [82], [106], [140]. The following example of this single-frequency disk method illustrates the gain and noise tradeoffs that an amplifier designer typically encounters. For this particular design, the gain G_T and noise figure F are functions of the input matching circuit only. At a single frequency, the input matching circuit corresponds to a point S_G in the unit disk of the complex plane: $|S_G| \leq 1$. Consequently, the gain $G_T(S_G)$ and the noise figure $F(S_G)$ are functions on the unit disk. A key result is that the *level sets* of the gain and noise figure are *circles.* That is, the set of complex numbers where the gain is constant is a circle in the unit disk. Likewise, the noise figure is constant on circles in the unit disk. Now an amplifier designer might want the gain to exceed 8 dB and the noise figure to fall below 2 dB. These inequalities result in the gain and noise *disks* plotted in Figure 0.2. The real and imaginary axes

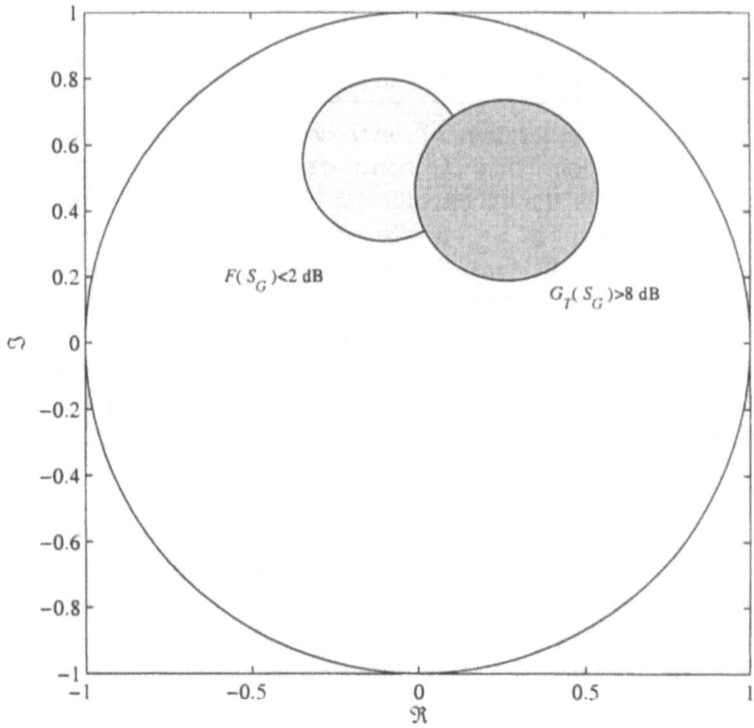

Fig. 0.2. Gain and noise disks in the unit disk at 4 GHz.

are labeled "ℜ" and "ℑ," respectively. Each point in the gain disk corresponds to a input matching circuit that delivers gain in excess of 8 dB. Each point in the noise disk corresponds to a input matching circuit that keeps the noise below 2 dB. Because the gain and noise disks intersect, the constraints are *feasible* and an input matching circuit may be extracted from the intersection. If the disks do not intersect, no matching circuit exists and the constraints are *not feasible.* In this case, the amplifier designer must relax the constraints and plot anew. However, these gain and noise performance bounds hold only at a single frequency.

The H^∞ Multidisk Method follows this engineering approach for wideband amplifier design. The designer specifies wideband gain, noise, and stability constraints. At each frequency, the corresponding gain, noise, and stability disks are computed. As a functions of frequency, these disks turn into gain, noise, and stability *tubes*. Figure 0.3 presents such a gain tube. The vertical slices contain a gain disk similar to Figure 0.2. As the frequency sweeps from 0 to 20 GHz, the gain disks sweep out the gain tube. The Amplifier Matching Problem admits a solution if and only if the intersection of the gain, noise, and stability tubes with H^∞ is nonempty.

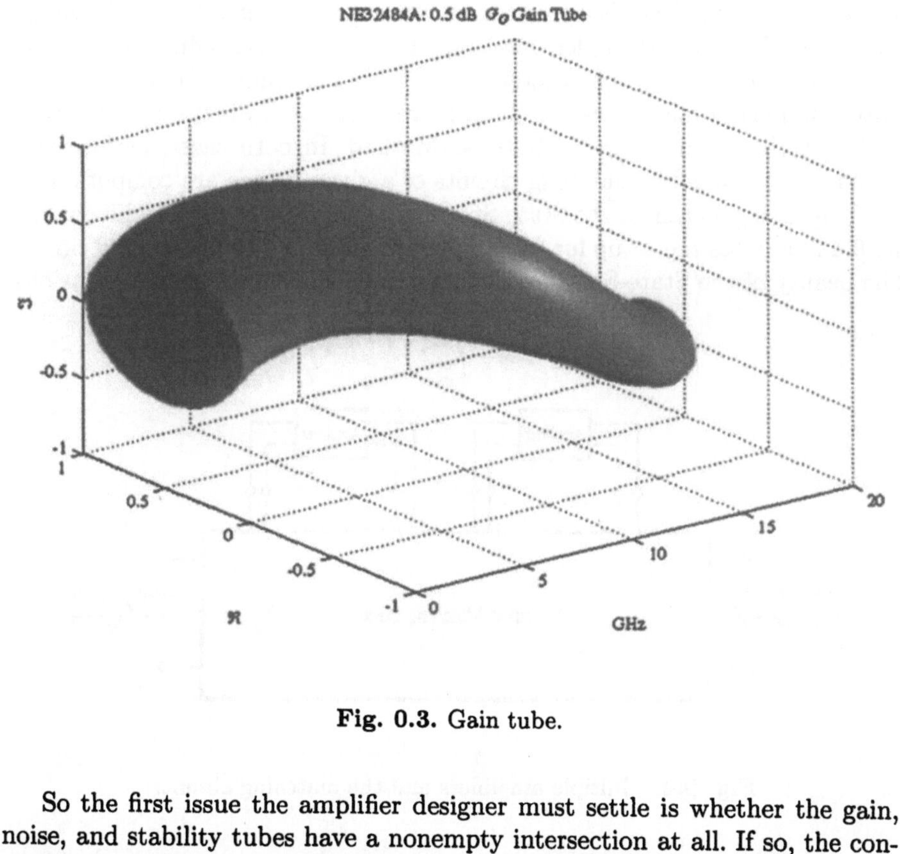

Fig. 0.3. Gain tube.

So the first issue the amplifier designer must settle is whether the gain, noise, and stability tubes have a nonempty intersection at all. If so, the constraints are said to be L^∞ *feasible*. If the constraints are L^∞ *feasible*, the second issue to settle is whether a matching circuit lies in this intersection. Because the matching circuits are H^∞ functions, the problem to determine if the nonempty intersection of the gain, noise, and stability tubes also intersect H^∞. If so, the constraints are said to be H^∞ *feasible*. Continuity then lets us assert that lumped matching circuits exist that can force the amplifier to meet the design constraints to arbitrary precision. Thus, the H^∞ Multidisk Method computes the best possible gain, noise, and stability trade-offs for the amplifier circuit of Figure 0.1. The problem is that the circuit topology constrains the performance and that these bounds do not have a circuit "size" associated with them. The State-Space Methods attack both problems.

Chapter 8 presents the State-Space Method for a single amplifier. This technique stands between the H^∞ solutions and the engineering approaches. The H^∞ solutions compute best possible performance bounds by optimizing over all possible matching circuits. The engineering approaches typically specify a matching circuit topology and optimize over the component values. The State-Space Method stands between these extremes by optimizing over all possible matching circuits *of a specified degree.* Thus, the number of components is fixed while optimization takes place over all possible circuit topologies. Critical for implementations, these topologies admit an efficient and numerically benign parameterization from the group of real orthogonal matrices. This parameterization gives more detail to both the H^∞ Multiobjective Method and the H^∞ Multidisk Method. In both cases, performance bounds for all possible matching circuits of a given degree are computable.

Chapter 9 generalizes the State-Space Method for multiple amplifiers. Figure 0.4 illustrates the setup for two amplifiers connected to a matching circuit. The beauty of the State-Space Method is that all such matching circuits of a

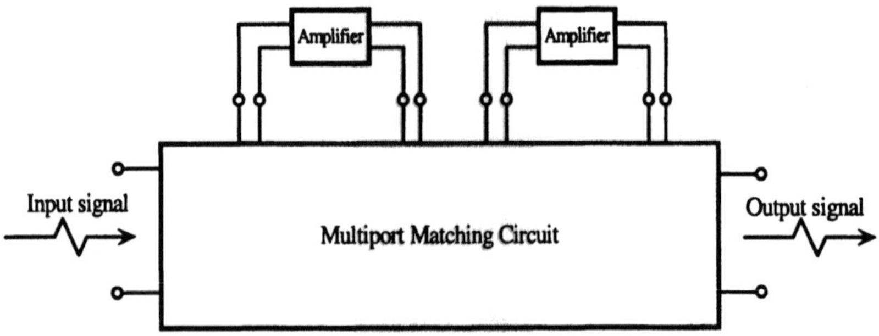

Fig. 0.4. Multiple amplifiers and the matching circuit.

fixed degree admit an easy parameterization. The difficulty is to generalize the noise computations. Not only does the input noise rattle through the matching circuit, with various paths arriving at each amplifier to be boosted in power and then injected back into the matching circuit, but each amplifier's noise is also injected into the matching circuit. The State-Space Method generalizes to handle this complex noise interaction. That is, the State-Space Method lets us optimize multiple amplifiers over all possible matching circuits of a given degree—including all possible feedback topologies. We find that these more general matching topologies offer more gain, less noise, and bring a substantial margin of stability.

In summary, the H^∞ Methods compute the best possible performance bounds attainable by any matching circuit with the dependence on degree captured by the State-Space Methods.

As we march through this analysis, many research questions arise. In the text, we make these questions explicit as they occur. Chapter 10 organizes these questions into a collection of research topics. The bulk of these questions concern representations of the matching circuits. The lumped matching circuits belong to the disk algebra. Distributed matching circuits, like the transmission line, are not in the disk algebra but still belong to H^∞. The Circuit-Scattering Correspondence for such matching circuits is open as is their state-space representation. More generally, we can ask if a similar representation exists for circuits containing amplifiers. However, the most pressing question is the construction of an optimal matching circuit. Considerable effort was expended to bring promising approaches into the purview of the applied mathematician. Briefly, the various classes of matching circuits correspond to unexplored classes of rational functions in several complex variables.

The vitality of a subject is the quality of the unexplored questions. For the mathematician, a small effort invested in circuit theory opens up wonderful research topics such as the orbits and Darlington's Theorem, the Circuit-Scattering Correspondence, and the Circuit Synthesis Problems. The payoff is an immediate application to the Amplifier Matching Problem. For the electrical engineer, the computation of the best possible performance limits begs to be linked to standard impedance matching programs. For both disciplines, generalizing the H^∞ theory to encompass multiple amplifiers awaits. In summary, this monograph is less about the amplifier matching problem and more about the research opportunities that lie at the interface of mathematics and electrical engineering.

Acknowledgments

This monograph was suggested by Dr. Reza Malek-Madani. Partial support was provided by Dr. Wen C. Masters the Office of Naval Research. Partial support from SPAWAR System Center is acknowledged with gratitude. The support of John Rockway of SPAWAR for the H^∞ approach and the collaboration with David Schwartz of SPAWAR laid the foundation for this work. Martin Jordan of SPAWAR proofread the first draft and his brave efforts are acknowledged with gratitude. Taylor Jordan assisted with the graphics. Rainer Pauli is thanked for sharing his views on circuit synthesis. Eldon Staggs of the Ansoft Corporation is thanked for his expertise and the amplifier data. Dennis Healy of DARPA and University of Maryland helped focus the exposition and improved the analysis. The diligence and expertise of Birkhäuser's staff: Tom Grasso, Seth Barnes, Tricia Manning, Elizabeth Loew, and their anonymous reviewer is most appreciated. Finally, the debt to J. W. Helton is profound. More that one mathematician I have met refers to Helton as "a great mathematician." This monograph is a simple, special case in the immense edifice of his creation.

Jeffery C. Allen
15 February 2004

Notation

$\mathbf{R}$	Real numbers
$\mathbf{R}_+$	Nonnegative real numbers
$\mathbf{C}$	Complex numbers
$\mathbf{C}_+$	Open right half plane
$\mathbf{D}$	Open unit disk
$\mathbf{T}$	Unit circle
$j\mathbf{R}$	Imaginary axis: $j = +\sqrt{-1}$
$\mathbf{R}^N$	Real N-space
$\mathbf{R}^{M\times N}$	$M \times N$ real matrices
$\mathbf{C}^N$	Complex N-space
$\mathbf{C}^{M\times N}$	$M \times N$ complex matrices
I_N	$N \times N$ identity matrix
0_N	$N \times N$ zero matrix
A^T	Transpose of matrix A
$\overline{A}$	Conjugate of matrix A
A^H	Hermitian or conjugate transpose of matrix A
$A \otimes B$	Tensor product of matrices A and B
$A \oplus B$	Direct sum of matrices A and B
$\mathcal{O}[N]$	Orthogonal Group: $U \in \mathbf{R}^{N\times N}$ and $U^TU = I_N$
$\mathcal{SO}[N]$	Special Orthogonal Group: $U \in \mathcal{O}[N]$ and $\det[U] = 1$

p	Complex frequency: $p = \sigma + j\omega \in \mathbf{C}_+$
ω	Radian frequency (radians per second)
f	Frequency (Hertz): $\omega = 2\pi f$
σ	Neper frequency (Nepers per second)
$\Re$	Real part: $\Re[\sigma + j\omega] = \sigma$
$\Im$	Imaginary part: $\Im[\sigma + j\omega] = \omega$
MHz	Mega-Hertz: 10^6 Hertz
GHz	Giga-Hertz: 10^9 Hertz
dB	Decibel
$Z(p)$	Impedance matrix
$Y(p)$	Admittance matrix
$S(p)$	Scattering matrix
$T(p)$	Chain matrix
$\Theta(p)$	Chain scattering matrix
$\|s\|_\infty$	Essential supremum of $\|s(z)\|$ on $\mathbf{T}$
$\|s\|_{-\infty}$	Essential infimum of $\|s(z)\|$ on $\mathbf{T}$
$L^\infty(\mathbf{T})$	Essentially bounded functions on the unit circle
$C(\mathbf{T})$	Continuous functions on the unit circle
$H^\infty(\mathbf{D})$	Hardy space on the unit disk
$\mathcal{A}(\mathbf{D})$	Disk algebra on the unit disk: $\mathcal{A}(\mathbf{D}) = H^\infty(\mathbf{D}) \cap C(\mathbf{T})$
$\|s\|_\infty$	Essential supremum of $\|s(j\omega)\|$ on $j\mathbf{R}$
$\|s\|_{-\infty}$	Essential infimum of $\|s(j\omega)\|$ on $j\mathbf{R}$
$L^\infty(j\mathbf{R})$	Essentially bounded functions on $j\mathbf{R}$
$C_0(j\mathbf{R})$	Continuous functions on $j\mathbf{R}$ that vanish at $\pm j\infty$
$H^\infty(\mathbf{C}_+)$	Hardy space on the right half plane
$\mathcal{A}_1(\mathbf{C}_+)$	Disk algebra on right half plane: $\mathcal{A}_1(\mathbf{C}_+) = H^\infty(\mathbf{C}_+) \cap 1 \dot{+} C_0(j\mathbf{R})$

$L^\infty(j\mathbf{R}, \mathbf{C}^{M\times N})$	Bounded matrix-valued functions on $j\mathbf{R}$
$H^\infty(\mathbf{C}_+, \mathbf{C}^{M\times N})$	Hardy space of matrix-valued function on $\mathbf{C}_+$
$BL^\infty(j\mathbf{R}, \mathbf{C}^{M\times N})$	Open unit ball of $L^\infty(j\mathbf{R}, \mathbf{C}^{M\times N})$: $\\|\phi\\|_\infty < 1$
$\overline{B}L^\infty(j\mathbf{R}, \mathbf{C}^{M\times N})$	Closed unit ball of $L^\infty(j\mathbf{R}, \mathbf{C}^{M\times N})$: $\\|\phi\\|_\infty \leq 1$
$\overline{D}(C,R)$	Closed disk with center C and radius R: $\phi \in \overline{D}(C,R) \Leftrightarrow (\phi - C)^H(\phi - C) \leq R$
$\Re L^\infty(j\mathbf{R}, \mathbf{C}^{M\times N})$	The real L^∞ functions: $\overline{\phi(\overline{j\omega})} = \phi(j\omega)$
$\Re\overline{B}H^\infty(j\mathbf{R}, \mathbf{C}^{M\times N})$	The closed unit ball of the real H^∞ functions
$U^+(N)$	Lossless scattering matrices: $S \in H^\infty(\mathbf{C}_+, \mathbf{C}^{N\times N})$ $S(j\omega)^H S(j\omega) = I_N$
$\gamma(X)$	Image of X under the function $\gamma : X \to Y$ $\gamma(X) = \{\gamma(\mathbf{x}) : \mathbf{x} \in X\}$
$[\gamma \leq \alpha]$	Sublevel set of the function $\gamma : X \to \mathbf{R}$: $[\gamma \leq \alpha] = \{\mathbf{x} \in X : \gamma(\mathbf{x}) \leq \alpha\}$
Λ_α	Lipschitz functions of order $\alpha \in (0,1]$ $\vert\phi(t_1) - \phi(t_2)\vert \leq A_\phi \vert t_1 - t_2\vert^\alpha$
$C^{n+\alpha}$	Continuous functions with $\phi^{(n)} \in \Lambda_\alpha$
C_ω	Dini-continuous functions
:=	definition operator
///	End-of-proof symbol
$\emptyset$	Empty set

Contents

List of Tables

List of Figures

H^∞ Engineering and Amplifier Optimization

1

Electric Circuits for Mathematicians

This chapter presents the required circuit theory. This low-key review is aimed at the applied mathematician with three goals in mind. First, that all the amplifier functions be carefully defined and accompanied by a physical meaning. Second, that the reader come away with a self-contained, computationally viable understanding of the scattering formalism, its connection with unitary and J-unitary systems, and the various linear-fractional transformations that underlie system theory. Third, after making such an investment of time and effort, the reader is equipped to assay the electrical engineering literature and develop their own circuits for analysis.

1.1 The Frequency Domain

Figure 1.1 shows a circuit containing an inductor and resistor being driven by a generator with voltage $V_G(t)$ as a function of time t. The generator forces a

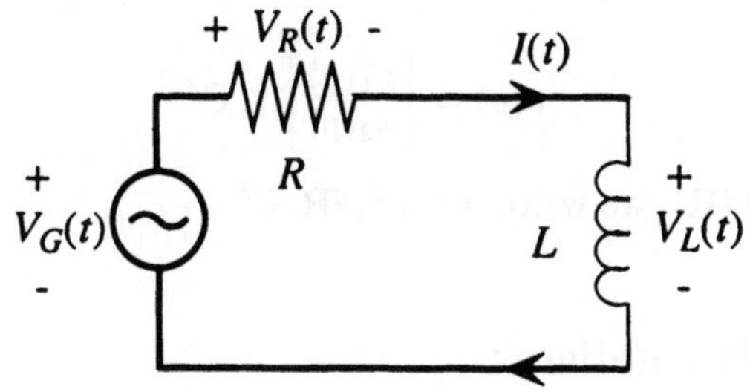

Fig. 1.1. A current loop.

current $I(t)$ to flow around the loop. The current induces voltages across the resistor $V_R(t) = RI(t)$ and the inductor

$$V_L(t) = L\frac{dI}{dt}(t).$$

The voltages around the loop sum to zero [8]:

$$-V_G(t) + RI(t) + L\frac{dI}{dt}(t) = 0. \tag{1.1}$$

Electrical engineers use the Laplace transform [8]

$$i(p) = \int_{-\infty}^{\infty} e^{-pt} I(t)dt,$$

with the *complex frequency* written as [57, Chapter 13]

$$p = \sigma + j\omega$$

to solve Equation 1.1:

$$v_G(p) = (R + pL)i(p).$$

Accordingly, we will work in the frequency domain exclusively. We also assume that $I(t)$ and $V(t)$ belong to $L^2(\mathbf{R})$

$$\int_{-\infty}^{\infty} |V(t)|^2 dt < \infty,$$

where $\mathbf{R}$ denotes the real line. Equivalently, $i(p)$ and $v(p)$ belong to $L^2(j\mathbf{R})$:

$$\int_{-\infty}^{\infty} |v(j\omega)|^2 d\omega < \infty.$$

When voltages are vector-valued with individual components in $L^2(j\mathbf{R})$, for example,

$$\mathbf{v}(p) = \begin{bmatrix} v_1(p) \\ v_2(p) \end{bmatrix} \in \mathbf{C}^2$$

with v_1, $v_2 \in L^2(j\mathbf{R})$, we write $\mathbf{v} \in L^2(j\mathbf{R}, \mathbf{C}^2)$.

1.2 N-Port Formalisms

Figure 1.2 represents an N-port that models a "black box." We don't know what is in the box but there are N pairs of wires physically attached to the unknown circuit. The use of the word "port" means that each pair of wires obeys a *conservation of current*—the current flowing into one wire of the pair equals the current flowing out of the other wire. We can imagine characterizing such a box by supplying current and voltage input signals of given frequency at the various ports and observing the current and voltages induced at the other

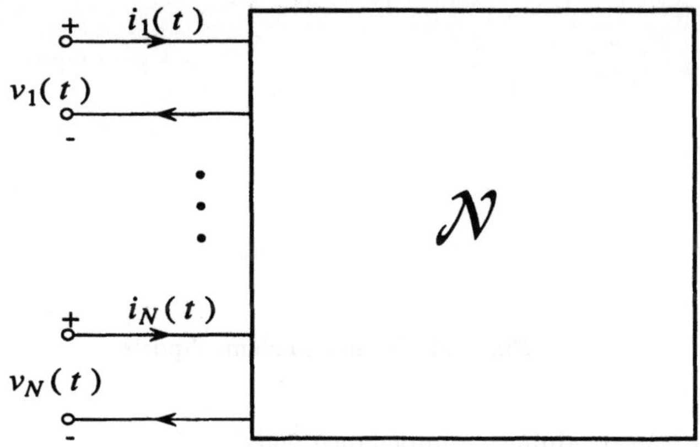

Fig. 1.2. The N-port.

ports. Mathematically, the N-port is defined as the collection $\mathcal{N}$ of voltage $\mathbf{v}(p)$ and current $\mathbf{i}(p)$ vectors that can appear on its ports [92], [73]:

$$\mathcal{N} \subseteq L^2(j\mathbf{R}, \mathbf{C}^N) \times L^2(j\mathbf{R}, \mathbf{C}^N).$$

If $\mathcal{N}$ is a linear subspace, then the N-port is called a *linear* N-port. Figure 1.3 presents the fundamental linear 1-ports and their voltage-current relationships constructed from the classic *lumped elements*:

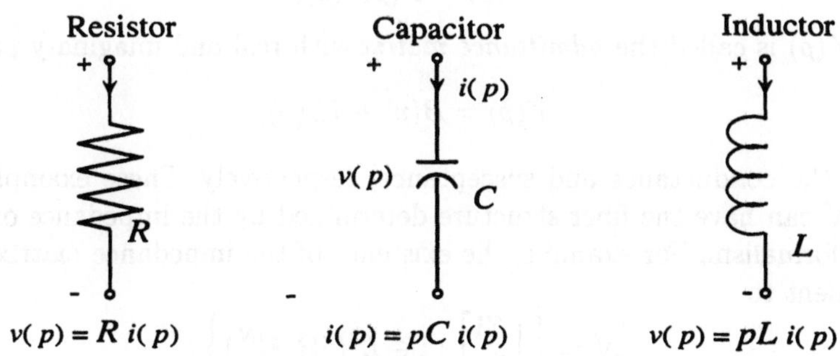

Fig. 1.3. The lumped elements.

By wiring together these 1-ports, we can construct multiports. For example, Figure 1.4 presents the series and shunt 2-ports. The series 2-port admits the admittance matrix

$$Y(p) = \frac{1}{z(p)} \begin{bmatrix} 1 & -1 \\ -1 & 1 \end{bmatrix}.$$

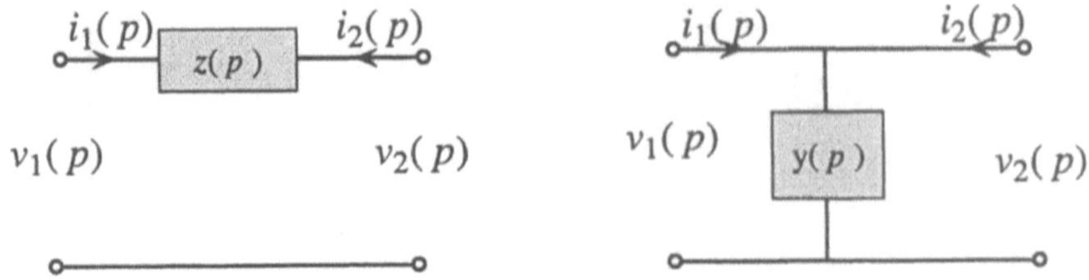

Fig. 1.4. Series and shunt 2-ports.

The shunt 2-port admits an impedance matrix

$$Z(p) = \frac{1}{y(p)} \begin{bmatrix} 1 & 1 \\ 1 & 1 \end{bmatrix}.$$

More generally, if the voltage and current are related as

$$\mathbf{v}(p) = Z(p)\mathbf{i}(p),$$

then $Z(p)$ is called the *impedance matrix* with real and imaginary parts

$$Z(p) = R(p) + jX(p)$$

called the resistance and reactance, respectively. If the voltage and current are related as

$$\mathbf{i}(p) = Y(p)\mathbf{v}(p),$$

then $Y(p)$ is called the *admittance matrix* with real and imaginary parts

$$Y(p) = B(p) + jG(p),$$

called the conductance and susceptance, respectively. These examples show that $\mathcal{N}$ can have the finer structure determined by the impedance or admittance formalism. For example, the existence of the impedance matrix $Z(p)$ is equivalent to

$$\mathcal{N} = \left\{ \begin{bmatrix} Z\mathbf{i} \\ \mathbf{i} \end{bmatrix} : \mathbf{i} \in L^2(j\mathbf{R}, \mathbf{C}^N) \right\},$$

or that $\mathcal{N}$ is the graph of the impedance matrix.

Figure 1.5 presents the basic 2-ports used to build multiports. These 2-ports are best represented in the chain formalism. The *chain matrix* $T(p)$ relates 2-port voltages and currents as

$$\begin{bmatrix} v_1 \\ i_1 \end{bmatrix} = \begin{bmatrix} t_{11}(p) & t_{12}(p) \\ t_{21}(p) & t_{22}(p) \end{bmatrix} \begin{bmatrix} v_2 \\ -i_2 \end{bmatrix}.$$

The ideal transformer has chain matrix [6, Eq. 2.4]:

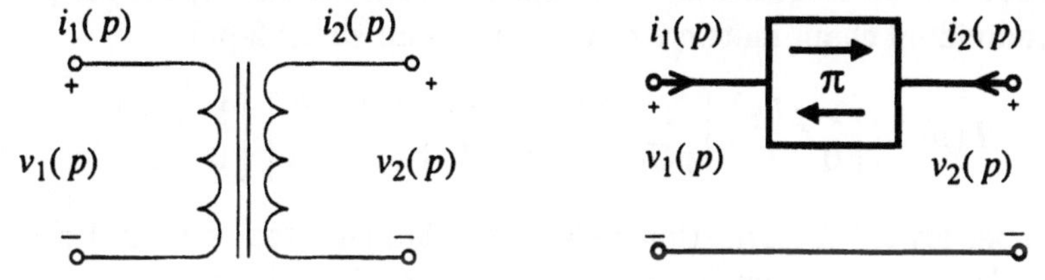

Fig. 1.5. The transformer and gyrator.

$$T_{\text{transformer}}(p) = \begin{bmatrix} n^{-1} & 0 \\ 0 & n \end{bmatrix}, \tag{1.2}$$

where n is the *turns ratio* of the windings on the transformer. The gyrator has chain matrix [6, Eq. 2.14]:

$$T_{\text{gyrator}}(p) = \begin{bmatrix} 0 & \alpha \\ \alpha^{-1} & 0 \end{bmatrix}.$$

The series and shunt 2-ports of Figure 1.4 have the chain matrices [130, Table 7-1]:

$$T_{\text{series}}(p) = \begin{bmatrix} 1 & z(p) \\ 0 & 1 \end{bmatrix}, \quad T_{\text{shunt}}(p) = \begin{bmatrix} 1 & 0 \\ y(p) & 1 \end{bmatrix}, \tag{1.3}$$

determined by the series impedance $z(p)$ and shunt admittance $y(p)$.

With these basic electrical objects in place, it is possible to connect these objects to produce useful circuits. For example, connecting the series and shunts in a *chain* produces a 2-port called a *ladder*. Ladders are classic matching 2-ports. Figure 1.6 shows a low-pass ladder. The inductors are wires at

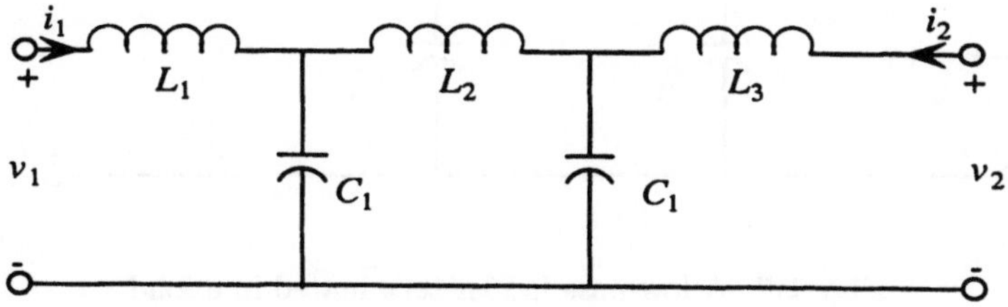

Fig. 1.6. A low-pass ladder.

low frequencies and open circuits at high frequencies; the capacitors are the converse. Consequently, the low-pass ladder attenuates the high frequencies and passes the low frequencies. The ladder's chain matrix $T(p)$ is the product of the individual chain matrices of the series and shunt 2-ports:

$$T(p) = \begin{bmatrix} 1 & pL_1 \\ 0 & 1 \end{bmatrix} \begin{bmatrix} 1 & 0 \\ pC_1 & 1 \end{bmatrix} \begin{bmatrix} 1 & pL_2 \\ 0 & 1 \end{bmatrix} \begin{bmatrix} 1 & 0 \\ pC_2 & 1 \end{bmatrix} \begin{bmatrix} 1 & pL_3 \\ 0 & 1 \end{bmatrix}. \quad (1.4)$$

The "upper-lower" factorization reflects the ladder's topology and choice of reactive elements. The natural question is to consider the converse:

> Given a chain matrix $T(p)$ of a ladder, how can we get this LU factorization?

This question opens the vast area of *circuit synthesis* [6], [8], [99], [47]:

> Given mathematical circuit data—such as a chain matrix $T(p)$, an impedance matrix $Z(p)$, a terminated impedance, or digital samples of these functions—how can one extract a useful circuit that best fits this data?

Basic to circuit synthesis and matching is the notion of an orbit.

1.3 Orbits

The ladder circuits, the 2-ports, and the N-ports are all supposed to "do something." Specifically, if a 2-port is terminated in a load, the resulting 1-port is the load viewed through the 2-port. An adroit choice of the 2-port transforms the load into a more useful form. For example, if the low-pass ladder terminates Port 2 in a load z_L as shown in Figure 1.7, the impedance

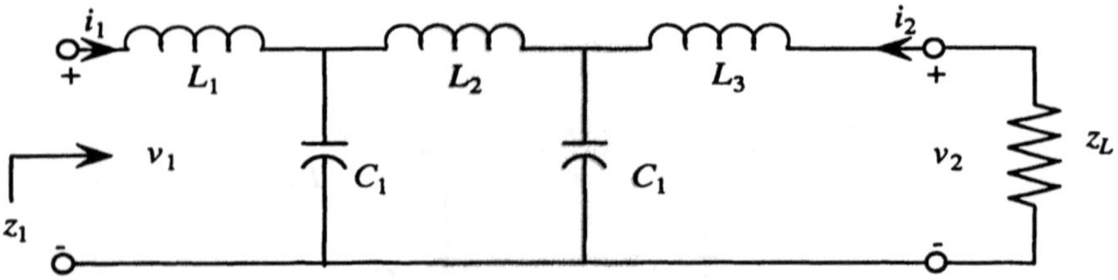

Fig. 1.7. A low-pass ladder terminated in a load.

looking into Port 1 is computed using the homogenous coordinates supplied by the chain matrix:

$$z_1 = \frac{v_1}{i_1} = \frac{t_{11}z_L + t_{12}}{t_{21}z_L + t_{22}} =: \mathcal{G}(T, z_L). \quad (1.5)$$

The chain matrix $T(p)$ is given by Equation 1.4. Let $\mathcal{T}$ denote a collection of such chain matrices determined by Equation 1.4. The corresponding impedances at Port 1 is denoted

$$\mathcal{G}(\mathcal{T}, z_L) := \{\mathcal{G}(T, z_L) : T \in \mathcal{T}\}$$

and called *orbit of the load z_L under the action of the low-pass ladders in $\mathcal{T}$.*

These orbits are fundamental for the matching problem. Even at this elementary level, mathematicians can raise some substantial questions regarding how these ladders themselves, or how the orbit of their load, sits in H^∞. Unfortunately, the impedance, the admittance, and the chain formalisms do not have good numerical properties. A basic problem is that the formalisms fail to exist for many N-ports:

- The transformer does not have an impedance matrix.
- The series 2-port does not have an impedance matrix.
- The shunt 2-port does not have an admittance matrix.

There are difficulties inherent in attempting the matching problem in a formalism where the basic objects under discussion fail to exist. In fact, much of the debate in electrical engineering in the 1960s focused on finding a formalism that guaranteed that every N-port had a representation as the graph of a linear operator. The *scattering formalism*, which has its roots in microwave theory [106] and deep connections to operator theory [73], satisfies this basic existence problem:

EXISTENCE OF THE SCATTERING MATRIX [6], [8], [73] *Any linear, causal, passive, time-invariant, solvable N-port $\mathcal{N}$ always admits a* scattering matrix $S \in \overline{B}H^\infty(\mathbb{C}_+, \mathbb{C}^{N\times N})$ *so that*

$$\mathcal{N} = \left\{ \begin{bmatrix} \mathbf{a} \\ S\mathbf{a} \end{bmatrix} : \mathbf{a} \in L^2(j\mathbf{R}, \mathbb{C}^N) \right\}.$$

The list of N-port qualifiers is discussed in depth in Appendix A. For all practical purposes, the reader is asked to accept as a working hypothesis that every circuit corresponds to a scattering matrix that belongs to H^∞. The remainder of this chapter reviews the scattering formalism. It is in the scattering formalism that provides the computational framework for the Amplifier Matching Problem. We make explicit the computational techniques using the scattering matrix, interconnections, power computations, and how the algebraic structure of the scattering matrix reveals the electrical circuit structure of its associated N-port.

1.4 The Scattering Matrix

Specializing to the 2-port in Figure 1.8, define the *incident signal* [6, Eq. 4.25a], [8, page 234]

$$\mathbf{a} = \frac{1}{2}\{R_0^{-1/2}\mathbf{v} + R_0^{1/2}\mathbf{i}\} \tag{1.6}$$

and the *reflected signal* [6, Eq. 4.25b], [8, page 234]

$$\mathbf{b} = \frac{1}{2}\{R_0^{-1/2}\mathbf{v} - R_0^{1/2}\mathbf{i}\}, \tag{1.7}$$

with respect to the normalizing[1] matrix

$$R_0 = \begin{bmatrix} r_1 & 0 \\ 0 & r_2 \end{bmatrix}.$$

The *scattering matrix* $S(p)$ maps the incident signal $\mathbf{a}(p)$ to the reflected

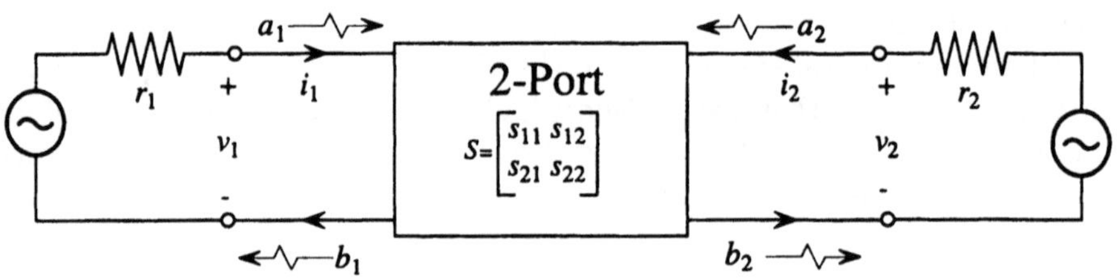

Fig. 1.8. The 2-port scattering formalism.

signal $\mathbf{b}(p)$:

$$\mathbf{b} = \begin{bmatrix} b_1 \\ b_2 \end{bmatrix} = \begin{bmatrix} s_{11} & s_{12} \\ s_{21} & s_{22} \end{bmatrix} \begin{bmatrix} a_1 \\ a_2 \end{bmatrix} = S\mathbf{a}.$$

The scattering matrix determines the *normalized impedance matrix* as [6, Eq. 4.59]:

$$\widetilde{Z} := R_0^{-1/2} Z R_0^{-1/2} = (I+S)(I-S)^{-1}.$$

To see this, invert Equations 1.6 and 1.7 and substitute into $\mathbf{v} = Z\mathbf{i}$. Conversely, if the N-port admits an impedance matrix, normalize and Cayley transform to get [6, Eq. 4.66]:

$$S = (\widetilde{Z} + I)^{-1}(\widetilde{Z} - I).$$

[1] Two accessible books on the scattering formalism are Baher [6] and Balabanian and Bickart [8]. Baher omits the factor of 1/2 but carries this rescaling into the power definitions. Most other books use the *power-wave normalization* [37]: $\mathbf{a} = R_0^{-1/2}\{\mathbf{v} + Z_0\mathbf{i}\}/2$, where the normalizing matrix $Z_0 = R_0 + jX_0$ is diagonal with diagonal resistance $R_0 > 0$ and reactance X_0.

Usually $R_0 = r_0 I$. The mathematical community typically sets $r_0 = 1$. Electrical engineers have endless arguments about normalizations [125] but typically take $r_0 = 50$ ohms, as we do also.

1.5 The Chain Scattering Matrix

Closely related to the scattering matrix is the *chain scattering matrix* Θ [56, page 148]:

$$\begin{bmatrix} b_1 \\ a_1 \end{bmatrix} = \Theta \begin{bmatrix} a_2 \\ b_2 \end{bmatrix} = \begin{bmatrix} \theta_{11} & \theta_{12} \\ \theta_{21} & \theta_{22} \end{bmatrix} \begin{bmatrix} a_2 \\ b_2 \end{bmatrix}.$$

When multiple 2-ports are connected in a chain, as in Figure 1.9, the chain scattering matrix is the product of the individual chain scattering matrices.

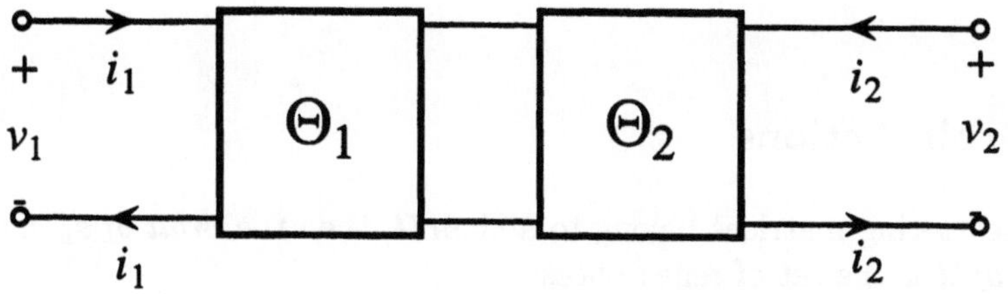

Fig. 1.9. The chain of the 2-ports has chain scattering matrix $\Theta = \Theta_1\Theta_2$.

The mappings between the scattering and chain scattering matrices are [56]:

$$S \mapsto s_{21}^{-1} \begin{bmatrix} -\det[S] & s_{11} \\ -s_{22} & 1 \end{bmatrix} = \Theta \mapsto \theta_{22}^{-1} \begin{bmatrix} \theta_{12} & \det[\Theta] \\ 1 & -\theta_{21} \end{bmatrix} = S. \tag{1.8}$$

Although every 2-port has a scattering matrix, it admits a chain scattering matrix only if s_{21} is invertible. The chain scattering matrix facilitates the scattering computations when 2-ports are terminated in passive or active loads.

1.6 Passive Terminations

In Figure 1.8, assume Port 2 is terminated with the load reflectance s_L only (i.e., there is no generator driving Port 2). Looking out from Port 2 gives the scattering from the load:

$$a_2 = s_L b_2. \tag{1.9}$$

The reflectance looking into Port 1 is computed from the chain-scattering matrix:

$$s_1 := \frac{b_1}{a_1} = \frac{\theta_{11}a_2 + \theta_{12}b_2}{\theta_{21}a_2 + \theta_{22}b_2} = \frac{\theta_{11}s_L + \theta_{12}}{\theta_{21}s_L + \theta_{22}} =: \mathcal{G}_1(\Theta, s_L). \tag{1.10}$$

Equation 1.8 also gives s_1 as the linear-fractional form of the scattering matrix:

$$s_1 = s_{11} + s_{12}s_L(1 - s_{22}s_L)^{-1}s_{21} =: \mathcal{F}_1(S, s_L). \tag{1.11}$$

Similarly, if Port 1 of the 2-port is terminated with the load reflectance s_G only (no generator driving Port 1), the reflectance looking into Port 2 is

$$\begin{aligned} s_2 = \mathcal{G}_2(\Theta, s_G) &:= \frac{\theta_{22}s_G + \theta_{21}}{\theta_{12}s_G + \theta_{11}} \\ &= s_{22} + s_{21}s_G(1 - s_{11}s_G)^{-1}s_{12} \\ &=: \mathcal{F}_2(S, s_G). \end{aligned}$$

For impedance matching, the engineer adroitly chooses 2-ports that push the load into a more useful reflectance. That is, a 2-port acts on the orbit of the load.

1.7 Orbit Actions

If the scattering matrices belong to a class $\mathcal{U}$, then the *orbit of* s_L *under the action of* $\mathcal{U}$ is the set of reflectances:

$$\mathcal{F}_1(\mathcal{U}, s_L) := \{\mathcal{F}_1(S, s_L) : S \in \mathcal{U}\}.$$

For example, the 2-port series inductor of Figure 1.4 has scattering and chain scattering matrices [56, Table 6.2]:

$$S(p) = \frac{1}{2 + Lp}\begin{bmatrix} Lp & 2 \\ 2 & Lp \end{bmatrix}, \quad \Theta(p) = \begin{bmatrix} 1 - Lp/2 & Lp/2 \\ -Lp/2 & 1 + Lp/2 \end{bmatrix}.$$

With $p = j$, the chain scattering matrix acts on $s_L \in \mathbf{D}$

$$\mathcal{G}(\Theta; s_L) = -\frac{\overline{a}}{a}\frac{s_L - a}{1 - \overline{a}s_L},$$

where $a = (1 + j2/(\omega L))^{-1}$. The Möbius group of symmetries of $\mathbf{D}$ consists of all maps $\mathbf{g} : \mathbf{D} \to \mathbf{D}$ [144]:

$$\mathbf{g}(s) = e^{j\theta}\frac{s - a}{1 - \overline{a}s},$$

where $a \in \mathbf{D}$ and $\theta \in \mathbf{R}$. Thus, the series inductor is a special case of a Möbius transform. The series inductor acts on unit disk as shown in Figure 1.10. Each panel shows the unit disk with the real and imaginary components labeled $\Re$ and $\Im$, respectively. The upper left panel shows the unit disk partitioned

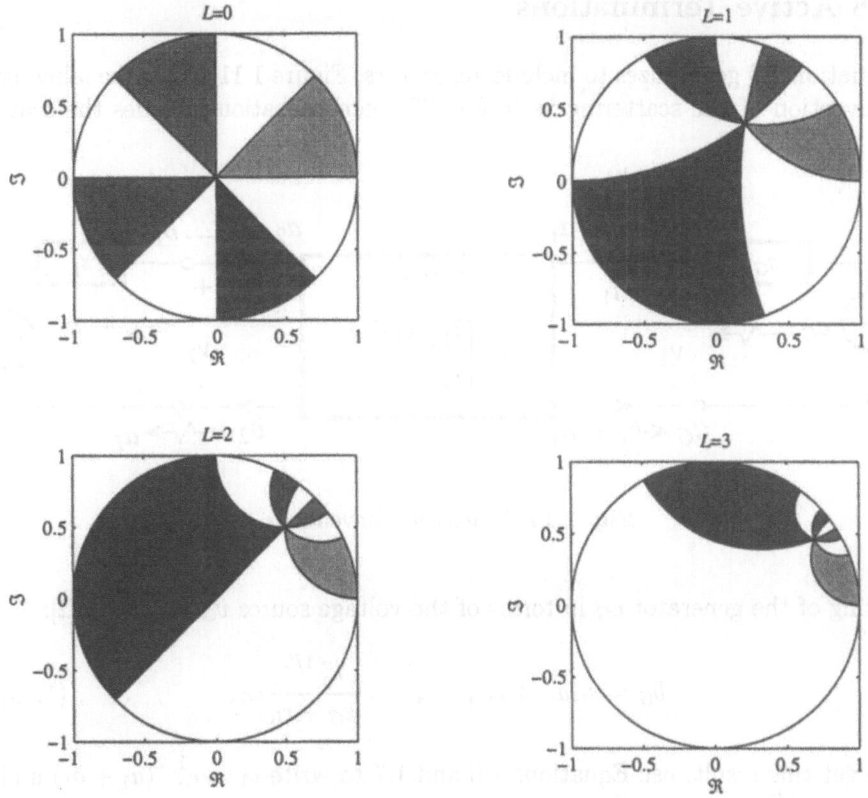

Fig. 1.10. Orbit actions of the series inductor.

into radial segments. As the inductance increases, the 2-port warps the radial pattern to the boundary.

The radial segments are geodesics of the *pseudo-hyperbolic*[2] metric on **D** [144, page 58]:

$$\rho(s_1, s_2) := \left| \frac{s_1 - s_2}{1 - \overline{s_1} s_2} \right| \quad (s_1, s_2 \in \mathbf{D}).$$

That ρ is invariant under the Möbius maps $\mathbf{g}$ is fundamental [144, page 58]:

$$\rho(\mathbf{g}(s_1), \mathbf{g}(s_2)) = \rho(s_1, s_2). \tag{1.12}$$

We use the pseudo-hyperbolic metric to understand the transducer power gain in Section 1.11.

[2] Also known as the Poincaré hyperbolic distance function.

1.8 Active Terminations

Equation 1.9 generalizes to include generators. Figure 1.11 shows the labeling convention of the scattering variables. The generalization includes the scat-

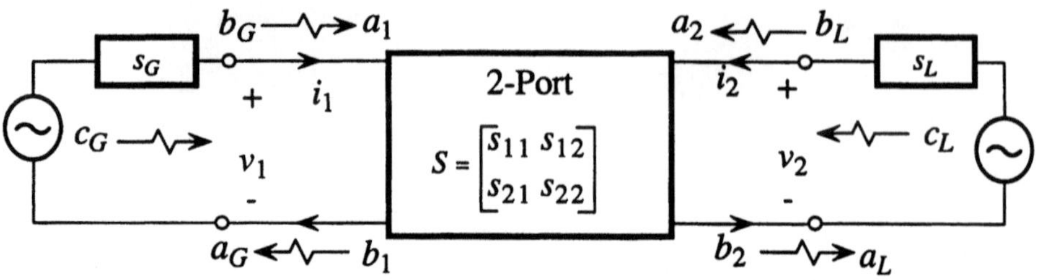

Fig. 1.11. Scattering conventions.

tering of the generator c_G in terms of the voltage source v_G [37, Eq. 3.2]:

$$b_G = s_G a_G + c_G; \quad c_G := \frac{r_0^{-1/2}}{z_G + r_0} v_G. \tag{1.13}$$

To get this result, use Equations 1.6 and 1.7 to write $v_1 = r_0^{1/2}(a_1 + b_1)$ and $i_1 = r_0^{-1/2}(a_1 - b_1)$. Substitute into the voltage drops $v_G = z_G i_1 + v_1$ of Figure 1.11 to get

$$c_G = \frac{r_0^{-1/2} v_G}{z_G + r_0} = a_1 - \frac{z_G - r_0}{z_G + r_0} b_1 = b_G - s_G a_G.$$

We can now analyze the setup in Figure 1.11. Equations 1.9 and 1.13 give

$$\mathbf{a} = \begin{bmatrix} a_1 \\ a_2 \end{bmatrix} = \begin{bmatrix} s_G & 0 \\ 0 & s_L \end{bmatrix} \begin{bmatrix} b_1 \\ b_2 \end{bmatrix} + \begin{bmatrix} c_G \\ c_L \end{bmatrix} =: S_X \mathbf{b} + \mathbf{c}_X.$$

Substitution into $\mathbf{b} = S\mathbf{a}$ and solving for the incident signal $\mathbf{a}$ completes the analysis of Figure 1.11:

$$\mathbf{a} = (I_2 - S_X S)^{-1} \mathbf{c}_X. \tag{1.14}$$

1.9 Power Formalisms

The complex power[3] delivered to an N-port is [8, page 241]:

[3] Baher uses [6, Eq. 2.17]: $W(p) = \mathbf{i}(p)^H \mathbf{v}(p)$. Gonzalez's power definitions [53] requires a 1/2.

$$W(p) := \mathbf{v}(p)^H \mathbf{i}(p).$$

Because $\mathbf{v}(p)$ has units volts-sec and $\mathbf{i}(p)$ has units amps-sec, $W(p)$ has units of Watts/Hz^2. The average power delivered to the N-port is [53, page 19]:

$$P_{\text{avg}} := \frac{1}{2}\Re[W] = \frac{1}{2}\{\mathbf{a}^H\mathbf{a} - \mathbf{b}^H\mathbf{b}\} = \frac{1}{2}\mathbf{a}^H\{I - S^H S\}\mathbf{a}. \tag{1.15}$$

If the N-port consumes power ($P_{\text{avg}} \geq 0$) for all its voltage and current pairs, the N-port is said to be passive. If the N-port consumes no power ($P_{\text{avg}} = 0$) for all its voltage and current pairs, the N-port is said to be lossless. If the N-port delivers power ($P_{\text{avg}} < 0$) to some of its voltage and current pairs, the N-port is declared to be active.

In terms of the N-port's scattering matrix, passivity and activity are characterized as follows [65]:

- Passive: $S^H(j\omega)S(j\omega) \leq I_N$ for all $\omega \in \mathbf{R}$.
- Lossless: $S^H(j\omega)S(j\omega) = I_N$ for all $\omega \in \mathbf{R}$.
- Active: $S^H(j\omega)S(j\omega) > I_N$ for some $\omega \in \mathbf{R}$.

Thus, a *lossless* N-port has a unitary scattering matrix. In terms of the impedance matrix $Z(p)$ of the N port [6]:

- Passive: $\Re[Z^H(j\omega)Z(j\omega)] \geq 0$ for all $\omega \in \mathbf{R}$.
- Lossless: $\Re[Z^H(j\omega)Z(j\omega)] = 0$ for all $\omega \in \mathbf{R}$.
- Active: $\Re[Z^H(j\omega)Z(j\omega)] < 0$ for some $\omega \in \mathbf{R}$.

Thus, a *negative resistor* is one model of an amplifier (Section 10.3). Finally, activity of a $2N$-port may be characterized by its chain scattering matrix $\Theta(p)$ using the signature matrix

$$J_N := \begin{bmatrix} I_N & 0 \\ 0 & -I_N \end{bmatrix}.$$

- Passive: $\Theta^H(j\omega)J_N\Theta(j\omega) \leq J_N$ for all $\omega \in \mathbf{R}$.
- Lossless: $\Theta^H(j\omega)J_N\Theta(j\omega) = J_N$ for all $\omega \in \mathbf{R}$.
- Active: $\Theta^H(j\omega)J_N\Theta(j\omega) > J_N$ for some $\omega \in \mathbf{R}$.

Thus, a *lossless* $2N$-port has a chain scattering matrix that is *J-unitary*. Electrical engineers employ a J-unitary approach [56]. A mathematical development of J-unitary operators with far-reaching applications is under sustained development by Helton, Joseph Ball, and their colleagues [65], [73], [11], [12], [13], [14]. Finally, to round out the taxonomy of the formalisms, Efimov and Potapov developed a substantial piece of the mathematical circuit analysis using the J_1-properties of the chain matrix $T(p)$ [45]:

- Passive: $T(j\omega)^H J_1 T(j\omega) \geq J_1, \quad J_1 = \begin{bmatrix} 0 & I_N \\ I_N & 0 \end{bmatrix}.$

1.10 Power Flows in the 2-Port

Specializing these notions of power to the 2-port of Figure 1.12 gives the following power flows:

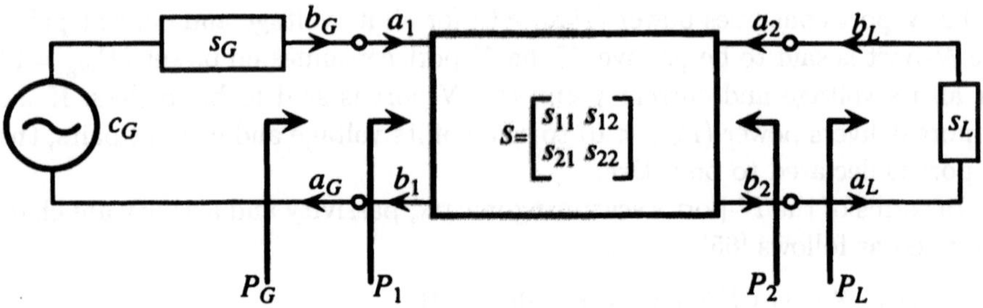

Fig. 1.12. Matching circuit and reflectances.

- The average power delivered to Port 1:

$$P_1 := \frac{1}{2}(|a_1|^2 - |b_1|^2) = \frac{|a_1|^2}{2}(1 - |s_1|^2).$$

- The average power delivered to Port 2:

$$P_2 := \frac{1}{2}(|a_2|^2 - |b_2|^2) = -P_L.$$

- The average power delivered to the load [53, Eq. 2.6.6]:

$$P_L := \frac{1}{2}(|a_L|^2 - |b_L|^2) = \frac{|b_2|^2}{2}(1 - |s_L|^2).$$

- The average power delivered by the generator:

$$P_G = \frac{1}{2}(|b_G|^2 - |a_G|^2).$$

To compute P_G, observe that Figure 1.12 gives $a_G = b_1$ and $b_G = a_1$. Substitute these and $b_1 = s_1 a_1$ into Equation 1.13 to get $c_G = (1 - s_G s_1)a_1$. Then

$$P_G = \frac{1}{2}(|a_1|^2 - |b_1|^2) = \frac{|a_1|^2}{2}(1 - |s_1|^2) = \frac{|c_G|^2}{2}\frac{1 - |s_1|^2}{|1 - s_G s_1|^2}. \qquad (1.16)$$

Lemma 1.10.1 *Assume the setup of Figure 1.12. There always holds*

$$P_2 = -P_L \quad \text{and} \quad P_G = P_1.$$

If the 2-port is lossless, $P_1 + P_2 = 0$.

1.11 Power Gains in the 2-Port

The matching network should map the generator's power to an output power that will be more useful at the load. The modification of power is generically called "gain." The matching problem requires gain computations, and we need the maximum power and mismatch definitions. The maximum power available from a generator is defined as the average power delivered by the generator to a conjugately matched load. Use Equation 1.16 to get [53, Eq. 2.6.7]:

$$P_{G,\max} := P_G|_{s_1=\overline{s_G}} = \frac{|c_G|^2}{2}(1-|s_G|^2)^{-1}.$$

The *source mismatch factor* is [53, Eq. 2.7.17]:

$$\frac{P_G}{P_{G,\max}} = \frac{(1-|s_G|^2)(1-|s_1|^2)}{|1-s_G s_1|^2}.$$

The maximum power available from the matching network is defined as the average power delivered from the network to a conjugately matched load [53, Eq. 2.6.19]:

$$P_{L,\max} := P_L|_{s_L=\overline{s_2}} := \frac{|b_2|_{s_L=\overline{s_2}}|^2}{2}(1-|s_2|^2).$$

The *load mismatch factor* is [53, Eq. 2.7.25]:

$$\frac{P_L}{P_{L,\max}} = \frac{(1-|s_L|^2)(1-|s_2|^2)}{|1-s_L s_2|^2}.$$

We need the following gains [53, page 213]:

- Transducer power gain

$$G_T := \frac{P_L}{P_{G,\max}} = \frac{\text{power delivered to the load}}{\text{maximum power available from the generator}}.$$

- Power gain or operating power gain

$$G_P := \frac{P_L}{P_1} = \frac{\text{power delivered to the load}}{\text{power delivered to the network}}.$$

- Available power gain

$$G_A := \frac{P_{L,\max}}{P_{G,\max}} = \frac{\text{maximum power available from the network}}{\text{maximum power available from the generator}}.$$

Lemma 1.11.1 *If the 2-port of Figure 1.12 is lossless and s_1, $s_G < 1$,*

$$G_T = \frac{(1-|s_G|^2)(1-|s_1|^2)}{|1-s_G s_1|^2}.$$

Proof:

$$G_T = \frac{P_L}{P_{G,\max}} \overset{\text{Lemma 1.10.1}}{=} \frac{-P_2}{P_{G,\max}} \overset{\text{lossless}}{=} \frac{P_1}{P_{G,\max}} \overset{\text{Lemma 1.10.1}}{=} \frac{P_G}{P_{G,\max}}.$$

///

The proof makes it clear that the equality holds because the power flowing into the lossless 2-port is the power flowing out of the 2-port, which is what one would expect from a lossless 2-port.

The transducer power gain is analyzed using the pseudo-hyperbolic metric of Equation 1.12—when the 2-port is lossless. Define the *power mismatch* between two passive reflectances s_1, s_2 as [66]:

$$\Delta P(s_1, s_2) := \left| \frac{\overline{s_1} - s_2}{1 - s_1 s_2} \right| = \rho(\overline{s}_1, s_2). \tag{1.17}$$

The power mismatch is the pseudo-hyperbolic distance between $\overline{s}_1$ and s_2 measured along their geodesic.

Lemma 1.11.2 *If the 2-port of Figure 1.12 is lossless, and* s_1, $s_G < 1$,

$$G_T(s_G, S, s_L) = 1 - \Delta P(s_G, s_1)^2 = 1 - \Delta P(s_G, \mathcal{F}(S, s_L))^2.$$

That is, *maximizing* G_T *is equivalent to minimizing the power mismatch.* Moreover, when the 2-port is lossless, we can match at either Port 1 or Port 2.

Lemma 1.11.3 [3] *If the 2-port of Figure 1.12 is lossless and* s_1, $s_G < 1$,

$$G_T(s_G, S, s_L) = 1 - \Delta P(s_G, \mathcal{F}_1(S, s_L))^2 = 1 - \Delta P(\mathcal{F}_2(S, s_G), s_L)^2.$$

The equalities do not hold if S is strictly passive. If $S^H S < I_2$, define the gains at Port 1 and 2 as

$$G_1(s_G, S, s_L) := 1 - \Delta P(s_G, \mathcal{F}_1(S, s_L))^2,$$

$$G_2(s_G, S, s_L) := 1 - \Delta P(\mathcal{F}_2(S, s_G), s_L)^2.$$

If S is only passive, we can only say

$$G_T \leq G_1, G_2.$$

To see this inequality, write G_1 and G_2 using the mismatch factors:

$$G_1(s_G, S, s_L) = 1 - \Delta P(s_G, s_1)^2 = \frac{P_G}{P_{G,\max}},$$

$$G_2(s_G, S, s_L) := 1 - \Delta P(s_2, s_L)^2 = \frac{P_L}{P_{L,\max}}.$$

If we believe that a passive 2-port forces both the available gain G_A and the power gain G_P not to amplify: G_A, $G_P \leq 1$, the inequalities $G_T \leq G_1$, G_2 are explained by the equalities:

$$G_T = \frac{P_L}{P_{G,\max}} = \frac{P_{L,\max}}{P_{G,\max}} \frac{P_L}{P_{L,\max}} = G_A G_2,$$

$$G_T = \frac{P_L}{P_{G,\max}} = \frac{P_1}{P_{G,\max}} \frac{P_L}{P_1} = G_P G_1.$$

Finally, when a general 2-port is used in Figure 1.12—be it lossless, passive, or active—setting $c_L = 0$ gives the specific solutions from Equation 1.14:

$$\begin{bmatrix} a_1 \\ a_2 \end{bmatrix} = \frac{c_G}{\det[I_2 - S_X S]} \begin{bmatrix} 1 - s_L s_{22} \\ s_L s_{21} \end{bmatrix},$$

and

$$\begin{bmatrix} b_1 \\ b_2 \end{bmatrix} = \frac{c_G}{\det[I_2 - S_X S]} \begin{bmatrix} (1 - s_L s_{22}) s_1 \\ s_{21} \end{bmatrix}.$$

Substitution into the definition of G_T produces

$$\begin{aligned} G_T &= |s_{21}|^2 \frac{|\det[I - S_X]|^2}{|\det[I - S_X S]|^2} \\ &= |s_{21}|^2 \frac{(1 - |S_{G,0}|^2)(1 - |S_{L,0}|^2)}{|(1 - S_{11} S_{G,0})(1 - S_{22} S_{L,0}) - S_{12} S_{21} S_{G,0} S_{L,0}|^2}. \end{aligned}$$

1.12 Modeling the Lumped, Lossless 2-Ports

The Amplifier Matching Problem consists of two components: the amplifier, which is given, and the matching circuits, which are tweaked to optimize the overall performance. For most of our analysis, the optimization requires us to search over classes of lumped, lossless 2-ports. Belevitch's Theorem makes precise the scattering matrix of any lumped, lossless 2-port and provides an excellent summary of the preceding concepts.

A *lumped, passive 2-port* contains only wires, transformers, gyrators, and the lumped elements (inductors, capacitors, resistors). In this case, its scattering matrix [99], [128]

$$S(p) = \begin{bmatrix} S_{11}(p) & S_{12}(p) \\ S_{21}(p) & S_{22}(p) \end{bmatrix}$$

is rational and analytic on the open right half plane $\mathbf{C}_+$ that is also real

$$S(p) = \overline{S(\overline{p})} \quad (p \in \mathbf{C}_+),$$

and a contraction:

$$S(p)^H S(p) \le \begin{bmatrix} 1 & 0 \\ 0 & 1 \end{bmatrix} \quad (p \in \mathbf{C}_+).$$

A *lumped, lossless 2-port* omits the resistors. As a consequence, its scattering matrix $S(p)$ is a real, analytic contraction on $\mathbf{C}_+$ that is also unitary on the boundary of $\mathbf{C}_+$:

$$S(j\omega)^H S(j\omega) = \begin{bmatrix} 1 & 0 \\ 0 & 1 \end{bmatrix}.$$

That is, $S(p)$ is an real, rational, inner function on $\mathbf{C}_+$.

A lumped, passive, reciprocal 2-port omits the gyrators. In this case, the scattering matrix $S(p)$ is a real, rational, analytic contraction on $\mathbf{C}_+$ that is symmetric:

$$S(p) = S(p)^T \quad (p \in \mathbf{C}_+).$$

Belevitch's Theorem asserts the existence of a scattering matrix $S(p)$ for each lumped, lossless 2-port such that $S(p)$ is a real, rational, inner function, and also obtains the *converse of these statements*.

BELEVITCH'S THEOREM [134], [6, pages 83–86] *A lumped, lossless 2-port $\mathcal{N}$ admits a scattering matrix*

$$S(p) = \frac{1}{g(p)} \begin{bmatrix} h(p) & f(p) \\ \epsilon f_*(p) & -\epsilon h_*(p) \end{bmatrix},$$

where $h_*(p) = h(-p)$ and (f, g, h) is a Belevitch triple:
B-1 $f(p)$, $g(p)$, and $h(p)$ are real polynomials; $\epsilon = \pm 1$,
B-2 $g(p)$ is strict Hurwitz (no zeros in the closed right half plane).
B-3 $g_*(p)g(p) = f_*(p)f(p) + h_*(p)h(p)$ for all $p \in \mathbf{C}$.
Conversely, any such $S(p)$ is the scattering matrix for some lumped, lossless 2-port $\mathcal{N}$. If the 2-port $\mathcal{N}$ is reciprocal

$$S(p) = \frac{1}{g(p)} \begin{bmatrix} h(p) & f(p) \\ f(p) & -\epsilon h_*(p) \end{bmatrix},$$

where either $f(p)$ is even and $\epsilon = -1$, or $f(p)$ is odd and $\epsilon = 1$. Conversely, any such $S(p)$ is the scattering matrix for some lumped, lossless, reciprocal 2-port $\mathcal{N}$.

Belevitch's Theorem is fundamental to filter design [134], circuit synthesis [17], and Darlington's Theorem [6] and is still a topic of research (i.e., lumped-distributed ladders generalize Belevitch's Theorem to several complex variables [114]). Moreover, Belevitch's Theorem characterizes several classes of 2-ports: low-pass, high-pass, mid-series, and mid-shunt ladders. Because we use these ladders for matching, we close this section with relevant ladder properties.

Figure 1.13 presents a generic ladder. If any one of the series arms is an open circuit ($z_k = \infty$) the ladder is an open circuit. Similarly, if any one of the shunt arms is a short circuit ($y_k = \infty$), the ladder is a short circuit. Thus, the poles of the series and shunt arms force *transmission zeros* of the ladder. Transmission zeros are the solutions of

$$0 = s_{21}(p) = \frac{f(p)}{g(p)}.$$

The finite transmission zeros are the zeros of $f(p)$. The remaining transmission zeros are at $p = \infty$.

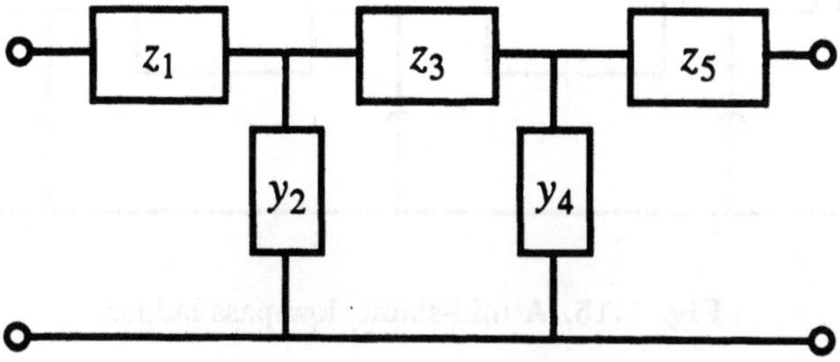

Fig. 1.13. A generic ladder.

Theorem 1.12.1 [119, page 166] *Let a ladder be constructed with series impedances z_k's and shunt admittances y_k's. Then the zeros of $f(p)$ are contained the finite poles of the z_k's and the y_k's.*

The low-pass ladder in Figure 1.6 has series inductors $z_k(p) = L_k p$ and shunt capacitors $y_k(p) = C_k p$. This forces all transmission zeros to $p = \infty$. Moreover, the low-pass ladder admits the scattering matrix *characterization* [6, page 121]:

$$f(p) = \text{constant}.$$

A high-pass ladder is illustrated in Figure 1.14. The shunt inductors $y_k(p) = (L_k p)^{-1}$ and the series capacitors $z_k(p) = (C_K p)^{-1}$ force all transmission zeros at $p = 0$. This condition also forces the converse. Thus, the high-pass ladder admits the scattering matrix characterization [6, page 122]:

$$f(p) = \text{constant} \times p^M,$$

where M is the degree of the polynomial $g(p)$.

Figures 1.15 and 1.16 illustrate the mid-shunt and mid-series ladders. The mid-shunt ladder has series impedance

$$z_k(p) = \frac{pL_k}{1 + p^2 L_k C_k}.$$

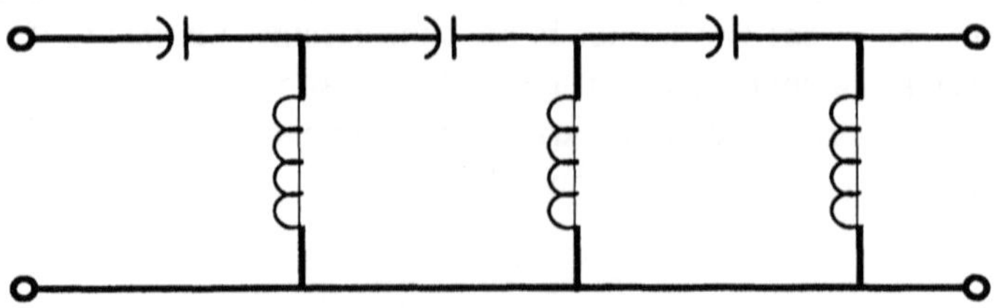

Fig. 1.14. A high-pass ladder.

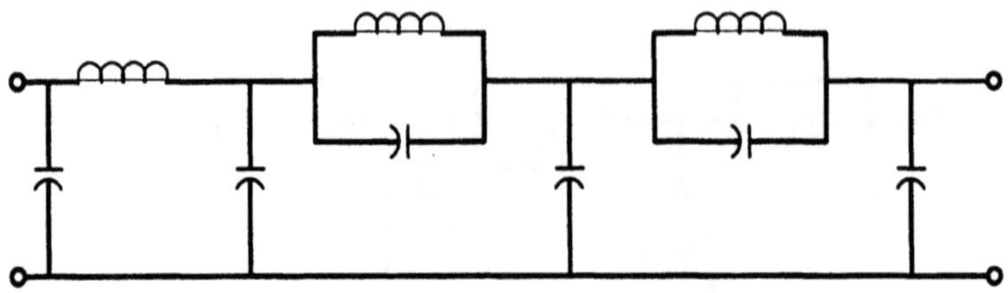

Fig. 1.15. A mid-shunt, low-pass ladder.

The mid-series ladder has shunt admittance

$$y_k(p) = \frac{pC_k}{1 + p^2 L_k C_k}.$$

Both cases force finite poles at $\pm\omega_k = \pm(L_k C_k)^{-1/2}$ so that

$$f(p) = \text{constant} \times \prod (p^2 + \omega_k^2).$$

The Fujisawa Theorems state the interlace condition that characterizes these ladders. Fujisawa's Theorems of 1955 are the theoretical *finale* of classical ladder theory [50], [6, pages 124-1-29], [134]. An impressive *cadenza* was per-

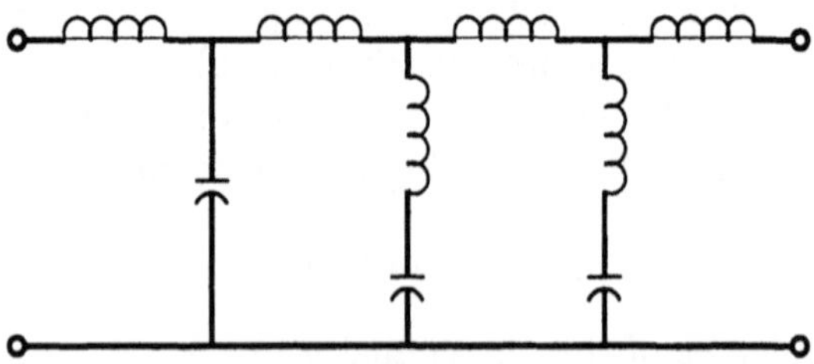

Fig. 1.16. A mid-series, low-pass ladder.

formed by Fialkow in 1979 [47]. Fialkow developed a general theory of parallel ladders. However, Fialkow's analysis has made little headway in the mainstream electrical engineering literature. Consequently, this rich body of parallel ladder theory awaits further exploration by applied mathematicians. Section 10.5 offers a general context for matching using such ladders.

1.13 Modeling the Small-Signal Amplifiers

The basic object under discussion is the amplifier. The physical examples we analyze are transistors. The H^{∞} theory only needs the scattering matrix of the transistor. The transistor's scattering matrix is obtained from the manufacturer's data sheets. These data sheets report the scattering matrix measured over a frequency band when the transistor is mounted in a biasing network. Figure 1.17 illustrates a typical biasing network where a Gallium-Arsenide Field-Effect Transistor (GaAs FET) is mounted in a common source configuration. Although the transistor is a nonlinear amplifier, the data sheets report

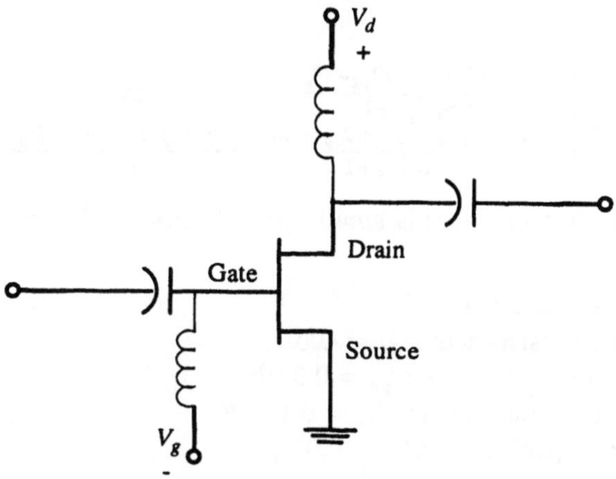

Fig. 1.17. Transistor and biasing network [106, pages 602–604] (courtesy of John Wiley).

the transistor's small-signal performance. That is, for a particular set of bias voltages and currents, and sufficiently small input, a linear scattering model may be assumed.

Figure 1.18 shows the small-signal equivalent circuit of the GaAs FET. This 2-port circuit includes both the FET and its biasing network. From Pozar [106, pages 602–603]:

> The dependent current generator $g_m V_c$ depends on the voltage across the gate-to-source capacitor C_{gs}, leading to a value of $|S_{21}| > 1$ under

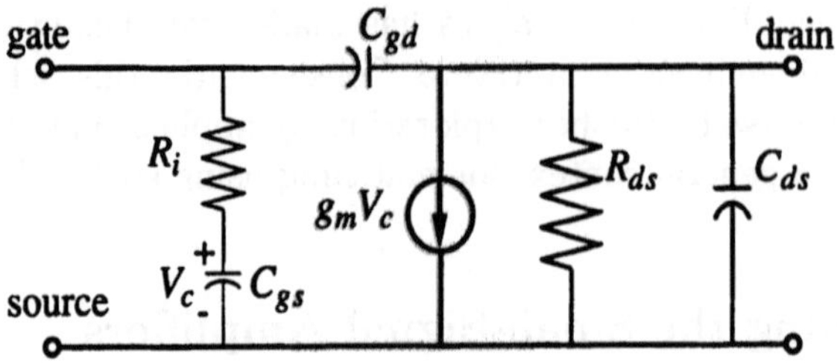

Fig. 1.18. GaAs FET small-signal equivalent circuit [106, pages 602–604] (courtesy of John Wiley).

normal operating conditions (where Port 1 is the gate and Port 2 is the drain). The reverse signal path, given by S_{12}, is due solely to the capacitance C_{gd}. As can be seen from the above data, this is typically a very small capacitor which can often be ignored in practice. In this case, $S_{12} = 0$, and the device is said to be unilateral.

Of the various 2-port formalisms for this amplifier, its admittance matrix is the simplest:

$$Y(p) = \begin{bmatrix} \frac{p(pC_{gd}R_iC_{gs}+C_{gd}+C_{gs})}{pR_iC_{gs}+1} & -pC_{gd} \\ -\frac{p^2C_{gd}R_iC_{gs}+pC_{gd}-g_m}{pR_iC_{gs}+1} & \frac{1+p(C_{ds}R_{ds}+C_{gd}R_{ds})}{R_{ds}} \end{bmatrix}.$$

The parameters that model this small-signal GaAs FET over 4–18 GHz are [106, page 602]:

- series gate resistance: $R_i = 7\Omega$,
- drain-to-source resistance: $R_{ds} = 400\Omega$,
- gate-to-source capacitance: $C_{gs} = 0.3$ pF,
- drain-to-source capacitance: $C_{ds} = 0.12$ pF,
- gate-to-drain capacitance: $C_{gs} = 0.01$ pF,
- transconductance: $g_m = 40$ mS.

Because the capacitor, resistor, and transconductance values are all positive, the admittance matrix is real and analytic on the open right half plane $\mathbf{C}_+$. Consequently, the scattering matrix [6, Eq. 4.67]

$$S(p) := \left(I + \widetilde{Y}(p)\right)^{-1}\left(I - \widetilde{Y}(p)\right),$$

is also real and analytic on $\mathbf{C}_+$. Here $\widetilde{Y}(p)$ denotes the normalized admittance matrix as in Section 1.4 [6, Eq. 4.64]:

$$\widetilde{Y}(p) = r_0 Y(p),$$

where $r_0 = 50$ ohms. Figure 1.19 plots $S(j\omega)$ as a function of $\omega = 2\pi f$ to reveal the active nature of this 2-port. To plot this matrix-valued function, the

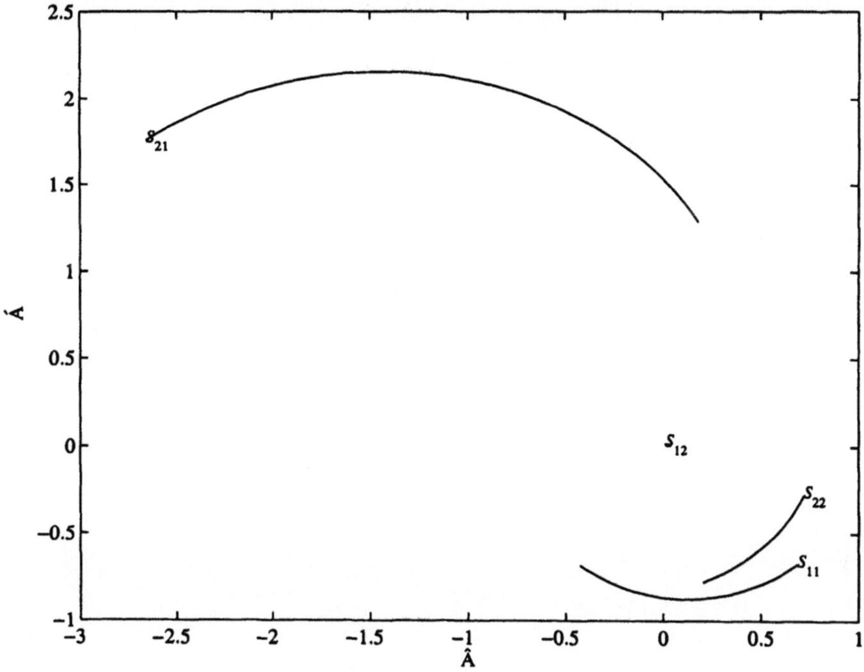

Fig. 1.19. Scattering matrix of a small-signal GaAs FET model over 4–18 GHz.

individual complex-valued functions $S_{11}(j\omega)$, $S_{21}(j\omega)$, $S_{12}(j\omega)$, and $S_{22}(j\omega)$ are plotted as curves in the complex plane $\mathbf{C}$. The label for each curve starts at the low frequency or 4 GHz position. The real axis is tagged with "$\Re$." The imaginary axis is tagged with "$\Im$." The plot shows that S_{12} is relatively small, thereby backing the claim that this amplifier is unilateral, which makes matching easy (see Chapter 2). The plot also shows that S_{11} and S_{22} both lie within the unit disk, which is good for stability, but that S_{11} is nearly unity, so that this amplifier has stability problems (see Chapter 2). Finally, it is interesting to see that this model is similar to the measured scattering matrices displayed in Chapter 6.

2

The Amplifier Matching Problem

Figure 2.1 shows an amplifier with scattering matrix[1]

$$S = \begin{bmatrix} S_{11} & S_{12} \\ S_{21} & S_{22} \end{bmatrix}$$

with Ports 1 and 2 terminated in the input and output matching circuits connecting a generator to a load. S_G denotes the reflectance of the input

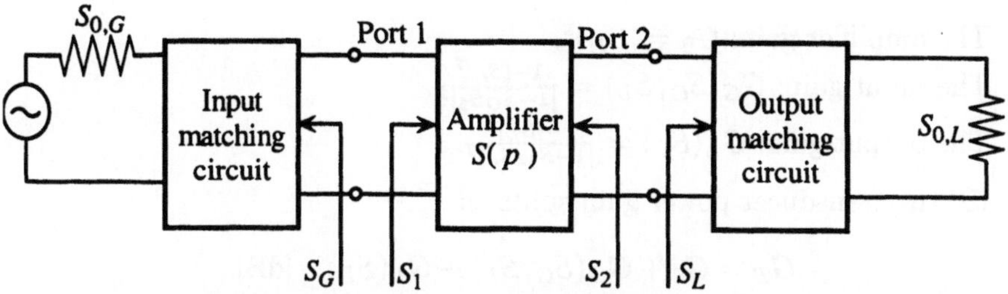

Fig. 2.1. Generic amplifier, matching circuits, and reflectances.

matching circuit terminated in the generator's reflectance $S_{0,G}$. S_L denotes the reflectance of the output matching circuit terminated in the load's reflectance $S_{0,L}$. S_1 denotes the reflectance looking into Port 1 of the amplifier with Port 2 terminated in S_L. Likewise, S_2 denotes the reflectance looking into Port 2 of the amplifier with Port 1 terminated in S_G. The Amplifier Matching Problem is to select input and output matching circuits that simultaneously

AMP-1 maximize the transducer power gain,
AMP-2 minimize the noise figure,
AMP-3 guarantee stability.

[1] All scattering variables are referenced to a real impedance Z_0.

These are competing objectives: increasing the gain increases the noise and decreases stability; asking for more stability decreases the gain and increases the noise. Consequently, the amplifier designer searches over collections of input and output matching circuits to find an acceptable *trade-off* between these amplifier functions (e.g., accepting a smaller gain to get less noise and better stability). The amplifier functions are introduced in the following sections.

2.1 Transducer Power Gain

From Pozar [106, page 610]:

> The most useful gain definition is the transducer power gain ... which accounts for both source and load mismatch.

The transducer power gain G_T is the ratio of the power P_L dissipated in the load S_L to the maximum power available P_{avs} from the source [106, Eq. 11.16]:

$$G_T = \frac{P_L}{P_{\text{avs}}} = |S_{21}|^2 \frac{1 - |S_G|^2}{|1 - S_G S_1|^2} \frac{1 - |S_L|^2}{|1 - S_{22} S_L|^2}.$$

The transducer power gain G_T factors as follows:

- The amplifier gain: $G_0 = |S_{21}|^2$.
- The input gain: $G_G(S_G, S_L) = \frac{1-|S_G|^2}{|1-S_G S_1|^2}$.
- The output gain: $G_L(S_L) = \frac{1-|S_L|^2}{|1-S_{22} S_L|^2}$.

In dB, the transducer power gain splits as

$$G_T = G_0 + G_G(S_G, S_L) + G_L(S_L) \quad \text{[dB]}.$$

Because the amplifier gain G_0 is fixed, the circuit designer looks for matching circuits to maximize G_G and G_L. Why is this splitting important? If the amplifier is approximately unilateral ($S_{12} \approx 0$) then the load is decoupled from the input reflectance so that

$$G_G(S_G, S_L) = G_G(S_G) = \frac{1 - |S_G|^2}{|1 - S_G S_{11}|^2}.$$

That is, $G_T(S_G, S_L)$ may be maximized by separately maximizing $G_G(S_G)$ and $G_L(S_L)$. Thus, a 4-port optimization problem is split into two independent 2-port problems.

2.2 Stability

Figure 2.1 shows that S_1 is the reflectance looking into Port 1 of the amplifier with Port 2 terminated in S_L [106, Eq. 11.6a]:

$$S_1 = S_{11} + S_{12}S_L(1 - S_{22}S_L)^{-1}S_{21}.$$

Likewise, S_2 is the reflectance looking into Port 2 of the amplifier with Port 1 terminated in S_G [106, Eq. 11.6b]:

$$S_2 = S_{22} + S_{21}S_G(1 - S_{11}S_G)^{-1}S_{12}.$$

When either $|S_1| > 1$ or $|S_2| > 1$, oscillations arise and the resulting circuit is not stable. Because these reflectances depend on the passive source and load reflectances, electrical engineers identify two types of stability that are associated with an amplifier [106, page 612]:

- An amplifier is called *unconditionally stable* provided $|S_1|$, $|S_2| < 1$ for any generator and load reflectances.
- An amplifier is called *conditionally stable* provided $|S_1|$, $|S_2| < 1$ for some generator and load reflectances.

An amplifier that is unconditionally stable permits us to drop stability (AMP-3) from the matching problem. There are at least two ways to make an amplifier unconditionally stable [106, page 616], [121, page 94]:

- Add resistive matching elements.
- Add feedback.

From Vendelin [121, page 94]:

> In practice, neither of these approaches is recommended, since other properties of the amplifier are usually degraded: gain, bandwith, noise figure, rf output power, and dc bias power.

Similar observations also arise in the noise figure.

2.3 Noise Figure

When a signal and the input noise are amplified by a noiseless amplifier, the signal-to-noise ratio (SNR) is unchanged. If the amplifier is noisy, noise is added to the output and the SNR is decreased. Let S_i/N_i denote the input SNR. Let S_o/N_o denote output SNR. The noise figure F measures this degradation of the SNR [106, Eq. 10.10], [140, Eq. 4-13]:

$$F = \frac{S_i/N_i}{S_o/N_o} \geq 1.$$

For a 2-port amplifier, the noise figure F is [106, Eq. 11.58]

$$F = F_{\text{min}} + \frac{4R_N}{Z_0}\frac{|S_G - S_{\text{opt}}|^2}{(1 - |S_G|^2)|1 + S_{\text{opt}}|^2},$$

where the following parameters are typically measured and reported on the amplifier's data sheet:

- F_{min} is the minimum noise figure of the amplifier when $S_G = S_{\text{opt}}$.
- R_N is the equivalent noise resistance of amplifier.
- S_G is the generator's reflectance presented to the amplifier.

Consequently, only the input matching circuit controls the noise figure. An excellent tutorial on noise is found in Leach [93, page 1516]:

> In any circuit containing resistors, capacitors, and inductors, only the resistors generate thermal noise.

Specific to the noise figure (NF) [93, page 1521]:

> The NF can be a very misleading specification. If an attempt is made to minimize an amplifier NF by adding resistors in either series or in parallel with the source, the SNR is always decreased. This is referred to as the *noise figure fallacy.*

To avoid inserting extra noise sources, we use lossless matching circuits—no resistors—although the theory will be developed to handle lossy matching circuits in Chapter 9.

3

H^∞ Tools for Electrical Engineers

This chapter reviews the necessary H^∞ theory. The presentation is aimed at electrical engineers with the goal of matching the mathematical results to a corresponding physical meaning. For example, a lumped, lossless 2-port admits a scattering matrix $S(p)$ in H^∞ that is real, rational, and lossless. The converse is true and the mapping from $S(p)$ to an actual circuit via the Foster or Cauer expansions is a time-honored torment of the undergraduate electrical engineers. Suppose now that $S(p)$ belongs to H^∞ and is lossless. Then $S(p)$ is rational if and only if $S(p)$ is continuous. Consequently, continuity is equivalent to the lossless 2-port being constructed from lumped elements.

This physical meaning of continuity stands in contrast to the observation that most microwave circuits use distributed elements. Consequently, the distributed 2-ports cannot have a continuous scattering matrix $S(p)$. An open question we develop in Chapter 4 asks how these distributed 2-ports reside in H^∞. We will see that this interplay between continuity and H^∞ underlies much of the analysis.

The fundamental tool of H^∞ engineering is *Nehari's Theorem.* Electrical engineers know the practical manifestation of Nehari's Theorem from wideband impedance matching. For example, a wideband matching technique is described as follows [140]: The load impedance z_L is plotted in the Smith chart. The engineer is to terminate this load with a cascade of lossless 2-ports. By repeatedly applying "shunt-stub/series-line cascades, a skilled designer using simulation software can see [the terminated impedance z_T] form into a fairly tight circle around $z = 1$." The appearance of this circle demonstrates that Nehari's Theorem is heuristically understood by microwave engineers. The radius of this circle determines the best possible matching attainable by any lossless 2-port. This matching limit may be computed by Nehari's Theorem as an eigenvalue problem. That is, the substantial problem of computing the performance limit of high-degree matching circuits is evaded by a standard matrix computation. And, in concordance with the underlying theme of continuity, we will see that continuity determines the existence of a matching 2-port. Thus, the attention paid to the continuous functions sets up the discussion for Nehari's Theorem, the existence of optimal matching 2-ports, and allows us to explore the relationship between the lumped and distributed 2-ports.

3.1 The Function Spaces

The real numbers are denoted by $\mathbf{R}$. The complex numbers are denoted by $\mathbf{C}$. The set of complex $M \times N$ matrices is denoted by $\mathbf{C}^{M\times N}$. The $N \times N$ identity matrix is denoted as I_N. The transpose of matrix A is denoted A^T and the conjugate transpose by A^H. Complex frequency is typically written $p = \sigma + j\omega$, where $j = \sqrt{-1}$ is used because i is used by electrical engineers for current. The open right half plane is denoted by $\mathbf{C}_+ := \{p \in \mathbf{C} : \Re[p] > 0\}$. The open unit disk is denoted by $\mathbf{D}$ and its unit circle by $\mathbf{T}$. The end-of-proof symbol is ///.

- $L^\infty(j\mathbf{R})$ denotes the class of Lebesgue-measurable functions defined on $j\mathbf{R}$ with norm $\|\phi\|_\infty := \text{ess.sup}\{|\phi(j\omega)| : \omega \in \mathbf{R}\}$.
- $C_0(j\mathbf{R})$ denotes the subspace of those continuous functions on $j\mathbf{R}$ that vanish at $\pm\infty$ with sup norm.
- $H^\infty(\mathbf{C}_+)$ is the Hardy space of functions bounded and analytic on $\mathbf{C}_+$ with norm $\|h\|_\infty := \sup\{|h(p)| : p \in \mathbf{C}_+\}$.

$H^\infty(\mathbf{C}_+)$ is identified with a subspace of $L^\infty(j\mathbf{R})$ whose elements are obtained by the pointwise limit $h(j\omega) = \lim_{\sigma\to 0} h(\sigma + j\omega)$ that converges almost everywhere [91, page 153]. Convergence in norm occurs if and only if the H^∞ function has continuous boundary values. Those H^∞ functions with continuous boundary values constitute the "disk algebra":

- $\mathcal{A}_1(\mathbf{C}_+) := 1 \dot{+} H^\infty(\mathbf{C}_+) \cap C_0(j\mathbf{R})$ denotes those continuous $H^\infty(\mathbf{C}_+)$ functions that are constant at infinity.

These spaces nest as

$$\mathcal{A}_1(\mathbf{C}_+) \subset H^\infty(\mathbf{C}_+) \subset L^\infty(j\mathbf{R}).$$

Tensoring with $\mathbf{C}^{M\times N}$ gives the corresponding matrix-valued functions [95]:

$$L^\infty(j\mathbf{R}, \mathbf{C}^{M\times N}) := L^\infty(j\mathbf{R}) \otimes \mathbf{C}^{M\times N}$$

with norm $\|\phi\|_\infty := \text{ess.sup}\{\|\phi(j\omega)\| : \omega \in \mathbf{R}\}$ induced by the matrix norm.

3.2 Unit Balls

The *open unit ball* of $L^\infty(j\mathbf{R}, \mathbf{C}^{M\times N})$ is denoted as

$$BL^\infty(j\mathbf{R}, \mathbf{C}^{M\times N}) := \left\{\phi \in L^\infty(j\mathbf{R}, \mathbf{C}^{M\times N}) : \|\phi\|_\infty < 1\right\}.$$

The *closed unit ball* of $L^\infty(j\mathbf{R}, \mathbf{C}^{M\times N})$ is denoted as

$$\overline{B}L^\infty(j\mathbf{R}, \mathbf{C}^{M\times N}) := \{\phi \in L^\infty(j\mathbf{R}, \mathbf{C}^{M\times N}) : \|\phi\|_\infty \le 1\}.$$

Likewise, the *open unit ball* of $H^\infty(\mathbf{C}_+, \mathbf{C}^{M\times N})$ is denoted as

$$BH^\infty(\mathbf{C}_+, \mathbf{C}^{M\times N}) := BL^\infty(\mathbf{C}_+, \mathbf{C}^{M\times N}) \cap H^\infty(\mathbf{C}_+, \mathbf{C}^{M\times M}).$$

Let $\mathbf{C}_\infty^N$ denote $\mathbf{C}^N$ equipped with the max norm:

$$\|\mathbf{x}\|_\infty = \max\{|x_n| : n = 1, \ldots, N\}.$$

Then $\phi \in L^\infty(j\mathbf{R}, \mathbf{C}_\infty^N)$ has norm

$$\begin{aligned}\|\phi\|_\infty &= \text{ess.sup}\{\max\{|\phi_n(j\omega)| : n = 1, \ldots, N\} : \omega \in \mathbf{R}\}\\ &= \max\{\|\phi_n\|_\infty : n = 1, \ldots, N\}.\end{aligned}$$

3.3 Real Functions

The impedance, admittance, and scattering functions that arise from physical circuits are known to electrical engineers as real functions *real*; $\phi \in L^\infty(j\mathbf{R}, \mathbf{C}^{M\times N})$ is called *real* provided

$$\overline{\phi(\overline{j\omega})} = \phi(j\omega) \quad \text{a.e.}$$

The collection of real L^∞ functions is denoted

$$\Re L^\infty(j\mathbf{R}, \mathbf{C}^{M\times N}) := \{\phi \in L^\infty(j\mathbf{R}, \mathbf{C}^{M\times N}) : \overline{\phi(\overline{j\omega})} = \phi(j\omega) \text{ a.e.}\}.$$

Specializing to the right half plane, $h \in H^\infty(\mathbf{C}_+, \mathbf{C}^{M\times N})$ is *real* provided

$$\overline{h(\overline{p})} = h(p) \quad (p \in \mathbf{C}_+).$$

A polynomial with real coefficients is the canonical example of a real function. The collection of real H^∞ functions in the closed unit ball is denoted

$$\Re\overline{B}H^\infty(\mathbf{C}_+, \mathbf{C}^{N\times N}).$$

3.4 Inner Functions

The matching circuits that arise in electrical engineering are typically *lossless* or have a scattering matrix that a mathematician would call *inner*. A function $S \in H^\infty(\mathbf{C}_+, \mathbf{C}^{N\times N})$ is called *inner* [58, page 68], [49, page 186], [116, page 190]

$$S(j\omega)^H S(j\omega) = I_N \quad \text{a.e.},$$

rather than the more general notion that $S(j\omega)$ is a partial isometry [108, page 94]. The class of *real inner* functions is denoted

$$U^+(N) := \{S \in \Re\overline{B}H^\infty(\mathbf{C}_+, \mathbf{C}^{N\times N}) : S(j\omega)^H S(j\omega) = I_N \quad \text{a.e.}\}.$$

Our lossless matching circuits are drawn from this class.

Lemma 3.4.1 *$U^+(N)$ is a closed subset of the boundary of $\Re\overline{B}H^\infty(\mathbf{C}_+, \mathbf{C}^{N\times N})$.*

Proof: It suffices to show closure. If $\{S_m\} \subset U^+(N)$ converges to $S \in H^\infty(\mathbf{C}_+, \mathbf{C}^{N\times N})$, then $S_m(j\omega) \to S(j\omega)$ a.e. so that

$$I_N = \lim_{m\to\infty} S_m(j\omega)^H S_m(j\omega) = S(j\omega)^H S(j\omega) \quad \text{a.e.}$$

Similar arguments show $S(j\omega)$ is real. ///

3.5 The Cayley Transform

Numerical computations are more conveniently placed in function spaces defined on the open unit disk $\mathbf{D}$ rather than on the open right half plane $\mathbf{C}_+$. The notation for the spaces on the disk follows the preceding nomenclature with the unit disk $\mathbf{D}$ replacing $\mathbf{C}_+$ and the unit circle $\mathbf{T}$ replacing $j\mathbf{R}$. $H^\infty(\mathbf{D})$ denotes the collection of analytic functions on the open unit disk with essentially bounded boundary values. $C(\mathbf{T})$ denotes the continuous functions on the unit circle. $\mathcal{A}(\mathbf{D})$ denotes the disk algebra

$$\mathcal{A}(\mathbf{D}) := H^\infty(\mathbf{D}) \cap C(\mathbf{T}).$$

$L^\infty(\mathbf{T})$ denotes the Lebesgue-measurable functions on the unit circle $\mathbf{T}$ with norm determined by the essential bound. A Cayley transform connects the function spaces on the right half plane to their counterparts on the disk.

Lemma 3.5.1 [59, page 99] *Let the Cayley transform $\mathbf{c} : \mathbf{C}_+ \to \mathbf{D}$*

$$\mathbf{c}(p) := \frac{p-1}{p+1}$$

extend to the composition operator $\mathbf{c} : L^\infty(\mathbf{T}) \to L^\infty(j\mathbf{R})$ as

$$h(p) := H \circ \mathbf{c}(p) \quad (p = j\omega).$$

Then $\mathbf{c}$ is an isometry mapping $\left\{\begin{array}{c} \mathcal{A}(\mathbf{D}) \\ H^\infty(\mathbf{D}) \\ C(\mathbf{T}) \\ L^\infty(\mathbf{T}) \end{array}\right\}$ onto $\left\{\begin{array}{c} \mathcal{A}_1(\mathbf{C}_+) \\ H^\infty(\mathbf{C}_+) \\ 1\dot{+}C_0(j\mathbf{R}) \\ L^\infty(j\mathbf{R}) \end{array}\right\}$.

3.6 Factoring H^∞ Functions

The *inner-outer* factorization of H^∞ functions is most conveniently developed on the unit disk and transplanted to the right half plane [81]. Let $\phi \in L^1(\mathbf{T})$ have the Fourier expansion

$$\phi(z) = \sum_{n=-\infty}^{\infty} \widehat{\phi}(n) z^n \quad (z = e^{j\theta})$$

with Fourier coefficients

$$\widehat{\phi}(n) := \int_{-\pi}^{\pi} e^{-jn\theta} \phi(e^{j\theta}) \frac{d\theta}{2\pi}.$$

For $1 \leq p \leq \infty$, define $H^p(\mathbf{D})$ as the subspace of $L^p(\mathbf{T})$ with vanishing negative Fourier coefficients [59, page 77]:

$$H^p(\mathbf{D}) := \{h \in L^p(\mathbf{T}) : \widehat{h}(n) = 0 \text{ for } n = -1, -2, \ldots\}.$$

Then $H^p(\mathbf{D})$ is a closed subspace of $L^p(\mathbf{T})$ and nest as [59, page 3]:

$$H^\infty(\mathbf{T}) \subset H^{p_2}(\mathbf{T}) \subset H^{p_1}(\mathbf{T}) \subset H^1(\mathbf{T}) \quad (1 \leq p_1 \leq p_2 \leq \infty)$$

Each $h \in H^p(\mathbf{D})$ admits an analytic extension on the open unit disk [59, page 77]:

$$h(z) = \sum_{n=0}^{\infty} \widehat{h}(n) z^n \quad (z = re^{j\theta}).$$

From the analytic extension, define $h_r(e^{j\theta}) := h(re^{j\theta})$ for $0 \leq r \leq 1$. For $r < 1$, h_r is continuous and analytic. As r increases to 1, h_r converges to h as in the L^p norm, provided $1 \leq p < \infty$. For $p = \infty$, h_r converges to h in the weak* topology (discussed in Section 3.7). If h_r does converge to h in the L^∞ norm, convergence is uniform and forces $h \in \mathcal{A}(\mathbf{D})$. The disk algebra $\mathcal{A}(\mathbf{D})$ is a strict subset of $H^\infty(\mathbf{D})$ in the norm topology and weak* dense in the weak* topology.

If ϕ is a positive, measurable function with $\log(\phi) \in L^1(\mathbf{T})$, the analytic function [110, page 370]:

$$q(z) = \exp\left(\int_{-\pi}^{\pi} \frac{e^{jt} + z}{e^{jt} - z} \log(\phi(e^{jt})) \frac{dt}{2\pi} \right) \quad (z \in \mathbf{D}),$$

is called an *outer function*. The magnitude of $q(z)$ matches ϕ [110, page 371]:

$$\lim_{r \to 1} |q_r(re^{j\theta})| = \phi(re^{j\theta}) \quad \text{a.e.}$$

and leads to the equivalence $\phi \in L^p(\mathbf{T}) \Leftrightarrow q \in H^p(\mathbf{D})$. We will call $q(z)$ a *spectral factor* of ϕ. Every $h \in H^\infty(\mathbf{D})$ admits an inner-outer factorization [110, pages 370–375]

$$h(z) = e^{j\theta_0} b(z) s(z) q(z),$$

where the outer function $q(z)$ is a spectral factor of $|h|$, the inner factor splits into the *Blaschke product* [110, page 333]

$$b(z) := z^k \prod_{n=1}^{\infty} \frac{z_n - z}{1 - \overline{z}_n z} \frac{\overline{z}_n}{z_n},$$

$z_n \neq 0$, $\sum(1 - |z_n|) < \infty$, and the *singular inner function*

$$s(z) = \exp\left(- \int_{-\pi}^{\pi} \frac{e^{jt} + z}{e^{jt} - z} d\mu(t) \right),$$

where μ is a finite, positive, Borel measure on $\mathbf{T}$ that is singular with respect to the Lebesgue measure.

Electrical engineers know that the Blaschke products correspond to lumped, lossless circuits while the singular inner functions can look like transmission lines. Electrical engineers also know the outer functions as *minimum-phase* functions and have been computing spectral factors for decades [132], [99], [135].

3.7 The Weak* Topology

The dual space of $L^\infty(j\mathbf{R})$ admits a representation as $L^\infty(j\mathbf{R})^* = L^1(j\mathbf{R})$, where $w \in L^1(j\mathbf{R})$ acts as the linear functional

$$\langle w, \phi \rangle := \int_{-\infty}^{\infty} w(j\omega)\phi(j\omega)d\omega.$$

The weak* topology on $L^\infty(j\mathbf{R})$ is the weakest vector topology that makes these functionals continuous. A weak* subbasis at $0 \in L^\infty(j\mathbf{R})$ is the collection of weak* open sets

$$O[w, \epsilon] := \{\phi \in L^\infty(j\mathbf{R}) : |\langle w, \phi \rangle| < \epsilon\},$$

where $\epsilon > 0$ and $w \in L^1(j\mathbf{R})$. Every weak* open set that contains $0 \in L^\infty(j\mathbf{R})$ is a union of finite intersections of these subbasic sets.

The unit ball $\overline{B}L^\infty(j\mathbf{R})$ is weak* compact by the Banach-Alaogalu Theorem [109, Theorem 3.15]. Compactness also holds for a distorted version of the unit ball, a fact that will have significant import for the optimization problems we consider later.

Lemma 3.7.1 *Let c, $r \in L^\infty(j\mathbf{R})$ with $r \geq 0$ define the* disk

$$\overline{D}(c, r) := \{\phi \in L^\infty(j\mathbf{R}) : |\phi - c| \leq r \text{ a.e.}\}.$$

Then $\overline{D}(c, r)$ is a closed, convex subset of $L^\infty(j\mathbf{R})$ that is also weak compact.*

Proof: Closure and convexity follow from pointwise closure and convexity. To prove weak* compactness, let $M_r : L^\infty(j\mathbf{R}) \to L^\infty(j\mathbf{R})$ be multiplication: $M_r\phi := r\phi$. Observe $\overline{D}(k, r) = k + M_r\overline{B}L^\infty(j\mathbf{R})$. Assume for now that M_r is weak* continuous. Then $M_r\overline{B}L^\infty(j\mathbf{R})$ is weak* compact, because $\overline{B}L^\infty(j\mathbf{R})$ is weak* compact, and the image of a compact set under a continuous function is compact. This forces $\overline{D}(k,r)$ to be weak* compact, provided M_r is weak* continuous. To see that M_r is weak* continuous, it suffices to show that M_r pulls subbasic sets back to subbasic sets. Let $\epsilon > 0$, $w \in L^1(j\mathbf{R})$. Then,

$$\begin{aligned}\psi \in M_r^{-1}(O[w,\epsilon]) &\iff M_r\psi \in O[w,\epsilon]\\ &\iff |\langle w, r\psi\rangle| < \epsilon\\ &\iff |\langle rw, \psi\rangle| < \epsilon\\ &\iff \psi \in O[rw,\epsilon],\end{aligned}$$

noting that $rw \in L^1(j\mathbf{R})$. ///

If K is a convex subset $L^\infty(j\mathbf{R})$, then K is closed $\Leftrightarrow$ K is weak* closed [41, page 422]. Because $H^\infty(\mathbf{C}_+)$ is a closed subspace of $L^\infty(\mathbf{C}_+)$, it is also weak* closed. Intersecting weak* closed $H^\infty(\mathbf{C}_+)$ with the weak* compact unit ball of $L^\infty(j\mathbf{R})$ forces $\overline{B}H^\infty(\mathbf{C}_+)$ to be weak* compact.

3.8 Weierstrass Theorem

The classic result that a continuous function on compact set attains its minimum is generalized by extending the notion of continuity and compactness.

Definition 3.8.1 [110, pages 38–39], [142, page 150] *Let γ be a real or extended-real function on a topological space X.*

- *γ is lower semicontinuous provided $\{x \in X : \gamma(x) \le \alpha\}$ is closed for every real α.*
- *γ is lower semicompact provided $\{x \in X : \gamma(x) \le \alpha\}$ is compact for every real α.*

If $\gamma : K \subseteq X \to \mathbf{R}$, any $x \in K$ that minimizes γ:

$$\gamma(x) = \inf\{\gamma(k) : k \in K\}$$

is called a *minimizer* of γ on K. The Weierstrass Theorem identifies properties of γ that produce minimizers.

Theorem 3.8.1 (Weierstrass) [142, page 152] *Let K be a nonempty subset of a topological space X. Let γ be a real or extended-real function defined on K. If either condition holds:*

- *γ is lower semicontinuous on the compact set K, or*
- *γ is lower semicompact,*

then $\inf\{\gamma(x) : x \in K\}$ *admits minimizers.*

3.9 Nehari's Theorem

The Toeplitz and Hankel operators are most conveniently defined on $L^2(\mathbf{T})$ using the Fourier basis $\{\exp(jn\theta)\}$. Let $\phi \in L^2(\mathbf{T})$ have the Fourier expansion

$$\phi(z) = \sum_{n=-\infty}^{\infty} \widehat{\phi}(n) z^n \qquad (z = e^{j\theta}).$$

Let P denote the orthogonal projection of $L^2(\mathbf{T})$ onto $H^2(\mathbf{D})$:

$$P(\phi) = \sum_{n=0}^{\infty} \widehat{\phi}(n) z^n.$$

The *Toeplitz operator* with symbol $\phi \in L^\infty(\mathbf{T})$ is the mapping $\mathcal{T}_\phi : H^2(\mathbf{D}) \to H^2(\mathbf{D})$

$$\mathcal{T}_\phi(h) := P(\phi h).$$

The *Hankel operator* with symbol $\phi \in L^\infty(\mathbf{T})$ is the mapping $\mathcal{H}_\phi : H^2(\mathbf{D}) \to H^2(\mathbf{D})$

$$\mathcal{H}_\phi(h) := U(I - P)(\phi h),$$

where $U : H^2(\mathbf{D})^\perp \to H^2(\mathbf{D})$ is the unitary "flipping" operator:

$$U(h; z) := z^{-1} h(z^{-1}).$$

These operators admit matrix representations with respect to the Fourier basis [140, page 173]:

$$\mathcal{T}_\phi = \begin{bmatrix} \widehat{\phi}(0) & \widehat{\phi}(1) & \widehat{\phi}(2) & \ddots \\ \widehat{\phi}(-1) & \widehat{\phi}(0) & \widehat{\phi}(1) & \ddots \\ \widehat{\phi}(-2) & \widehat{\phi}(-1) & \widehat{\phi}(0) & \ddots \\ \ddots & \ddots & \ddots & \ddots \end{bmatrix}$$

and [140, page 191]

$$\mathcal{H}_\phi = \begin{bmatrix} \widehat{\phi}(-1) & \widehat{\phi}(-2) & \widehat{\phi}(-3) & \cdots \\ \widehat{\phi}(-2) & \widehat{\phi}(-3) & \widehat{\phi}(-4) & \cdots \\ \widehat{\phi}(-3) & \widehat{\phi}(-4) & \widehat{\phi}(-5) & \cdots \\ \vdots & \vdots & \vdots & \end{bmatrix}.$$

We will need the following computations from Douglas [38, pages 177–184] and Helton [66, page 42].

Lemma 3.9.1 *Let $\phi \in H^\infty(\mathbf{C}_+)$ and $\psi \in L^\infty(j\mathbf{R})$. Define*

$$\check{\phi}(e^{j\theta}) := \phi(e^{-j\theta}).$$

(a) $\mathcal{T}_\psi^* = \mathcal{T}_{\overline{\psi}}$ *and* $\mathcal{H}_\psi^* = \mathcal{H}_{\overline{\psi}}$.
(b) $\mathcal{T}_\psi \mathcal{T}_\phi = \mathcal{T}_{\psi\phi}$.
(c) If ϕ *is outer,* $\mathcal{T}_\psi^{-1} = \mathcal{T}_{\psi^{-1}}$.
(d) $\mathcal{H}_\psi \mathcal{T}_\phi = \mathcal{H}_{\psi\phi}$.
(e) $\mathcal{T}_{\check{\phi}} \mathcal{H}_\psi = \mathcal{H}_{\psi\phi}$.

Throughout this book, the *quotient norm* or norm on a quotient space is used [109, page 29–31]. If X is a Banach space containing a subspace K, the quotient space X/K consists of the elements $x + K$ equipped with the norm

$$\|x - K\| := \inf\{\|x - k\| : k \in K\}.$$

That is, $\|x - K\|$ is the distance between x and K. For example, the norm on $L^\infty(\mathbf{T})/H^\infty(\mathbf{D})$ is

$$\|\phi - H^\infty(\mathbf{D})\|_\infty := \inf\{\|\phi - h\|_\infty : h \in H^\infty(\mathbf{D})\}$$

and measures the distance between ϕ and $H^\infty(\mathbf{D})$. The following version of Nehari's Theorem emphasizes existence, computation, and uniqueness of best approximations in terms of the quotient norms.

Theorem 3.9.1 (Nehari) [140], [105] *If* $\phi \in L^\infty(\mathbf{T})$, *then* ϕ *admits best approximations from* $H^\infty(\mathbf{D})$ *as follows:*

N-1 $\|\phi - H^\infty(\mathbf{D})\|_\infty = \|\mathcal{H}_\phi\|$.
N-2 $\|\phi - \{H^\infty(\mathbf{D}) + C(\mathbf{T})\}\|_\infty = \|\mathcal{H}_\phi\|_e$.
N-3 If $\|\mathcal{H}_\phi\|_e < \|\mathcal{H}_\phi\|$ *then best approximations are unique.*

The operator norm is

$$\|\mathcal{H}_\phi\| := \sup\{\|\mathcal{H}_\phi h\|_\infty : h \in \overline{B}H^\infty(\mathbf{D})\}$$

and the essential norm is

$$\|\mathcal{H}_\phi\|_e := \inf\{\|\mathcal{H}_\phi - K\| : K \text{ is a compact operator}\}.$$

Nehari's Theorem

- computes the distance from ϕ to $H^\infty(\mathbf{D})$,
- observes that H^∞ minimizers exist,
- links the unicity of these minimizers to the continuity of ϕ.

However, the physical matching circuits, such as the lumped, lossless 2-ports, force us to minimize from the disk algebra. Because the disk algebra $\mathcal{A}(\mathbf{D})$ is a strictly proper subset of $H^\infty(\mathbf{D})$, the inequality follows:

$$\|\phi - \mathcal{A}(\mathbf{D})\|_\infty \geq \|\phi - H^\infty(\mathbf{D})\|_\infty = \|\mathcal{H}_\phi\|.$$

Equality holds whenever ϕ is sufficiently smooth. Continuity supplies the simplest result for equality The following result is adapted from Koosis [91, pages 193–195] and Hintzman [78], [79]).

Theorem 3.9.2 *If* $\phi \in 1 \dotplus C_0(j\mathbf{R})$,

$$\|\phi - \mathcal{A}_1(\mathbf{C}_+)\|_\infty = \|\phi - H^\infty(\mathbf{C}_+)\|_\infty$$

and there is exactly one $h \in H^\infty(\mathbf{C}_+)$ *such that*

$$\|\phi - \mathcal{A}_1(\mathbf{C}_+)\|_\infty = |\phi - h(j\omega)| \quad \text{a.e.}$$

Theorem 3.9.2 computes the distance between ϕ and the disk algebra and offers a unique best approximant from H^∞. However, continuity alone cannot force the best approximant into the disk algebra.

Example 3.9.1 (Hintzman's Counterexample) [79] Let u and v denote the real and imaginary parts of the analytic function:

$$u(z) + jv(z) := -j \sum_{n=2}^{\infty} \frac{z^n}{n \log(n)}.$$

Construct

$$g(e^{j\theta}) = e^{-jv(e^{i\theta})}(e^{-j\theta} - e^{-u(e^{j\theta})}).$$

Then g belongs to $C(\mathbf{T})$ but does **not** have a best approximation from $\mathcal{A}(\mathbf{D})$.

Figure 3.1 plots $g(e^{i\theta})$ approximated by the finite sum with N replacing ∞. Even with a finite number of terms, the figure shows at least one point that is not smooth. To get existence in the disk algebra requires more than continuity. If $\phi : \mathbf{R} \to \mathbf{C}$ has period 2π, the *modulus of continuity* of ϕ is the function [42, page 71]:

$$\omega(\phi; t) := \sup\{|\phi(t_1) - \phi(t_2)| : t_1, t_2 \in \mathbf{R}, |t_1 - t_2| \le t\}.$$

Let Λ_α denote those functions that satisfy a *Lipschitz condition of order* $\alpha \in (0, 1]$:

$$|\phi(t_1) - \phi(t_2)| \le A|t_1 - t_2|^\alpha.$$

Let $C^{n+\alpha}$ denote those functions with $\phi^{(n)} \in \Lambda_\alpha$ [24]. Let C_ω denote those functions that are *Dini-continuous*:

$$\int_0^\epsilon \omega(\phi; t) t^{-1} dt < \infty,$$

for some $\epsilon > 0$. A sufficient condition for a function $\phi(t)$ to be Dini-continuous is that $|\phi'(t)|$ be bounded [51, section IV.2]. Carleson and Jacobs have an amazing paper that addresses best approximation from the disk algebra [24].

Theorem 3.9.3 (Carleson and Jacobs) [24] *If* $\phi \in L^\infty(\mathbf{T})$, *there always exists a best approximation* $h \in H^\infty(\mathbf{D})$:

$$\|\phi - h\|_\infty = \|\phi - H^\infty(\mathbf{D})\|_\infty.$$

If $\phi \in C(\mathbf{T})$, *then the best approximation is unique. Moreover,*

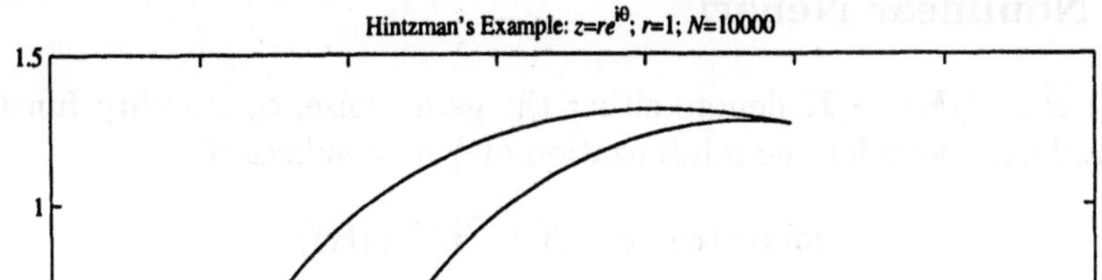

Fig. 3.1. Hintzman's continuous function that does not have a best approximation from $\mathcal{A}(\mathbf{D})$.

(a) If $\phi \in C_\omega$ *then* $h \in C_\omega$.
(b) If $\phi^{(n)} \in C_\omega$ *then* $h^{(n)} \in C_\omega$.
(c) If $0 < \alpha < 1$ *and* $\phi \in \Lambda_\alpha$ *then* $h \in \Lambda_\alpha$.
(d) If $0 < \alpha < 1$, $n \in N$, *and* $\phi \in C^{n+\alpha}$ *then* $h \in C^{n+\alpha}$.

As noted by Carleson and Jacobs [24]: "the function-theoretic proofs ... are all of a local character, and so all the results can easily be carried over to any region which has in each case a sufficiently regular boundary." Provided we can guarantee smoothness across ∞, Theorem 3.9.3 carries over to the right half plane using the Cayley transform $\mathbf{c}(p) = (p-1)(p+1)^{-1}$.

Corollary 3.9.1 *If* $\phi \in 1 \dot{+} C_0(j\mathbf{R})$, *the best approximation*

$$\|\phi - h\|_\infty = \|\phi - H^\infty(\mathbf{C}_+)\|_\infty$$

exists and is unique. Moreover, if $\phi \circ \mathbf{c}^{-1} \in C_\omega$, *then* $h \circ \mathbf{c}^{-1} \in C_\omega$ *so that*

$$\|\phi - h\|_\infty = \|\phi - H^\infty(\mathbf{C}_+)\|_\infty = \|\phi - \mathcal{A}_1(\mathbf{C}_+)\|_\infty.$$

Informally, the smoothness of the target function ϕ is invariant under the *best approximation operator* of H^∞.

3.10 Nonlinear Nehari

Let $\gamma : \overline{B}L^\infty(j\mathbf{R}) \to \mathbf{R}$ denote either the gain, noise, or stability functions. Standard questions for the minimization of γ on a subset K

$$\min\{\gamma(\phi) : \phi \in K \subseteq \overline{B}L^\infty(j\mathbf{R})\}$$

are the following:

- existence of minimizers,
- uniqueness of minimizers,
- characterization of minimizers.

A *sublevel set* of γ is denoted as

$$[\gamma \le \alpha] := \{\phi \in \overline{B}L^\infty(j\mathbf{R}) : \gamma(\phi) \le \alpha\}.$$

It was Helton's observation that many functions in electrical engineering and control theory have sublevel sets that are disks [74]. That is, there is a *center function* c_α and a *radius function* r_α in L^∞ such that

$$[\gamma \le \alpha] = \{\phi \in L^\infty(j\mathbf{R}) : |\phi(j\omega) - c_\alpha(j\omega)| \le r_\alpha(j\omega) \text{ a.e.}\} = \overline{D}(c_\alpha, r_\alpha).$$

The Weierstrass Theorem demonstrates that minimizers exist for such a γ.

Corollary 3.10.1 *Let $\gamma : \mathcal{D} \subseteq L^\infty(j\mathbf{R}) \to \mathbf{R}_+$ where $\mathcal{D}$ is nonempty and open. Assume γ has sublevel sets that are closed disks: for all $\alpha \ge 0$, there exist c_α, $r_\alpha \in L^\infty(j\mathbf{R})$ such that*

$$[\gamma \le \alpha] = \overline{D}(c_\alpha, r_\alpha).$$

Assume $K \subset \mathcal{D}$ is closed, bounded, and convex. Then γ admits minimizers in K.

Proof: Because K is convex, K is closed $\Leftrightarrow$ K is weak* closed [41, Theorem V.13.11.13]. The boundedness of K coupled with the weak* closure of K forces K to be weak* compact [41, Corollary V.4.2.3]. By Lemma 3.7.1, the sublevel sets of γ are weak* compact so that γ is weak* lower semicontinuous. By the Weierstrass Theorem, γ admits minimizers in K. ///

As a special case, we get the classic result that $L^\infty(j\mathbf{R})$ admits best $H^\infty(\mathbb{C}_+)$ approximations.

Corollary 3.10.2 [91, page 197] *If $\phi \in L^\infty(j\mathbf{R})$ then there is an $h \in H^\infty(\mathbb{C}_+)$ such that $\|\phi - H^\infty(\mathbb{C}_+)\|_\infty = \|\phi - h\|_\infty$.*

The disk structure also couples with Nehari's Theorem to characterize minimizers. Helton provided the fundamental link between disks and operators.

Theorem 3.10.1 [66, Theorem 4.2] *Let C, P, $R \in L^\infty(\mathbf{T}, \mathbf{C}^{N\times N})$. Assume P and R are uniformly strictly positive. Define the disk*

$$\overline{D}(C,R,P) := \{\Phi \in L^\infty(\mathbf{T}, \mathbf{C}^{N\times N}) : (\Phi - C)P^2(\Phi - C)^H \leq R^2\}$$

and set $\check{R}(j\omega) := R(-j\omega)$. Then

$$\emptyset \neq \overline{D}(C,R,P) \cap H^\infty(\mathbf{D}, \mathbf{C}^{N\times N}) \iff \mathcal{H}_C \mathcal{T}_{P^{-2}}^{-1} \mathcal{H}_C^* \leq \mathcal{T}_{\check{R}^2}.$$

For certain classes of problems, such as lossless impedance matching and our Amplifier Matching Problem, the sublevel sets of γ are contained in $\overline{B}L^\infty(j\mathbf{R})$. Consequently, the unit ball constraint may be ignored and Theorem 3.10.1 directly applies:

$$\overline{D}(c_\alpha, r_\alpha) \cap \overline{B}H^\infty(\mathbf{C}_+) = \overline{D}(c_\alpha, r_\alpha) \cap H^\infty(\mathbf{C}_+).$$

The following result applies Theorem 3.10.1 to the disk theory under this stronger assumption.

Corollary 3.10.3 *Let $\gamma : \overline{B}L^\infty(j\mathbf{R}) \to \mathbf{R}$. Assume γ has sublevel sets that are disks contained in $\overline{B}L^\infty(j\mathbf{R})$:*

$$[\gamma \leq \alpha] = \overline{D}(c_\alpha, r_\alpha) \subseteq \overline{B}L^\infty(j\mathbf{R}).$$

Let $C_\alpha := c_\alpha \circ \mathbf{c}^{-1}$ and $R_\alpha = r_\alpha \circ \mathbf{c}^{-1}$ where $\mathbf{c}$ is the Cayley transform of Lemma 3.5.1. Assume R_α is strictly uniformly positive with spectral factor $Q_\alpha \in H^\infty(\mathbf{D})$. Then the following are equivalent:

(a) $\overline{D}(c_\alpha, r_\alpha) \cap \overline{B}H^\infty(\mathbf{C}_+) \neq \emptyset$.
(b) $\mathcal{H}_{C_\alpha} \mathcal{H}_{C_\alpha}^* \leq \mathcal{T}_{\check{R}_\alpha^2}$.
(c) $\|Q_\alpha^{-1} C_\alpha - H^\infty(\mathbf{D})\|_\infty \leq 1$.

Proof: By Theorem 3.10.1, all that is needed is to prove (a)⇔(c). If (a) is true, there exists an $H \in \overline{B}H^\infty$ such that $|H - C_\alpha| \leq R_\alpha = |Q_\alpha|$ a.e. Because $R_\alpha > 0$ on $\mathbf{T}$, we may divide by $|Q_\alpha|$ to get $|Q_\alpha^{-1}H - Q_\alpha^{-1}C_\alpha| \leq 1$ a.e., or that (c) must be true. Conversely, if (c) is true, there exists an $G \in H^\infty$ such that $|G - Q_\alpha^{-1}C_\alpha| \leq 1$ a.e. Because Q_α is outer, $H = Q_\alpha G \in H^\infty$ and $|H - C_\alpha| \leq R_\alpha$ a.e. Then $H \in \overline{D}(C_\alpha, R_\alpha) \cap H^\infty(\mathbf{C})$. Because $\overline{D}(C_\alpha, R_\alpha)$ is contained in the unit ball of $L^\infty(\mathbf{T})$, (a) must hold. ///

The eigenvalue test of Corollary 3.10.3(b) permits a nice graphical display. Let $\lambda_{\inf}(\alpha)$ denote the smallest "eigenvalue" of $\mathcal{T}_{\check{R}_\alpha^2} - \mathcal{H}_{C_\alpha}\mathcal{H}_{C_\alpha}^*$. The plots in Chapter 7 illustrate that $\lambda_{\inf}(\alpha)$ is a decreasing function of α that crosses zero at the minimum of γ on $\overline{B}H^\infty(\mathbf{C}_+)$. Because γ has sublevel sets that are weak* closed disks, it is lower semicontinuous in the weak* topology and H^∞ minimizers exist.

Corollary 3.10.4 *Let $\gamma : \overline{B}L^\infty(j\mathbf{R}) \to \mathbf{R}$. Assume γ has sublevel sets that are disks contained in $\overline{B}L^\infty(j\mathbf{R})$:*

$$[\gamma \le \alpha] = \overline{D}(c_\alpha, r_\alpha) \subseteq \overline{B}L^\infty(j\mathbf{R}).$$

Then γ has minimizers $h_{\min} \in \overline{B}H^\infty(\mathbf{C}_+)$:

$$\gamma_{\overline{B}H^\infty} := \min\{\gamma(h) : h \in \overline{B}H^\infty(\mathbf{C}_+)\}.$$

Let $c_{\min}$ and $r_{\min}$ denote the corresponding center and radius functions in $L^\infty(j\mathbf{R})$. Let $C_\alpha := c_\alpha \circ \mathbf{c}^{-1}$ and $R_\alpha = r_\alpha \circ \mathbf{c}^{-1}$ where $\mathbf{c}$ is the Cayley transform of Lemma 3.5.1. Assume R_α is strictly uniformly positive with spectral factor $Q_{\min}$. Then the following are equivalent:

Min-1 $\overline{D}(c_{\min}, r_{\min}) \cap \overline{B}H^\infty \neq \emptyset$.
Min-2 $0 = \lambda_{\inf}(\gamma_{\overline{B}H^\infty})$.
Min-3 $\|Q_{\min}^{-1}C_{\min} - H^\infty(\mathbf{D})\|_\infty = 1$.

Moreover, if $Q_{\min}^{-1}C_{\min} \in C(\mathbf{T})$, then the minimizer $h_{\min}$ is unique.

Proof:
Min-1⇒Min-3: If the inequality were strict, $|C_{\min} - H| < R_{\min}$ a.e. for some $H \in H^\infty(\mathbf{D})$. Then $h = H \circ \mathbf{c}$ belongs to $H^\infty(\mathbf{C}_+)$ and drops γ below its minimum: $\gamma(h) < \gamma_{\overline{B}H^\infty}$. This contradiction forces equality at the minimum.
Min-3⇒Min-1: Corollary 3.10.3.
Min-1⇒Min-2: Theorem 3.10.1 forces $\mathcal{H}_{C_{\min}}\mathcal{H}_{C_{\min}}^* \le T_{R_{\min}^2}$ or $0 \le \lambda_{\inf}(\gamma_{\overline{B}H^\infty})$. By Lemma 3.9.1 or Helton [66, page 42], the operator inequality is equivalent to $1 \ge \|\mathcal{H}_{Q_{\min}^{-1}C_{\min}}\|$. By Nehari's Theorem,

$$1 \ge \|\mathcal{H}_{Q_{\min}^{-1}C_{\min}}\| = \|Q_{\min}^{-1}C_{\min} - H^\infty(\mathbf{D})\|_\infty = 1,$$

where the equivalence of Min-1 and Min-3 gives the last equality. Thus, the inequality must be an equality.
Min-2⇒Min-1: $0 = \lambda_{\inf}(\gamma_{\overline{B}H^\infty})$ forces $1 = \|\mathcal{H}_{Q_{\min}^{-1}C_{\min}}\|$. By Nehari's Theorem,

$$1 = \|Q_{\min}^{-1}C_{\min} - H^\infty(\mathbf{D})\|_\infty,$$

or that $\overline{D}(C_{\min}, R_{\min}) \cap H^\infty(\mathbf{D}) \neq \emptyset$. ///

For minimization in the disk algebra, the inclusion of the disk algebra in H^∞ forces

$$\gamma_{\overline{B}H^\infty} \le \gamma_{\overline{B}A} := \inf\{\gamma(h) : h \in \overline{B}\mathcal{A}_1(\mathbf{C}_+)\}.$$

Under smoothness and continuity conditions, equality between the disk algebra and H^∞ minima is possible.

Corollary 3.10.5 *In addition to the assumptions of Corollary 3.10.4, assume that $Q_{\min}^{-1}C_{\min}$ is Dini-continuous. Then*

$$\gamma_{\overline{B}H^\infty} = \gamma_{\overline{B}A^\infty} = \min\{\gamma(h) : h \in \overline{B}\mathcal{A}_1(\mathbf{C}_+)\}$$

and the minimum is unique.

Proof: By Corollary 3.10.4, there is a unique minimizer $H_{\min} \in H^\infty(\mathbf{D})$

$$1 = \|Q_{\min}^{-1} C_{\min} - H^\infty(\mathbf{D})\|_\infty = \|Q_{\min}^{-1} C_{\min} - H_{\min}\|_\infty .$$

By Corollary 3.9.1, Dini-continuity forces $H_{\min}$ to be Dini-continuous or $h_{\min} = H \circ \mathbf{c} \in \mathcal{A}_1(\mathbf{C}_+)$. Thus, the inclusion of the H^∞ minimizer in the disk algebra forces $\gamma_{\overline{B}H^\infty} = \gamma_{\overline{B}\mathcal{A}^\infty}$. Uniqueness of the disk algebra minimizer follows from the unicity of the H^∞ minimizer. ///

If $R_{\min}$ is continuously differentiable and uniformly positive, $Q_{\min}$ is also continuously differentiable, and so is $Q_{\min}^{-1}$. If $C_{\min}$ is continously differentiable, so is $Q_{\min}^{-1} C_{\min}$. By the comments preceding Theorem 3.9.3, $Q_{\min}^{-1} C_{\min}$ is then Dini-continuous. Consequently, if the center and radius functions are continously differentiable and the radius function uniformly positive at the minimum value of γ, the unique minimizer of γ is in the disk algebra. These unicity arguments will be used in the later chapters.

3.11 Disks with Strict Inequalities

This section remarks upon disks defined by strict inequalities:

$$D(c, r) := \{\phi \in L^\infty(j\mathbf{R}) : |\phi(j\omega) - c(j\omega)| < r(j\omega) \text{ a.e.}\}.$$

First, $D(c, r)$ need not be open. For example, $D(0, 1)$ contains the open unit ball and is contained in its closure:

$$BL^\infty(j\mathbf{R}) \subset D(0, 1) \subset \overline{B}L^\infty(j\mathbf{R}).$$

However,

$$\phi(j\omega) := \frac{\omega}{1 + |\omega|}$$

belongs to $D(0, 1)$ but with $\|\phi\|_\infty = 1$, there is no neighborhood of ϕ that is contained in the open unit ball.

Second, consider what the strict inequalities mean for those $\gamma : L^\infty(j\mathbf{R}) \to \mathbf{R}$ that are continuous with sublevel sets

$$[\gamma \leq \alpha] = \overline{D}(c_\alpha, r_\alpha).$$

We cannot claim that $[\gamma < \alpha]$ is $D(c_\alpha, r_\alpha)$. Instead, $[\gamma < \alpha]$ is an *open set* contained by $D(c_\alpha, r_\alpha)$. In this regard, the following result gives us some control of the strict inequality.

Theorem 3.11.1 *Let* c, $r \in L^\infty(j\mathbf{R})$. *Assume* $r^{-1} \in L^\infty(j\mathbf{R})$. *Let* V *be any nonempty open subset of* $L^\infty(j\mathbf{R})$ *such that* $V \subseteq D(c, r)$. *For any* $\phi \in V$,

$$\|r^{-1}(\phi - c)\|_\infty < 1.$$

Proof: For any $\phi \in V$, the openness of V implies there is an $\epsilon > 0$ such that

$$\phi + \epsilon BL^\infty(j\mathbf{R}) \subset V.$$

Consider the particular element of the open ball:

$$\Delta\phi := \epsilon' \times \operatorname{sgn}(\phi - c)\frac{r}{\|r\|_\infty},$$

where $0 < \epsilon' < \epsilon$ and

$$\operatorname{sgn}(z) := \begin{cases} z/|z| & z \neq 0 \\ 0 & z = 0 \end{cases}.$$

Then $\phi + \Delta\phi \in D(c, r)$ so that

$$r > |\phi + \Delta\phi - c| = |\phi - c| + \epsilon'\frac{r}{\|r\|_\infty} \quad \text{a.e.}$$

Divide by r and take the norm to get

$$1 \geq \|r^{-1}(\phi - c)\|_\infty + \epsilon'\|r\|_\infty^{-1},$$

or that $1 > \|r^{-1}(\phi - c)\|_\infty$. To complete the argument, we need to demonstrate that the preceding argument is not vacuous. That is, $D(c, r)$ does indeed contain an open set. Because r does not "pinch off," $0 < \|r\|_{-\infty}$. Choose any $0 < \eta < \|r\|_{-\infty}$. For any $\phi \in BL^\infty(j\mathbf{R})$

$$\|(\eta\phi + c) - c\|_\infty \leq \eta < r \quad \text{a.e.}$$

Thus, the open set $c + \eta BL^\infty(j\mathbf{R})$ is contained in $D(c, r)$. ///

3.12 The Real Constraint

Because the physical matching circuits have real scattering matrices, we are forced to optimize under the real constraint: $\phi \in L^\infty(j\mathbf{R})$ is real provided

$$\phi_*(j\omega) := \overline{\phi(\overline{j\omega})} = \phi(j\omega) \quad \text{a.e.}$$

For example, a rational scattering function $S(p)$ is real if and only if it is the ratio of two polynomials with real coefficients. If $X \subseteq L^\infty(j\mathbf{R})$, denote

$$\Re X := \{\phi \in X : \phi = \phi_*\}.$$

More generally, $X \subseteq L^\infty(j\mathbf{R})$ is *real invariant* provided

$$X_* := \{\phi_* : \phi \in X\} \subseteq X.$$

Many optimization problems are invariant with respect to real symmetry. For example, set inclusion forces the inequality

$$\inf\{\|\phi - h\|_\infty : h \in \Re H^\infty(\mathbf{C}_+)\} \geq \|\phi - H^\infty(\mathbf{C}_+)\|_\infty.$$

Equality is obtained provided ϕ is also real:

$$\phi = \phi_* \quad \Longrightarrow \quad \|\phi - \Re H^\infty(\mathbf{C}_+)\|_\infty = \|\phi - H^\infty(\mathbf{C}_+)\|_\infty.$$

The invariance of $\Re$ under the best approximation operator is characteristic of its general invariance for optimization problems. It is a measure of Helton's prowess that he revealed such a real invariance for nonlinear minimization [74].

Theorem 3.12.1 *Let $\Gamma : j\mathbf{R}\times\mathbf{C} \to \mathbf{R}_+$ be continuous. Define $\gamma : L^\infty(j\mathbf{R}) \to \mathbf{R}_+ \cup \infty$ by*

$$\gamma(\phi) := \text{ess.sup}\{\Gamma(j\omega, \phi(j\omega)) : \omega \in \mathbf{R}\}.$$

Then γ is well defined. Assume Γ has the following properties:

(a) Γ is real symmetric: $\Gamma(j\omega, z) = \Gamma(-j\omega, \overline{z})$ for all $\omega \in \mathbf{R}$ and all $z \in \mathbf{C}$.
(b) Γ has convex sublevel sets: $\{z \in \mathbf{C} : \Gamma(j\omega, z) \leq c\}$ is convex for all $\omega \in \mathbf{R}$, and all $c \in \mathbf{R}_+$.

Let $X \subseteq L^\infty(j\mathbf{R})$ be nonempty. Assume γ is finite for at least one element of X. Assume that X is convex and real-invariant:

$$X_* = \{\phi_* : \phi \in X\} \subseteq X.$$

Then

$$\inf\{\gamma(\phi) : \phi \in X\} = \inf\{\gamma(\phi) : \phi \in \Re X\} < \infty.$$

Proof: To see that γ is well defined, let $\mathcal{M}(\mathbf{R})$ denote the complex-valued measurable functions on $j\mathbf{R}$. Let id denote the identity map on $j\mathbf{R}$: $\text{id}(j\omega) = j\omega$. Define $\widetilde{\Gamma} : \mathcal{M}(j\mathbf{R}) \to \mathcal{M}(j\mathbf{R})$ by

$$\widetilde{\Gamma}[\phi] = \Gamma \circ (\text{id} \times \phi).$$

The continuity of Γ and the measurability of $\text{id}\times\phi$ force $\widetilde{\Gamma}[\phi]$ to be measurable [110, Theorem 1.7]. Thus, the essential supremum makes sense as an extended real number [110, page 67]. In particular, $\gamma(\phi) = \|\widetilde{\Gamma}[\phi]\|_\infty$. Observe that $\Re X$ is nonempty. Indeed, X is nonempty, so there is a $\phi \in X$. Because X is real invariant, $\phi_* \in X$. Because X is convex, $\phi + \phi_* \in X$, which also belongs to $\Re X$. Thus, $\Re X$ is nonempty. Because $\Re X$ is real invariant, it also is a subset of X. This inclusion $\Re X \subseteq X$ forces the inequality:

$$\begin{aligned}\gamma_X &:= \inf\{\gamma(\phi) : \phi \in X\}\\ &\leq \inf\{\gamma(\phi) : \phi \in \Re X\}\\ &=: \gamma_{\Re X} < \infty.\end{aligned}$$

The assumption that γ is finite forces both infima to be finite. Suppose now that $\phi \in X$. Real invariance forces $\phi_* \in X$. By Assumption (a),

$$\begin{aligned}\gamma(\phi_*) &= \text{ess.sup}\{\Gamma(j\omega, \overline{\phi(-j\omega)} : \omega \in \mathbf{R}\}\\ &= \text{ess.sup}\{\Gamma(-j\omega, \phi(-j\omega) : \omega \in \mathbf{R}\}\\ &= \gamma(\phi)\end{aligned}$$

or that

$$\phi_*(j\omega) \in \{z \in \mathbf{C} : \Gamma(j\omega, z) \le \gamma(h)\} =: S(j\omega, \gamma(\phi)) \quad \text{a.e.}$$

Let $\{\phi_m\} \subset X$ be a minimizing sequence: $\gamma(\phi_m) \to \gamma_X$. Because $\phi_m(j\omega)$ and $\phi_{m,*}(j\omega)$ belong to $S(j\omega, \gamma(\phi_m))$ a.e., Assumption (b) forces the convex combination into the sublevel set:

$$\frac{1}{2}(\phi_m(j\omega) + \phi_{m,*}(j\omega)) \in S(j\omega, \gamma(\phi_m)) \quad \text{a.e.}$$

This inclusion forces $\gamma((\phi_m + \phi_{m,*})/2) \le \gamma(\phi_m)$. Because $(\phi_m + \phi_{m,*})/2$ also belongs to $\Re X$, it follows that $\gamma_{\Re X} \le \gamma_X$. ///

Generally speaking, if the minimizing sets are real invariant and convex, the real constraint may be dropped.

3.13 Inner Functions in the Disk Algebra

A lossless matching N-port has a scattering matrix S that belongs to $U^+(N)$. That is, each $S \in U^+(N)$ is a real function, $S(p) = \overline{S(\overline{p})}$ for $p \in \mathbf{C}_+$ and inner:

$$S(j\omega)^H S(j\omega) = I_N \quad \text{a.e.}$$

Inner functions exhibit a fascinating behavior at the boundary.

Theorem 3.13.1 [34] *Let $E = \{e^{j\theta_m} : m = 1, 2, \ldots\} \subset \mathbf{T}$. Let $\{K_m\}$ be a sequence of nonempty, closed, connected subsets of $\overline{D}$. There exists a singular inner function*

$$S(z) = \exp\left(-\int_{-\pi}^{\pi} \frac{e^{jt} + z}{e^{jt} - z} d\mu(t)\right),$$

where the singular measure μ is supported on E, such that each K_m is the cluster set of $S(re^{j\theta_m})$ as $r \to 1$.

That is, inner functions are set-valued interpolators. In contrast to this boundary behavior, a lumped, lossless N-port forces its scattering function S to be rational. The following result shows that a rational scattering function S must be continuous.

Lemma 3.13.1 *If S belongs to $H^\infty(\mathbf{C}_+, \mathbf{C}^{N\times N})$ and is rational, then S is continuous or $S \in \mathcal{A}_1(\mathbf{C}_+, \mathbf{C}^{N\times N})$.*

Proof: Let $S(p) = H(p)/g(p)$ where $g(p)$ is a real polynomial

$$g(p) = g_0 + g_1 p + \ldots + g_L p^L,$$

that is strict Hurwitz (zeros only in open left half plane $\mathbf{C}_-$) and $H(p)$ is a real $N \times N$ polynomial

$$H(p) = H_0 + H_1 p + \ldots + H_M p^M.$$

Boundedness forces $L \geq M$. Then,

$$\frac{H(p)}{g(p)} = \frac{H_0 + \ldots + H_M p^M}{g_0 + \ldots + g_L p^L} \overset{p\to\infty}{\to} \begin{cases} 0 & L > M \\ H_N/g_N & L = M \end{cases}.$$

Thus, $H(p)/g(p)$ is continuous across $p = \pm j\infty$. Thus, $S(p)$ is continuous at $\pm j\infty$. ///

If S is a scalar-valued inner function, then the converse to Lemma 3.13.1 is true.

Lemma 3.13.2 *If $S \in \mathcal{A}_1(\mathbf{C}_+)$ is inner, then S is rational.*

This result for scalar-valued inner functions generalizes to matrix-valued inner functions. The demonstration requires a review of the matrix-valued H^∞ theory. For $a \in \mathbf{D}$, define the elementary Blaschke factor [87, Equation 4.2]:

$$b_a(z) := \begin{cases} \frac{|a|}{a}\frac{a-z}{1-\bar{a}z} & a \neq 0 \\ z & a = 0 \end{cases}.$$

To get a matrix-valued version, let $P \in \mathbf{C}^{N\times N}$ be an orthogonal projection: $P^2 = P$ and $P^H = P$. The *Blaschke-Potapov elementary factor* associated with a and P is [87, Equation 4.4]:

$$B_{a,P}(z) := I_M + (b_a(z) - 1)P.$$

There are several ways to see that $B_{a,P}$ is inner. Let U be a unitary matrix that diagonalizes P:

$$U^H P U = \begin{bmatrix} I_K & 0 \\ 0 & 0 \end{bmatrix}.$$

Then U shows that $B_{a,P}$ is really a diagonal of inner functions:

$$U^H B_{a,P}(z) U = \begin{bmatrix} b_a(z) I_K & 0 \\ 0 & I_{M-K} \end{bmatrix}.$$

From this, we get the following [87, Equation 4.5]:

$$\det[B_{a,P}(z)] = b_a(z)^{\mathrm{rank}[P]}.$$

Definition 3.13.1 [87, pages 320–321] *The function $B : \mathbf{D} \to \mathbf{C}^{N\times N}$ is called* a left Blaschke-Potapov product *if either B is a constant unitary matrix or there exists a unitary matrix U, a sequence of orthogonal projection matrices $\{P_k : k \in \mathcal{K}\}$, and a sequence $\{z_k : k \in \mathcal{K}\} \subset \mathbf{D}$ such that*

$$\sum_{k\in\mathcal{K}} (1 - |z_k|)\mathrm{trace}[P_k] < \infty$$

and the representation

$$B(z) = \left\{ \overrightarrow{\prod_{k\in\mathcal{K}}} B_{z_k,P_k}(z) \right\} U$$

holds.

Definition 3.13.2 [87, pages 319] *Let $S \in H^\infty(\mathbf{D}, \mathbf{C}^{N\times N})$ be an inner function. S is called* singular *if and only if* $\det[S(z)] \neq 0$ *for all $z \in \mathbf{D}$.*

Theorem 3.13.2 [87, Theorem 4.1] *Let $S \in H^\infty(\mathbf{D}, \mathbf{C}^{N\times N})$ be an inner function. There exists a left Blaschke-Potapov product and a $\mathbf{C}^{N\times N}$-valued singular inner function Ξ such that*

$$S = B\Xi.$$

Moreover, the representation is unique up to a unitary matrix U. If

$$S = B_1\Xi_1 = B_2\Xi_2,$$

then $B_2 = B_1 U$ and $\Xi_2 = U^H \Xi_1$.

Critical for our purposes is that the determinant maps these matrix-valued generalizations of the Blaschke and singular functions to their scalar-valued counterparts.

Theorem 3.13.3 [87, Theorem 4.2] *Let $S \in \overline{B}H^\infty(\mathbf{D}, \mathbf{C}^{N\times N})$.*

(a) $\det[S] \in \overline{B}H^\infty(\mathbf{D})$.
(b) S is inner if and only if $\det[S]$ *is inner.*
(c) S is singular if and only if $\det[S]$ *is singular.*

With these results in place, Lemma 3.13.2 admits the following matrix-valued generalization.

Corollary 3.13.1 *Let $S \in H^\infty(\mathbf{C}_+, \mathbf{C}^{N\times N})$ be an inner function. The following are equivalent:*

(a) $S \in \mathcal{A}_1(\mathbf{C}_+, \mathbf{C}^{N\times N})$.
(b) S is rational.

Proof:
(a⇒b) Lemma 3.5.1 and Assumption (a) shows that $W = S \circ \mathbf{c}^{-1}$ is a continuous inner function in $\mathcal{A}(\mathbf{D}, \mathbf{C}^{N\times N})$. Theorem 3.13.2 shows that $W = B\Xi$ for a left Blaschke-Potapov product B and singular Ξ. Observe that $\det[W] = \det[B]\det[\Xi]$. If W is inner, then $\det[W]$ is inner by Theorem 3.13.3(a). Because W is continuous, $\det[W]$ is continuous and Lemma 3.13.2 forces $\det[W]$ to be rational. Therefore, $\det[W]$ cannot admit the singular factor $\det[\Xi]$. Consequently, W cannot have a singular factor by Theorem 3.13.3(c). Because $\det[W]$ is rational and

$$\det[W] = \det[B] = \prod b_{z_k}^{\mathrm{rank}[P_k]},$$

we see that B must be a *finite* left Blaschke-Potapov product. Consequently, $S = W \circ \mathbf{c}$ is rational. Finally, this shows that S is rational.
(b⇒a) Lemma 3.13.1 puts S in $\mathcal{A}_1(\mathbf{C}_+, \mathbf{C}^{N\times N})$. ///

4

Lossless N-Ports

Figure 4.1 illustrates the Amplifier Matching Problem of Chapter 2. Given an amplifier, an amplifier designer is to find input and output matching circuits that maximize the gain, minimize the noise, and guarantee stability. The

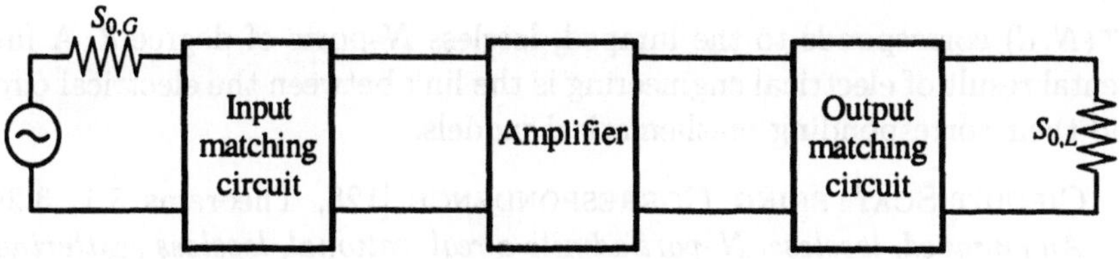

Fig. 4.1. Single amplifier and matching circuits.

matching circuits are the lossless 2-ports of Chapter 1. The scattering matrices of the lossless 2-ports belong to the following set:

$$U^+(2) := \{S \in \Re\overline{B}H^\infty(\mathbf{C}_+, \mathbf{C}^{2\times 2}) : S(j\omega)^H S(j\omega) = I_2 \ \text{a.e.}\}.$$

See the Preface and Chapter 3 for notation. Thus, amplifier matching is an optimization problem on $U^+(2) \times U^+(2)$.

Figure 4.2 shows a generalization to lossless 4-ports. Ports 1 and 2 correspond to the input and output ports. The amplifier is terminated into Ports 3 and 4. In this case, amplifier matching is an optimization problem over the lossless 4-ports or over the set

$$U^+(4) := \{S \in \Re\overline{B}H^\infty(\mathbf{C}_+, \mathbf{C}^{4\times 4}) : S(j\omega)^H S(j\omega) = I_4 \ \text{a.e.}\}.$$

Similarly, we can attach two amplifiers in a lossless 6-port, leaving two ports for input and output. In this case, amplifier matching is an optimization problem over $U^+(6)$. Thus, the lossless N-ports are the fundamental objects of amplifier optimization and their scattering matrices are the subject of this chapter.

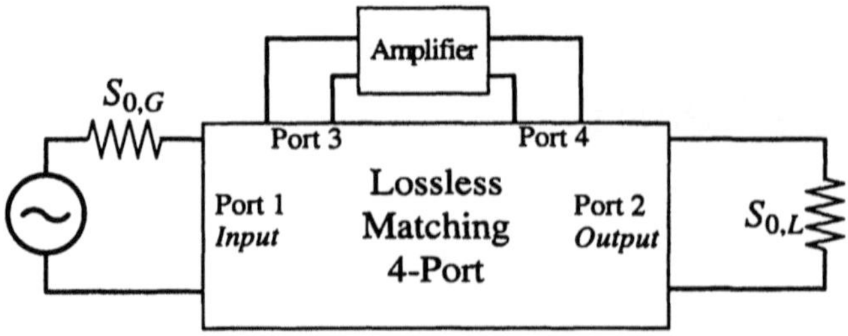

Fig. 4.2. General amplifier matching.

4.1 The Circuit-Scattering Correspondence

This chapter describes the scattering matrices that represent the lossless N-ports. These scattering matrices nest as follows:

$$U^+(N,d) \subset U^+(N,\infty) \subset U^+(N) \subset \Re\overline{B}H^\infty(\mathbf{C}_+, \mathbf{C}^{N\times N}).$$

$U^+(N,d)$ corresponds to the lumped, lossless N-ports of degree d. A fundamental result of electrical engineering is the link between the electrical circuits and their corresponding mathematical models.

> CIRCUIT-SCATTERING CORRESPONDENCE [128, Theorems 3.1, 3.2]
> *Any lumped, lossless, N-port admits a real, rational, lossless scattering matrix $S(p)$. Conversely, any real, rational, lossless scattering matrix $S(p)$ can be mapped to a lumped, lossless N-port.*

Thus, there is an exact correspondence between a lumped, lossless N-port and a scattering matrix $S(p)$ that belongs to $U^+(N,d)$.

Question 1. Does this correspondence extend beyond the lumped, lossless N-ports?

We first consider $U^+(N,d)$ as a subset of $U^+(N)$. What are the limits of all the lumped, lossless N-ports? Define

$$U^+(N,\infty) := \overline{\bigcup_{d\geq 0} U^+(N,d)},$$

where the overline denotes the closure in L^∞. The closure does not pick up any new N-ports. The reason is that any scattering matrix in $U^+(N,\infty)$—as the uniform limit of continuous scattering matrices—must also be continuous on $j\mathbf{R}$. We will see that any scattering matrix in $U^+(N,\infty)$ must be rational or, equivalently, the corresponding N-port must be lumped (Corollary 3.13.1).

Only by considering distributed elements do we get new N-ports, such as the forthcoming model of a transmission line. The transmission line has a scattering matrix in $U^+(2)$ but not in $U^+(2,\infty)$ There are several open questions on how various classes of lumped-distributed N-ports reside in $U^+(N)$.

Consequently, the circuit-scattering correspondence leads us to the following distinctions. First, when we refer to a lossless scattering matrix $S(p)$, the only requirement is that $S(p)$ be a real inner function: $S \in U^+(N)$. Second, when we refer to an N-port, we shall assume that an electrical circuit, however impractical, does exist. One of the great topics of electrical engineering is the mapping between N-ports and their corresponding representation by a scattering matrix. Depending on how one defines an N-port, the claim is that every linear, time-invariant, causal, passive, solvable N-port admits a scattering matrix [128], [73], [92], and Appendix A.

4.2 $U^+(N, d)$

Figure 4.3 illustrates the state-space representation of a passive, lumped N-port. The figure shows that by pulling the d reactive elements and the r lossy

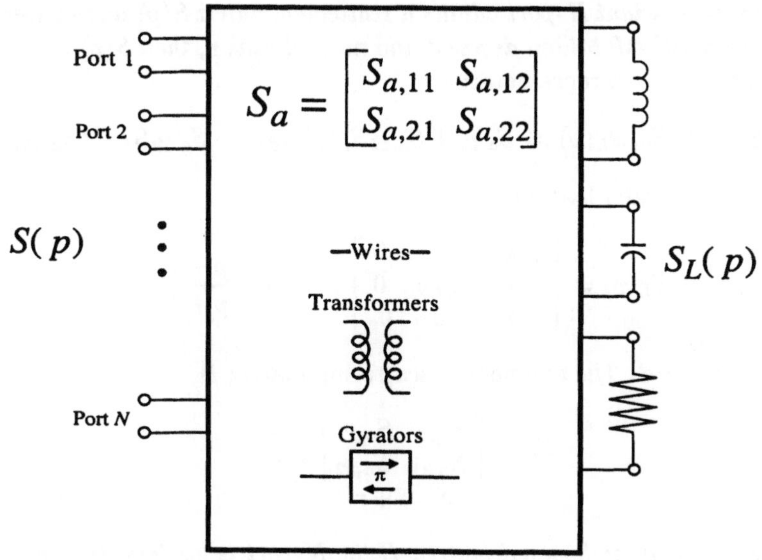

Fig. 4.3. State-space representation of a passive, lumped N-port containing d reactive elements and r resistors.

elements into the *augmented load* $S_L(p)$, what remains is an $(N + d + r)$-port containing only wires, transformers, and gyrators. This *constant* $(N + d + r)$-port has *constant* scattering matrix S_a called the *augmented scattering* matrix. The N-port is obtained by looking into Ports 1, 2, ..., N of the augmented scattering matrix S_a while its remaining ports are terminated in the augmented load $S_L(p)$. That is, $S(p)$ is the image of the augmented load $S_L(p)$ viewed through the augmented scattering matrix S_a. Following the supporting definitions, Theorem 4.2.1 gives the precise statement of this *state-space representation.*

Definition 4.2.1 *Let* $S \in \Re\overline{B}H^\infty(\mathbf{C}_+, \mathbf{C}^{N\times N})$ *and be rational [128, page 91]:*

- *The normal rank*[1] *of* $S(p)$ *is defined as*
$$r[S(p)] := \text{rank}[I_N - S(-p)^T S(p)].$$
- $S(p)$ *is symmetric provided* $S(p) = S(p)^T$.
- *Let the* K *distinct poles of* $S(p)$ *be denoted as* p_k. *Let* $\deg(S(p);p_k)$ *denote the largest order to which* $p = p_k$ *appears in any minor of* $S(p)$. *The Smith-McMillan degree of* $S(p)$ *is*
$$\deg_{\text{SM}}[S(p)] := \sum_{k=1}^{K} \deg(S(p);p_k).$$

Theorem 4.2.1 (State-Space) [128, pages 90–93] *Every lumped, passive, casual, time-invariant* N*-port admits a scattering matrix* $S(p)$ *and conversely. If* $S(p)$ *has Smith-McMillan degree* d *and normal rank* r*, then* $S(p)$ *admits the following* state-space representation*:*

$$S(p) = \mathcal{F}(S_a, S_L; p) := S_{a,11} + S_{a,12}S_L(p)(I_{d+r} - S_{a,22}S_L)^{-1}S_{a,21},$$

where the augmented load *is*

$$S_L(p) = \begin{bmatrix} qI_{N_L} & 0 & 0 \\ 0 & -qI_{N_C} & 0 \\ 0 & 0 & 0_r \end{bmatrix}, \quad \left(q = \frac{p-1}{p+1}\right),$$

and $N_L + N_C = d$. *The* augmented scattering matrix *is*

$$S_a = \begin{bmatrix} S_{a,11} & S_{a,12} \\ S_{a,21} & S_{a,22} \end{bmatrix} \begin{matrix} N \\ d+r \end{matrix}$$
$$\qquad\quad N \quad d+r$$

is a constant, real, orthogonal matrix. If the N*-port is lossless, then* $r = 0$. *If the* N*-port is reciprocal, then* S_a *is symmetric:* $S_a^T = S_a$.

The power of the state-space representation is that the structure of the scattering matrix encodes the structure of the N-port [128, page 91], [99]:

- Any N-port that has $S(p)$ as its scattering matrix contains at least $r[S(p)]$ resistors. Moreover, of all the N-ports that have $S(p)$ as their scattering matrix, there exists at least one having exactly $r[S(p)]$ resistors.

[1] Newcomb [99, page 126] uses the qualifier "normal" to compute the rank of $S(p)$ when p is taken to be a variable. Wohlers [128] applies normal to the rank of $I - S(p)^T S(p)$.

- Let N_L and N_C denote the number of inductors and capacitors in an N-port. For every N-port with scattering matrix $S(p)$, $N_L + N_C \geq \deg_{SM}[S(p)]$. Moreover, of all the N-ports that have $S(p)$ as their scattering matrix, there exists at least one N-port having exactly this many inductors and capacitors.
- The number of gyrators in any N-port must always exceed

$$g[S(p)] := \frac{1}{2}\text{rank}[S(p) - S(p)^T].$$

Wohlers [128, page 92] remarks that only in the case that $g[S(p)] = 0$ or when $S(p)$ is symmetric can an N-port can be found that is reciprocal or contains no gyrators.

This representation also provides a numerically efficient parameterization of the scattering matrices corresponding to the lumped, passive N-ports independent of any circuit topology [2], [3]. This state-space representation is applied to the general amplifier designs in Chapters 8 and 9.

Finally, we observe that Theorem 4.2.1 efficiently delineates the class of lumped, lossless N-ports:

$$U^+(N,d) := \{S \in U^+(N) : \deg_{SM}[S(p)] \leq d\}.$$

By efficiently, we mean that $U^+(N,d)$ is compact.

Theorem 4.2.2 *Let $d \geq 0$. $U^+(N,d)$ is a compact subset of $\mathcal{A}_1(\mathbf{C}_+, \mathbf{C}^{N\times N})$.*

Proof: Let $C(\mathbf{T}, \mathbf{C}^{N\times N})$ denote the continuous functions on the unit circle $\mathbf{T}$. Let $\mathcal{R}_M^L$ denote those rational functions $g^{-1}(q)H(q)$ in $C(\mathbf{T}, \mathbf{C}^{N\times N})$ where $g(q)$ and $H(q)$ are polynomials with degrees $\partial[g] \leq M$ and $\partial[H] \leq L$. The Existence Theorem [29, page 154] shows that $\mathcal{R}_M^L$ is a boundedly compact subset of $C(\mathbf{T}, \mathbf{C}^{N\times N})$. Lemma 3.5.1 shows the Cayley transform preserves compactness. Thus, $\mathcal{R}_M^L \circ \mathbf{c}$ is a boundedly compact subset of $1 \dotplus C(j\mathbf{R}, \mathbf{C}^{N\times N})$. By Lemma 3.4.1, $U^+(N)$ is a closed subset of $L^\infty(j\mathbf{R}, \mathbf{C}^{N\times N})$. The intersection of a closed and bounded set with a boundedly compact set is compact. Thus, $U^+(N) \cap \mathcal{R}_M^L \circ \mathbf{c}$ is a compact subset of $1 \dotplus C(j\mathbf{R}, \mathbf{C}^{N\times N})$. We claim that $U^+(N,d) = U^+(N) \cap \mathcal{R}_d^d \circ \mathbf{c}$. Observe that $\mathcal{R}_d^d \circ \mathbf{c}$ consists of all rational functions with the degree of the numerator and denominator not exceeding d and that are also continuous on $j\mathbf{R}$, including the point at infinity. If $S \in U^+(N) \cap \mathcal{R}_d^d \circ \mathbf{c}$,then $\deg_{SM}[S] \leq d$. This forces S into $U^+(N,d)$. Consequently, $U^+(N,d) \supseteq U^+(N) \cap \mathcal{R}_d^d \circ \mathbf{c}$. For the converse, suppose $S \in U^+(N,d)$. By Corollary 3.13.1, $S \in \mathcal{A}_1(\mathbf{C}_+, \mathbf{C}^{N\times N})$ and thus forces S into $\mathcal{R}_d^d \circ \mathbf{c}$. Thus, $U^+(N,d) \subseteq U^+(N) \cap \mathcal{R}_d^d \circ \mathbf{c}$ and equality must hold. Thus, $U^+(N,d)$ is compact. ///

The forthcoming chapters demonstrate that the amplifier functions are continuous on $U^+(N,d)$. Continuity and compactness force the existence of optimal solutions to the Amplifier Matching Problem on $U^+(N,d)$.

4.3 $U^+(2, \infty)$

A natural generalization drops the constraint on the number of reactive elements in the N-port and asks: *What is the matching set that is obtained as* $\deg_{\mathrm{SM}}[S(p)] \to \infty$*?* Define

$$U^+(N, \infty) := \overline{\bigcup_{d \geq 0} U^+(N, d)},$$

where the overline denotes the closure in L^∞. The physical meaning of $U^+(N, \infty)$ is that it contains the scattering matrices of all lumped, lossless 2-ports. It is worthwhile to ask if the closure has picked up additional circuits. The following result answers negatively.

Theorem 4.3.1 $U^+(N, \infty) = \bigcup_{d \geq 0} U^+(N, d) = U^+(N) \cap \mathcal{A}_1(\mathbf{C}_+)$.

Proof: Let $\{S_m\} \subset \bigcup_{d \geq 0} U^+(N, d)$ converge to $S \in H^\infty(\mathbf{C}_+, \mathbf{C}^{N \times N})$. By Corollary 3.13.1, $\{S_m\} \subset \mathcal{A}_1(\mathbf{C}_+, \mathbf{C}^{N \times N})$ so that $\{S_m\}$ converges uniformly to S. This forces S into $\mathcal{A}_1(\mathbf{C}_+, \mathbf{C}^{N \times N})$. By Corollary 3.13.1, S must be rational. That is, S must belong to some $U^+(N, d)$. ///

The uniform transmission line or unit element (UE) is an excellent example of a nonrational scattering matrix.

Example 4.3.1 (Uniform Transmission Line) *Figure 4.4 illustrates a uniform, lossless transmission line of* characteristic impedance Z_c *and* commensurate *length l. Such a 2-port is called a unit element (UE) and has the chain*

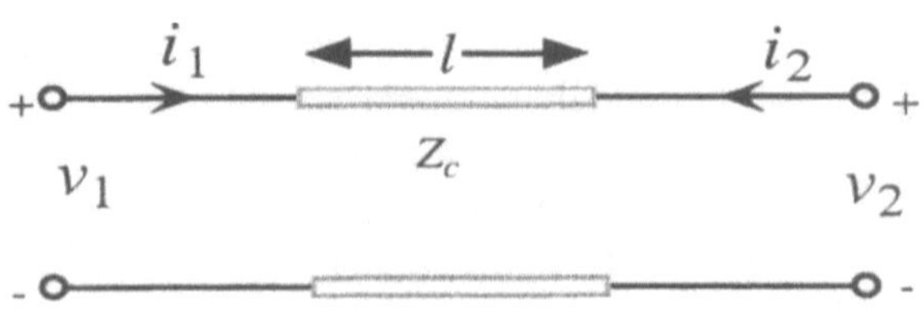

Fig. 4.4. The unit element transmission line.

matrix [6, Equation 8.1]

$$T_{\mathrm{UE}}(p) = \begin{bmatrix} \cosh(\tau p) & Z_c \sinh(\tau p) \\ Y_c \sinh(\tau p) & \cosh(\tau p) \end{bmatrix},$$

where τ is the commensurate one-way delay $\tau = l/c$ determined by the speed of propagation c. The corresponding impedance matrix is

$$Z_{\mathrm{UE}}(p) = \frac{Z_c}{\sinh(\tau p)} \begin{bmatrix} \cosh(\tau p) & 1 \\ 1 & \cosh(\tau p) \end{bmatrix}.$$

The scattering matrix **normalized to** Z_c *is*

$$S_{\mathrm{UE}}(p) = (Z(p) + Z_c I_2)^{-1}(Z(p) - Z_c I_2) = \begin{bmatrix} 0 & e^{-\tau p} \\ e^{-\tau p} & 0 \end{bmatrix}.$$

The transmission line gives rise to two observations. First, $S_{\mathrm{UE}}(j\omega)$ oscillates out to $\pm\infty$, so $S_{\mathrm{UE}}(j\omega)$ cannot be continuous across $\pm\infty$. Consequently, $U^+(2,\infty)$ cannot contain such a transmission line. Second, a physical transmission line *cannot* behave like this near $\pm\infty$. Many electrical engineering books mention only in passing that their models are applicable only for a given frequency band. For example, one rarely sees much discussion that the standard models for the inductor and capacitor are essentially low-frequency models. This holds true even for the standard model of wire. One cannot shine a light in one end of a 100-foot length of copper wire and expect much out of the other end. These model limitations notwithstanding, the circuit-scattering correspondence will be developed using these standard models. The transmission line on the disk is

$$S_{\mathrm{UE}} \circ \mathbf{c}^{-1}(z) = \begin{bmatrix} 0 & \exp\left(-\tau\frac{1+z}{1-z}\right) \\ \exp\left(-\tau\frac{1+z}{1-z}\right) & 0 \end{bmatrix}$$

and is recognizable as the simplest singular inner function analytic on $\mathbf{C}\setminus\{1\}$ [81, pages 66–69]. Figure 4.5 shows the essential singularity of the real part of the (1,2) element of $S_{\mathrm{UE}} \circ \mathbf{c}^{-1}(z)$ as z tends toward the boundary of the unit circle. Thus, it is natural to ask for a characterization of distributed or lumped-distributed N-ports as elements of $U^+(2)$.

4.4 $U^+(N)$

The definition of $U^+(N)$ is strictly mathematical:

$$U^+(N) := \{S \in \Re H^\infty(\mathbf{C}_+, \mathbf{C}^{N\times N}) : S(j\omega)^H S(j\omega) = I_N \text{ a.e.}\}.$$

Given the correspondence between lumped, lossless N-ports and $U^+(N,d)$, it is natural to ask the same question for $U^+(N)$.

Question 2. Does every element in $U^+(N)$ correspond to a lossless N-port?

The preceding sections established that $U^+(N,\infty)$ is a closed subset of $U^+(N)$ that consists of all real, rational inner functions. Physically, $U^+(N,\infty)$ models all the lumped, lossless N-ports—but not the transmission line. It is natural to wonder what subclass of $U^+(2)$ contains the lumped 2-ports and the transmission line. More precisely,

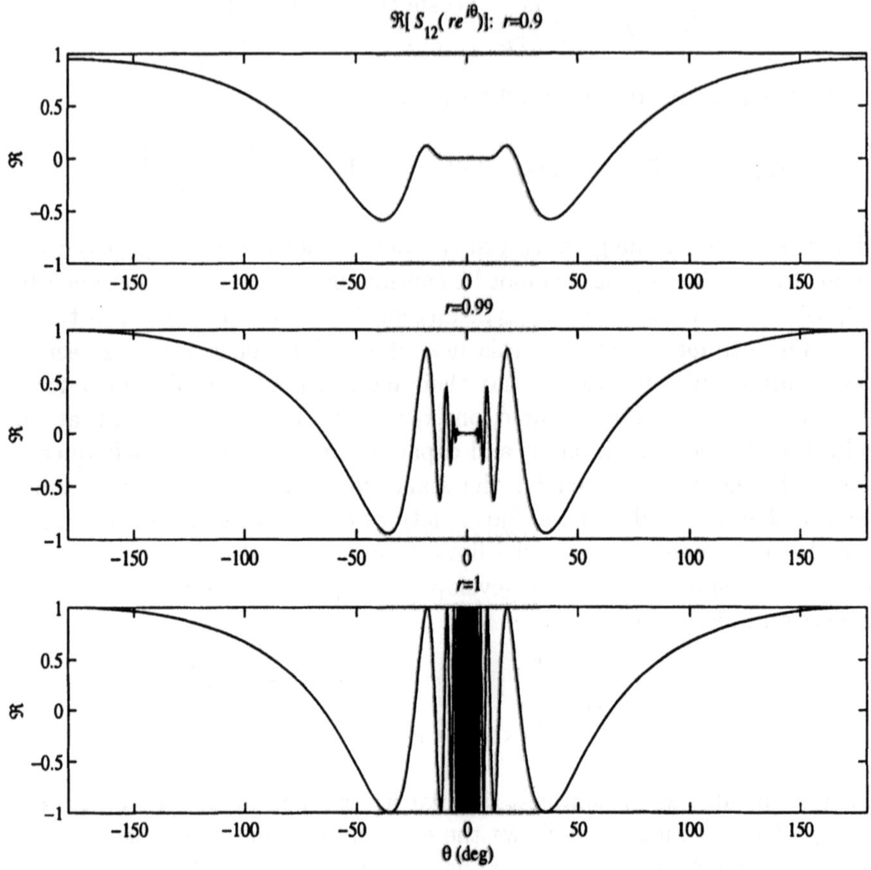

Fig. 4.5. Behavior of $\Re[S_{\mathrm{UE},12} \circ \mathbf{c}^{-1}(z)]$ for $z = re^{i\theta}$ as $r \to 1$.

- What constitutes a lumped-distributed network?
- How do we recognize its scattering matrix?

Wohlers [128, pages 168–172] addressed the first question. Figure 4.6 shows the state-space parameterization of the lossless, lumped-distributed N-ports consisting of N_L inductors, N_C capacitors, and N_U uniform transmission lines or unit elements (UE). The augmented scattering matrix S_a models the non-reactive multiport. The augmented load S_L models the reactive elements.

Let $U_\tau^+(N, N_L, N_C, N_U)$ denote the class of scattering matrices that represent such an N-port with the added restriction that all the UEs have the same electrical length τ. Wohlers [128, pages 168–172] claims that scattering matrices in $U_\tau^+(N, N_L, N_C, N_U)$ exist and have the form

$$S(p) = \mathcal{F}(S_a, S_L; p) = S_{a,11} + S_{a,12}S_L(p)(I_d - S_{a,22}S_L(p))^{-1}S_{a,21},$$

where the augmented scattering matrix

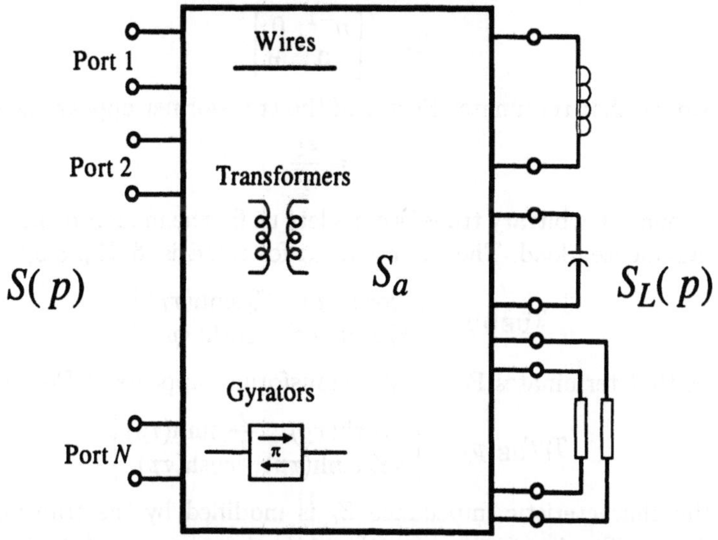

Fig. 4.6. State-space representation of the lossless, lumped-distributed N-ports.

$$S_a = \begin{bmatrix} S_{a,11} & S_{a,12} \\ S_{a,21} & S_{a,22} \end{bmatrix}$$

models a network of wires, transformers, and gyrators and is a constant, real, orthogonal matrix. The reactive elements are modeled by the augmented load

$$S_L(p) = \begin{bmatrix} qI_{N_L} & & 0 \\ 0 & -qI_{N_C} & 0 \\ 0 & 0 & I_{N_U} \otimes \begin{bmatrix} 0 & e^{-\tau p} \\ e^{-\tau p} & 0 \end{bmatrix} \end{bmatrix}.$$

This decomposition assumes the inductors and capacitors are normalized to unit resistance, the UEs all have the same electrical length, and the UEs are normalized to the characteristic impedance Z_{c,n_u}.

There exists some work characterizing these scattering matrices. However, Koga's extentions, which are reported in Wohlers [128, page 173], are false according to Choi [30]. Choi's results raise the question whether $U_\tau^+(N, N_L, N_C, N_U)$ is indeed parameterized by Figure 4.6, or that a Circuit-Scattering Correspondence exists for lumped-distributed N-ports. For our immediate use, we will simply accept that the state-space parameterization maps into—not onto—the scattering matrices of $U_\tau^+(N, N_L, N_C, N_U)$.

Finally, there is an important difference between the lumped representation and the lumped-distributed representation that has practical consequences for the numerical implementation. In the lumped multiports, the transformers sweep over all inductor and capacitor values. The transformer has the chain matrix [6, page 20]:

$$T_t = \begin{bmatrix} n^{-1} & 0 \\ 0 & n \end{bmatrix}.$$

Any load z_L that terminates Port 2 of the transformer appears at Port 1 as

$$z_1 = \frac{z_L}{n^2}.$$

The presence of arbitrary transformers lets us fix the inductors and capacitors in the augmented load. The chain matrix for a UE is [6, Eq. 8.5]:

$$T_{\mathrm{UE}}(p) = \begin{bmatrix} \cosh(\tau p) & Z_c \sinh(\tau p) \\ Y_c \sinh(\tau p) & \cosh(\tau p) \end{bmatrix}.$$

Any UE that terminates Port 2 of a transformer appears at Port 1 as

$$T_t T_{\mathrm{UE}}(p) = \begin{bmatrix} \cosh(\tau p) & \frac{Z_c}{n} \sinh(\tau p) \\ n Y_c \sinh(\tau p) & \cosh(\tau p) \end{bmatrix}.$$

Only the characteristic impedance Z_c is modified by the transformer—not the delay τ. The implication for the state-space representation is that the augmented multiport can only parameterize the characteristic impedance Z_c of each UE.

4.5 The Mathematics of Cascade Loading

Given the technical difficulties with mapping out the Circuit-Scattering Correspondence, it is appropriate to close with Helton's deep contribution to the mathematics of cascade loading. Helton asked [61]:

When does a cascade load have a meaning?

That is, when does the 1-port obtained by terminating a 2-port in a load have a scattering matrix? Formally, if the 2-port has scattering matrix $S(p)$ and the 1-port load has scattering matrix $S_L(p)$, when does the formal cascade

$$S_1(p) = \mathcal{F}_1(S, S_L; p) = S_{11}(p) + S_{12} S_L(p)(1 - S_{22}(p) S_L(p))^{-1} S_{21}(p)$$

make mathematical and physical sense? When the scattering matrices are rational functions, Theorem 4.2.1 and the Circuit-Scattering Correspondence provide an answer. Less clear is when distributed elements are involved. Returning to the basic circuit analysis, distributed 1-ports and 2-ports are modeled by a collection of *partial differential equations.* In this context of partial differential operators, the problem of connecting the 1-port to the 2-port requires a careful treatment of the domains. In general, such a cascade need not make sense and Helton constructs such an example. However, when a density condition is satisfied, the cascade does make sense and inherits properties common to both the 1-port and the 2-port. Consequently, Helton's cascade analysis sets out the mathematical framework for the analysis of the lumped-distributed N-ports.

5

The H^∞ Framework

The preceding chapters set forth the mathematical and engineering background in considerable detail. In contrast, this chapter formulates the Amplifier Matching Problem at a fairly general level. This high-level interlude examines the major features of the matching problem and methods of solution. As such, this chapter maps out the general topography of the matching problem to keep the reader oriented while plowing through the thickets of technical details.

5.1 N-Ports

The collection of all scattering matrices that represent the lossless N-ports is denoted

$$U^+(N) := \{S \in \Re\overline{B}H^\infty(\mathbf{C}_+, \mathbf{C}^{N\times N}) : S(j\omega)^H S(j\omega) = I_N \text{ a.e.}\}$$

and discussed in Chapter 4. H^∞ arises because every bounded, linear, time-invariant, causal, solvable N-port has a scattering matrix $S(p)$ that belongs to $H^\infty(\mathbf{C}_+, \mathbf{C}^{N\times N})$ [73], [99], [128]. Thus, all the 1- and 2-ports in Chapter 1, including the amplifier itself, have scattering matrices that belong to H^∞. Table 5.1 summarizes the nomenclature. As the table shows, the boundary values of a scattering matrix determines whether it is passive, lossless, or active [92, page 272]. The amplifier is active while the matching N-ports are typically lossless. Thus,

The lossless N-ports of $U^+(N)$ are "objects of discussion" of the matching problem.

In Chapters 8 and 9, a general amplifier design is developed using lossless N-ports with $N > 2$. For the classic matching problem discussed in this chapter, the amplifier designer looks for the best amplifier performance obtainable from the lossless 2-ports in $U^+(2)$.

Table 5.1. N-port nomenclature and scattering matrix properties.

N-Port	Scattering Matrix
Linear, time-invariant, causal, bounded	$S \in H^\infty(\mathbf{C}_+, \mathbf{C}^{N\times N})$
Real	$S(p) = \overline{S(\overline{p})}$
Passive	$\|S\|_\infty \leq 1$
Lossless	$S(j\omega)^H S(j\omega) = I_N$ a.e.
Active	$\|S\|_\infty > 1$

5.2 The Amplifier Matching Problem

As Figure 5.1 shows an amplifier—an active 2-port—cascaded with input and output matching 2-ports. The generator delivers the input signal S_i, which is corrupted by the additive input noise N_i to the input 2-port. The signal and

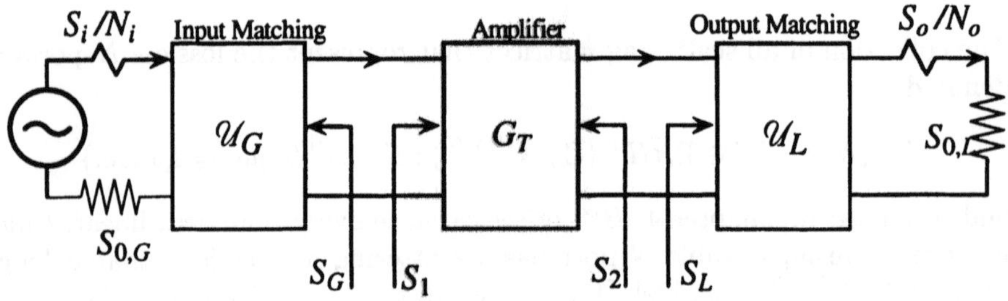

Fig. 5.1. Amplifier, input and output matching 2-port classes, reflectances, signals and noise.

noise are filtered by the input 2-port, amplified by the amplifier that also adds its own self-noise, and then filtered by the output 2-port to give the output signal S_o and the output noise N_o.

The performance of this amplifier circuit is a tradeoff of the signal gain to the increase in noise. The amplifier designer selects the input and output matching 2-ports from specified classes of lossless 2-ports:

- $\mathcal{U}_G \subseteq U^+(2)$ models the input matching 2-ports.
- $\mathcal{U}_L \subseteq U^+(2)$ models the output matching 2-ports.

The amplifier designer searches over the input and output matching 2-ports to maximize performance. If a scattering matrix is selected from the input matching circuits,

$$S_G = \begin{bmatrix} S_{G,11} & S_{G,12} \\ S_{G,21} & S_{G,22} \end{bmatrix} \in \mathcal{U}_G,$$

the reflectance looking into Port 2 of the input matching circuit with Port 1 terminated in $S_{G,0}$ is

$$s_G = S_{G,22} + S_{G,21}S_{G,0}(1 - S_{G,11}S_{G,0})^{-1}S_{G,12} =: \mathcal{F}_2(S_G, S_{G,0}),$$

as discussed in Chapter 1. The collection of all such reflectances is called *the orbit of the generator*:

$$\mathcal{F}_2(\mathcal{U}_G, S_{G,0}) := \{\mathcal{F}_2(S_G, S_{G,0}) : S_G \in \mathcal{U}_G\}.$$

Likewise, if a scattering matrix is selected from the output matching circuits,

$$S_L = \begin{bmatrix} S_{L,11} & S_{L,12} \\ S_{L,21} & S_{L,22} \end{bmatrix} \in \mathcal{U}_L,$$

the reflectance looking into Port 1 of the output matching circuit with Port 2 terminated in $S_{L,0}$ is

$$s_L = S_{L,11} + S_{L,12}S_{L,0}(1 - S_{L,22}S_{L,0})^{-1}S_{L,21} =: \mathcal{F}_1(S_L, S_{L,0}).$$

The collection of all such reflectances is called *the orbit of the load*:

$$\mathcal{F}_1(\mathcal{U}_L, S_{L,0}) := \{\mathcal{F}_1(S_L, S_{L,0}) : S_L \in \mathcal{U}_L\}.$$

The amplifier functions of Chapter 2 are computed from these s_G's and s_L's.
The noise figure is

$$F(s_G) = F_{\min} + \frac{4R_N}{Z_0}\frac{|s_G - s_{\text{opt}}|^2}{(1-|s_G|^2)|1+s_{\text{opt}}|^2}.$$

With the dependence on frequency made explicit,

$$F(s_G; j\omega) = F_{\min}(j\omega) + \frac{4R_N(j\omega)}{Z_0}\frac{|s_G(j\omega) - s_{\text{opt}}(j\omega)|^2}{(1-|s_G(j\omega)|^2)|1+s_{\text{opt}}(j\omega)|^2}.$$

The worst noise figure at any frequency is

$$\|F(s_G)\|_\infty := \text{ess.sup}\{|F(s_G; j\omega)| : \omega \in \mathbf{R}\}.$$

The stability reflectances are

$$s_1(s_L) = S_{11} + S_{12}s_L(1 - S_{22}s_L)^{-1}S_{21},$$

$$s_2(s_G) = S_{22} + S_{21}s_G(1 - S_{11}s_G)^{-1}S_{12}.$$

The worst loss of stability at any frequency is

$$\|S_1(S_L)\|_\infty := \text{ess.sup}\{|S_1(S_L;j\omega)| : \omega \in \mathbf{R}\},$$

$$\|S_2(S_G)\|_\infty := \text{ess.sup}\{|S_2(S_G;j\omega)| : \omega \in \mathbf{R}\}.$$

The transducer power gain is

$$G_T(S_G, S_L) = |S_{21}|^2 \frac{1-|S_G|^2}{|1-S_G S_1(S_L)|^2} \frac{1-|S_L|^2}{|1-S_{22}S_L|^2}.$$

The smallest transducer power gain at any frequency is

$$\|G_T(S_G, S_L)\|_{-\infty} := \text{ess.inf}\{|G_T(S_G, S_L; j\omega)| : \omega \in \mathbf{R}\}.$$

The amplifier designer can optimize these functions over

- the matching circuits $\mathcal{U}_G \times \mathcal{U}_L$, or
- the orbits $\mathcal{F}_2(\mathcal{U}_G, S_{G,0}) \times \mathcal{F}_1(\mathcal{U}_L, S_{L,0})$.

Thus, the Amplifier Matching Problem can be stated as follows:

AMPLIFIER MATCHING PROBLEM. Let $\mathcal{U}_G \times \mathcal{U}_L \subseteq U^+(2) \times U^+(2)$ be given. Find $S_G \in \mathcal{F}_2(\mathcal{U}_G, S_{G,0})$ and $S_L \in \mathcal{F}_1(\mathcal{U}_L, S_{L,0})$ to simultaneously tradeoff the competing objectives:
AMP-1 Maximize the transducer power gain $\|G_T(S_G, S_L)\|_{-\infty}$.
AMP-2 Minimize the noise figure $\|F(S_G)\|_\infty$.
AMP-3 Guarantee stability: $\|S_1(S_L)\|_\infty$, $\|S_2(S_G)\|_\infty \leq 1$.

5.3 H^∞ Engineering Approaches to Amplifier Matching

The preceding section closed with the general statement of the Amplifier Matching Problem. This section lays out general H^∞ approaches that will be explored in the subsequent chapters. The amplifier designer can optimize the amplifier's performance by searching over

- the matching circuits or,
- the orbits.

The matching-circuit approach specifies the topology and elements of the matching circuits. Then S_G and S_L are parameterized by the element values of the matching circuits. Optimization over these element values is the multiobjective optimization approach of Chapter 6. A more general matching-circuit approach drops the topology and element values to optimize over the lumped 2-ports $U^+(2,d)$ or the lumped-distributed 2-ports $U^+_\tau(2, N_L, N_C, N_U)$ of Chapter 4. This multiobjective optimization is developed in Chapters 8 and 9.

Optimization over the orbits leads to the H^∞ engineering approach developed in Chapter 7. From Chapter 4,

$$U^+(2,\infty) = \bigcup_{d\geq 0} U^+(2,d)$$

contains all the lumped, lossless 2-ports. When the reflectances of the generator and the load are zero, Darlington's Theorem identifies the orbit generated by $U^+(2,\infty)$ as a dense subset in disk algebra.

Theorem 5.3.1 (Darlington) [6, page 98] *Let the reflectance of the generator and load be zero:* $S_{G,0} = 0$, $S_{L,0} = 0$. *Both*

$$\mathcal{F}_2(U^+(2,\infty),0) \quad \text{and} \quad \mathcal{F}_1(U^+(2,\infty),0)$$

are strictly dense subsets of $\Re\overline{B}\mathcal{A}_1(\mathbf{C}_+)$.

That is, the S_G's and S_L's available[1] to the amplifier designer matching from the lumped, lossless 2-ports is a subset that is dense in the real, closed unit ball of the disk algebra. Consequently, the amplifier's performance obtained by lumped, lossless matching is bounded by the amplifier's performance obtained by matching over the disk algebra:

AMPLIFIER MATCHING OVER THE DISK ALGEBRA. Assume $S_{G,0} = S_{L,0} = 0$. Find S_G, $S_L \in \Re\overline{B}\mathcal{A}_1(\mathbf{C}_+)$ to simultaneously tradeoff the competing objectives:
AMP-1 Maximize the transducer power gain: $\|G_T(S_G, S_L)\|_{-\infty}$.
AMP-2 Minimize the noise figure $\|F(S_G)\|_\infty$.
AMP-3 Guarantee stability: $\|S_1(S_L)\|_\infty$, $\|S_2(S_G)\|_\infty \leq 1$.

Thus, the *matching circuits have dropped away while the resulting optima represent the best possible performance that the amplifier could ever attain.* However, all this nomenclature would be a baroque exercise in re-naming honest engineering functions were it not for the computational power embedded in the H^∞ approach. For example, the noise figure F is a function of S_G only. Minimizing the noise figure admits the following bounds:

$$\begin{aligned}
&\inf\{\|F(S_G)\|_\infty : S_G \in \mathcal{F}_2(U^+(2,d), S_{G,0} = 0)\} \\
&\geq \inf\{\|F(S_G)\|_\infty : S_G \in \mathcal{F}_2(U^+(2,\infty), S_{G,0} = 0)\} \\
&\geq \inf\{\|F(S_G)\|_\infty : S_G \in \Re\overline{B}\mathcal{A}_1(\mathbf{C}_+)\} \\
&\geq \inf\{\|F(S_G)\|_\infty : S_G \in \overline{B}H^\infty(\mathbf{C}_+)\}.
\end{aligned}$$

Consequently, the amplifier's performance admits the following bound:

AMPLIFIER MATCHING OVER H^∞. Assume $S_{G,0} = S_{L,0} = 0$. Find S_G, $S_L \in \overline{B}H^\infty(\mathbf{C}_+)$ to simultaneously tradeoff the competing objectives:

[1] See Arov [5] and Douglas and Helton [39], [40] for the limits of Darlington synthesis.

AMP-1 Maximize the transducer power gain: $\|G_T(S_G, S_L)\|_{-\infty}$.
AMP-2 Minimize the noise figure $\|F(S_G)\|_\infty$.
AMP-3 Guarantee stability: $\|S_1(S_L)\|_\infty$, $\|S_2(S_G)\|_\infty \leq 1$.

Under sufficient smoothness conditions, the H^∞ optima are also optima for the disk algebra. That is, the amplifier's performance over all the lumped, lossless matching circuits is bounded by the H^∞ bound, this bound is tight; and, most importantly, this bound is *computable by Nehari's Theorem.* This completes the high-level conversion of the Amplifier Matching Problem into the various H^∞ optimization problems.

This H^∞ computation has practical design utility for the amplifier designer. Typically, the amplifier designer is working from the amplifier's data sheet. The data sheet lists the amplifier's scattering matrix and noise figure measurements. The H^∞ computations maps these measurements into gain and noise bounds. These bounds are the best that can be obtained by *any* matching circuits. And knowing the global optimum of any optimization gives the amplifier designer confidence in the design process. For example, suppose that the amplifier designer computed that the best gain was 9 dB and the smallest noise figure was 1.2 dB. Suppose that a simple matching circuit was found that delivered 8.5 dB of gain at a noise figure of 1.4 dB. Knowing the global optima lets the amplifier designer assess the quality of this near-optimal design. However, there are technical problems lurking in the details of the H^∞ computation.

First, Nehari's Theorem optimizes over all of $H^\infty(\mathbf{C}_+)$ whereas the amplifier functions are optimized over $\overline{B}H^\infty(\mathbf{C}_+)$. The obvious solution—adding the additional objective $\|S_G\|_\infty$, $\|S_L\|_\infty \leq 1$—is confounded by the amplifier functions not being defined on all of H^∞. Handling the domains of the amplifier functions occupies a significant portion of the analysis of Chapter 7.

Second, Nehari's Theorem optimizes functions whose sublevel sets are disks. Even when the amplifier functions belong to this class, all the Amplifier Matching Problems use the word "simultaneously." This is equivalent to intersecting multiple sublevel sets and asking if that intersection contains any function in $H^\infty(\mathbf{C}_+)$. Currently, there is no simple (i.e., eigenvalue) solution to this problem. As a simple-minded solution, Chapter 7 starts by optimizing each amplifier function separately. This produces separate H^∞ performance bounds that are useful for numerical multiobjective optimizers. In particular, we can relax these bounds to suboptimal bounds and ask if the resulting sublevel sets have a nonempty intersection. If the intersection is nonempty, we can fit a disk in the intersection and apply Nehari's Theorem to see if the disk contains an H^∞ element. If the intersection of the disk with H^∞ is nonempty, matching circuits exists that can realize these suboptimal bounds. This iterative design process is developed in Chapter 7.

Third, the amplifier functions are built from sampled data whereas Nehari's Theorem requires knowledge of the function on the entire $j\mathbf{R}$ frequency axis. Currently, we use splines to "fill in" the amplifier functions for our numerical computations. The larger problem is the reconstruction of a function on the frequency axis from a finite number of noisy samples. This reconstruction problem is a major topic in electrical engineering. The most interesting observation is that extrapolation off the frequency band (i.e., extending the function to $\pm j\infty$) has a significant effect on the H^∞ computations. Ignoring this effect is part of the larger "Fundamental Mistake of H^∞ Control" developed by Helton [74, Section 10.4]. Chapter 10 makes explicit the research opportunities in this practical implementation of Nehari's Theorem.

Finally, it is interesting to compare the matching-circuit approach with the orbit approach. This comparison is arises in Chapters 8 and 9, when the Amplifier Matching Problem is extended to multiple amplifiers. The matching-circuit approach readily generalizes because of the State-Space representation of Chapter 4. However, the orbit approach and the corresponding generalization of Darlington's Theorem—one that would bring Nehari's Theorem to bear on matching multiple amplifiers—remain open problems. Chapter 10 offers further development of these research topics,

5.4 Why Lossless Matching?

As a final high-level look at the matching problem, we ask

Why match using the lossless 2-ports?

Recall that the scattering matrix $S(p)$ of a lossless 2-port is unitary:

$$S(j\omega)^H S(j\omega) = \begin{bmatrix} 1 & 0 \\ 0 & 1 \end{bmatrix}$$

almost everywhere. A much larger class are the passive 2-ports:

$$S(j\omega)^H S(j\omega) \leq \begin{bmatrix} 1 & 0 \\ 0 & 1 \end{bmatrix}.$$

Why restrict the amplifier matching circuits to the lossless 2-ports? Section 2.2 listed the engineering objections to lossy matching—gain is not improved but noise is increased. However, these objections are apparently empirical. What does a full-scale search over all the passive 2-ports yield? As part of an answer, Ball and Helton obtained a fundamental mathematical result showing that gain is not improved by lossy matching [10]:

Can one build a better amplifier (more gain) with passive connecting circuity than with lossless connecting circuity? This paper [10] proves that the answer is no.

However, the actual wideband amplifiers in Chapter 6 force us to return to this issue. These wideband amplifiers are barely stable. Moreover, such thin stability margins have been observed in several other wideband amplifiers. In all cases, boosting the gain or lowering noise further reduces the stability margin. In Chapter 9, stability is controlled by extending the class of lossless matching circuits to include feedback topologies, as illustrated in Figure 4.2. The stability problem of wideband amplifiers leads to a multiobjective Ball-Helton question: Given a wideband *unstable* amplifier, can one build a better amplifier (stable, more gain, less noise) with passive circuity than with lossless circuity? In Chapter 10, we revisit this question again when considering passive matching.

6

Amplifier Matching Examples

The Amplifier Matching Problem is a problem of multiobjective trade-offs. As Figure 6.1 illustrates, the goal is to amplify the input signal S_i to a larger

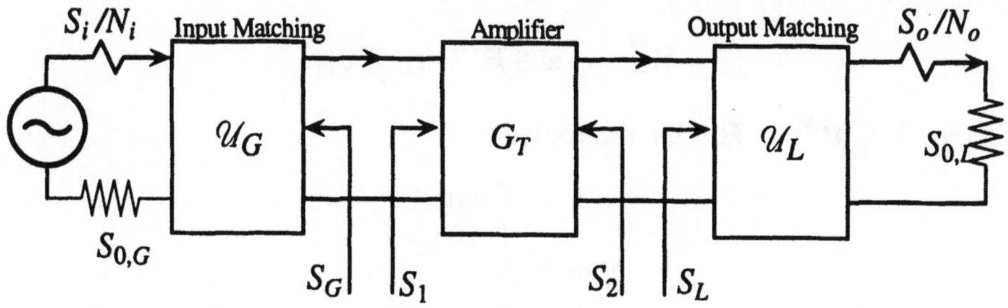

Fig. 6.1. Amplifier, classes of matching 2-ports, signals, noises, and reflectances.

output signal S_o, not to amplify the input noise N_i to a much larger output noise N_0, and not to have the entire circuit become unstable and blow up. To achieve these three competing objectives, the amplifier designer optimizes over classes of input and output matching circuits to simultaneously tradeoff

AMP-1 Maximizing the transducer power gain G_T.
AMP-2 Minimizing the noise figure F.
AMP-3 Guaranteeing stability: $|S_1| \leq 1$, $|S_2| \leq 1$.

Optimizing all three objectives simultaneously is a problem in multiobjective optimization. Section 6.1 reviews the general concepts of multiobjective optimization. Section 6.2 specializes a multiobjective minimizer—*Goal Attainment Method*—to the Amplifier Matching Problem. Section 6.3 applies the Goal Attainment Method to amplifier matching using a variety of matching circuits to acquaint us with the matching problem. Matching is done at a single frequency (i.e., narrowband) so stability problems are not immediately apparent. Section 6.4 undertakes the matching of *wideband* amplifiers using common matching circuits. These amplifiers will be our canonical examples. The matching circuits trade off gain and noise as before but have thin stability margins. We will contend with this delicate stability margin throughout this monograph. Section 6.5 highlights recent research on multiobjective optimization. With these examples and ideas in place, Section 6.6 sets out the H^∞ multiobjective approach to wideband amplifier matching and several research topics.

6.1 Multiobjective Optimization

Introduce the *partial order* on $\mathbf{R}^N$ by declaring

$$\mathbf{u} \leq \mathbf{v} \iff \mathbf{v} - \mathbf{u} \in \mathbf{R}_+^N,$$

where $\mathbf{R}_+^N$ denotes the closed positive orthant

$$\mathbf{R}_+^N := \{\mathbf{x} \in \mathbf{R}^N : x_n \geq 0\}.$$

Let $\gamma : X \subseteq \mathbf{R}^M \to \mathbf{R}^N$ be a mapping

$$\gamma(\mathbf{x}) := \begin{bmatrix} \gamma_1(\mathbf{x}) \\ \gamma_2(\mathbf{x}) \\ \vdots \\ \gamma_M(\mathbf{x}) \end{bmatrix}.$$

Each component γ_m is called an *objective function* so that γ is called a *multiobjective function.* We want to solve the vector-valued minimization of γ on X. The notion of a "minimizer" has been nicely generalized by Boyd and Vandenberghe [19, page 20]. Denote the *image of X under γ by*

$$\gamma(X) := \{\gamma(\mathbf{x}) : \mathbf{x} \in X\}.$$

Any $\gamma(\mathbf{x}) \in \gamma(X)$ is called a *minimum element* of $\gamma(X)$ provided

$$\gamma(\mathbf{x}) \leq \gamma(\mathbf{y})$$

for all $\mathbf{y} \in X$. In terms of the image of X under γ, we denote the preceding condition as follows:

$$\gamma(\mathbf{x}) \leq \gamma(X).$$

Figure 6.2 shows that $\gamma(\mathbf{x})$ is a minimum element is equivalent to

$$\gamma(X) \subseteq \gamma(\mathbf{x}) + \mathbf{R}_+^N.$$

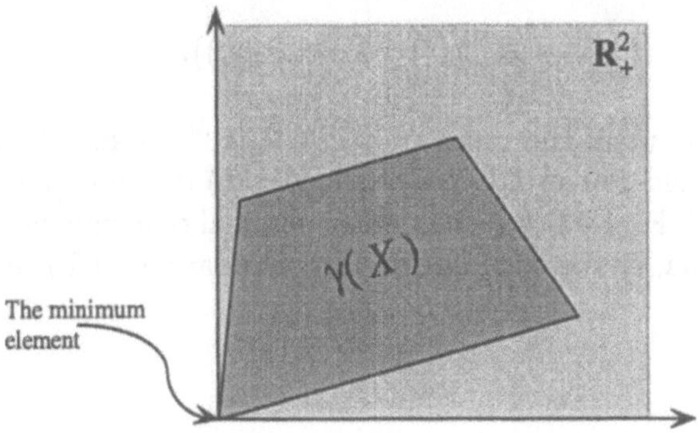

Fig. 6.2. The minimum element of $\gamma(X)$.

Not all sets admit a minimum element. More commonly, we look for the *minimal elements* as illustrated in Figure 6.3. Any $\gamma(\mathbf{x}) \in \gamma(X)$ is a minimal

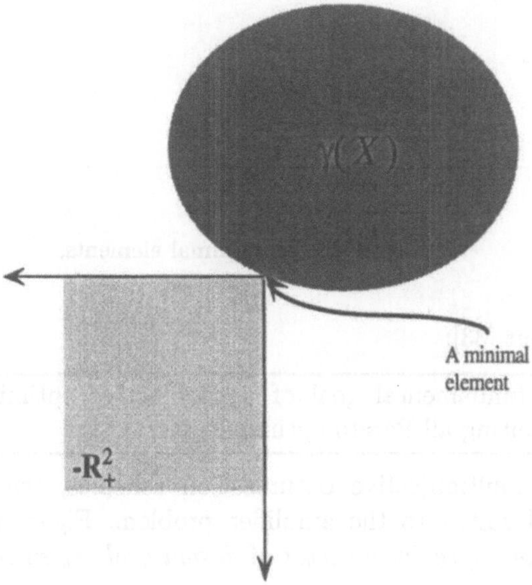

Fig. 6.3. A minimal element of $\gamma(X)$.

element of $\gamma(X)$ provided [19, page 21]

$$\gamma(\mathbf{y}) \leq \gamma(\mathbf{x}) \implies \gamma(\mathbf{y}) = \gamma(\mathbf{x}).$$

Figure 6.3 shows this is equivalent[1] to

$$(\gamma(\mathbf{x}) - \mathbf{R}_+^N) \bigcap \gamma(X) = \{\gamma(\mathbf{x})\}.$$

These definitions occur in the range of $\gamma : X \subseteq \mathbf{R}^M \to \mathbf{R}^N$. In *domain* of γ, any $\mathbf{x} \in X$ is called *Pareto optimal* provided $\gamma(\mathbf{x})$ is a minimal element of $\gamma(X)$ [19, page 102]. Figure 6.4 illustrates all minimal elements, or the images of the Pareto optima, as the dark line on the boundary of $\gamma(X)$. As Das and

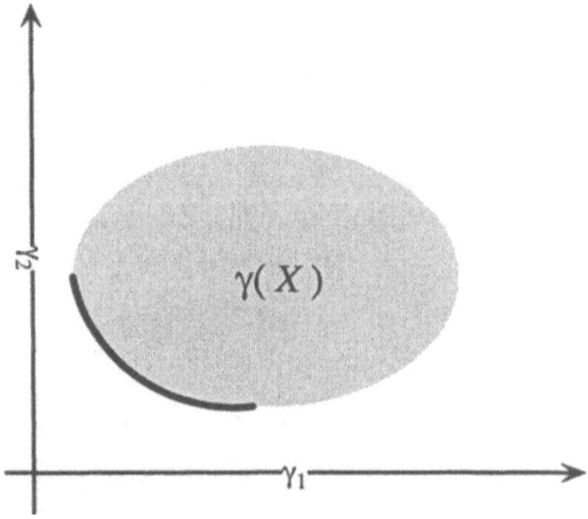

Fig. 6.4. The minimal elements.

Dennis point out [33]:

> The fundamental goal of mutiobjective optimization is computing all Pareto optima.

Of the many multiobjective optimization schemes, the *Goal Attainment Method* is well suited to the amplifier problem. Figure 6.5 illustrates the method. The user specifies a vector of *design goals* γ_u such that

$$\gamma_u \leq \gamma(X)$$

and a vector of non-negative *weights* $\mathbf{w}$. The minimizer attempts to shoot from γ_u along the direction of the weight vector $\mathbf{w}$ and hit the boundary of $\gamma(X)$. The stopping point, if it exists, may be a minimal element of $\gamma(X)$.

[1] More restrictive is the notion of *weak minimizers* [27]: $(\gamma(\mathbf{x}) - \mathrm{int}[\mathbf{R}_+^N]) \cap \gamma(X) = \emptyset$.

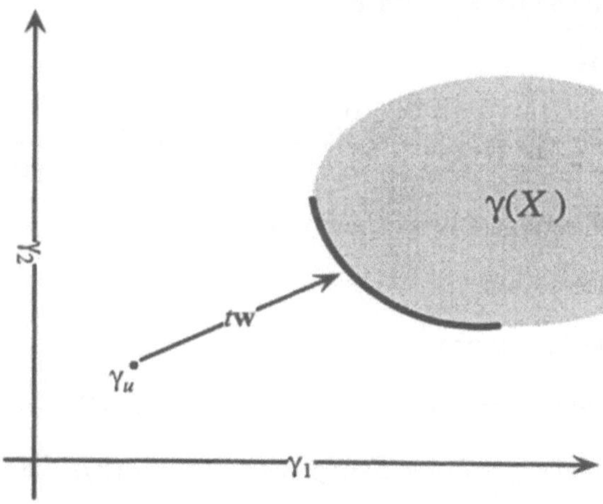

Fig. 6.5. The Goal Attainment Method.

THE GOAL ATTAINMENT METHOD [21]. Given $\gamma : X \subseteq \mathbf{R}^M \to \mathbf{R}^N$. Select a design goal $\gamma_u \leq \gamma(X)$. Select a weight vector $\mathbf{w} \in \mathbf{R}^N_+$.

$$\text{minimize}\{t \in \mathbf{R}\}$$

subject to $\mathbf{x} \in X$ and

$$\gamma(\mathbf{x}) - t\mathbf{w} \leq \gamma_u.$$

6.2 Multiobjective Optimization for Amplifiers

To apply the Goal Attainment Method to the amplifier matching, we need the circuit background of Chapter 1, the amplifier functions of Chapter 2, the class of matching circuits of Chapter 4, and the wideband norms of Chapter 5. In what follows, the matching circuit topologies are fixed. Only the values of the reactive elements vary. Let $\mathbf{x} \in \mathbf{R}^M$ be a vector parameterizing these element values. Realistic constraints are $\mathbf{x}_l \leq \mathbf{x} \leq \mathbf{x}_u$. Denote the parameterized reflectances as $S_G(\mathbf{x})$ and $S_L(\mathbf{x})$.

AMP-1 Maximize the transducer power gain.

The transducer power gain is

$$G_T(S_G(\mathbf{x}), S_L(\mathbf{x})) = |S_{21}|^2 \frac{1 - |S_G(\mathbf{x})|^2}{|1 - S_G(\mathbf{x})S_1(S_L(\mathbf{x}))|^2} \frac{1 - |S_L(\mathbf{x})|^2}{|1 - S_{22}S_L(\mathbf{x})|^2}.$$

The smallest transducer power gain at any frequency is $\|G_T(S_G(\mathbf{x}), S_L(\mathbf{x}))\|_{-\infty}$.

AMP-2 Minimize the noise figure F.

The noise figure is

$$F(S_G(\mathbf{x})) = F_{\min} + \frac{4R_N}{Z_0}\frac{|S_G(\mathbf{x}) - S_{\text{opt}}|^2}{(1-|S_G(\mathbf{x})|^2)|1+S_{\text{opt}}|^2}.$$

The worst noise figure at any frequency is $\|F(S_G(\mathbf{x}))\|_\infty$.

AMP-3 Guarantee stability.

The stability reflectances are

$$S_1(S_L(\mathbf{x})) = S_{11} + S_{12}S_L(\mathbf{x})(1 - S_{22}S_L(\mathbf{x}))^{-1}S_{21},$$

$$S_2(S_G(\mathbf{x})) = S_{22} + S_{21}S_G(\mathbf{x})(1 - S_{11}S_G(\mathbf{x}))^{-1}S_{12}.$$

The worst stability at any frequency is $\|S_1(S_L(\mathbf{x}))\|_\infty$ and $\|S_2(S_G(\mathbf{x}))\|_\infty$.

The Amplifier Matching Problem is the computation of the Pareto optima of

$$\gamma(\mathbf{x}) = \begin{bmatrix} -\|G_T(S_G(\mathbf{x}), S_L(\mathbf{x}))\|_{-\infty} \\ \|F(S_G(\mathbf{x}))\|_\infty \end{bmatrix},$$

where $\mathbf{x} \in \mathbf{R}^M$ belongs to the constraint set

$$\left\{\mathbf{x} \in \mathbf{R}^M : \begin{bmatrix} \|S_1(S_L(\mathbf{x}))\|_\infty \\ \|S_2(S_G(\mathbf{x}))\|_\infty \end{bmatrix} \le \begin{bmatrix} 1 \\ 1 \end{bmatrix}; \mathbf{x}_l \le \mathbf{x} \le \mathbf{x}_u \right\}.$$

More generally, the amplifier designer specifies gain and noise goals:

$$G_T \ge G_{T,u},$$

$$F \le F_u,$$

and stability constraints:

$$|S_1| \le S_{1,u},$$

$$|S_2| \le S_{2,u}.$$

The Goal Attainment Method lets us pull the stability functions out of the constraint set by setting associated weights to zero. Consequently, the matching problem may be implemented as

$$\min\{t \in \mathbf{R}\}$$

subject to $\mathbf{x} \in [\mathbf{x}_l, \mathbf{x}_u]$ and

$$\gamma(\mathbf{x}) - t\mathbf{w} = \begin{bmatrix} -\|G_T(S_G(\mathbf{x}), S_L(\mathbf{x}))\|_{-\infty} \\ \|F(S_G(\mathbf{x}))\|_\infty \\ \|S_1(S_L(\mathbf{x}))\|_\infty \\ \|S_2(S_G(\mathbf{x}))\|_\infty \end{bmatrix} - t\begin{bmatrix} w_G \\ w_F \\ 0 \\ 0 \end{bmatrix} \le \begin{bmatrix} -G_{T,u} \\ F_u \\ S_{1,u} \\ S_{2,u} \end{bmatrix} = \gamma_u.$$

That is, the Goal Attainment Method simultaneously maximizes gain and minimizes noise while enforcing stability. The following sections use this method to match a variety of amplifiers.

These examples build intuition and acquaint us with the problems faced by our amplifier designer and the limitations of sampled data. Regarding the last point, the gain, noise, and stability functions are measured at only a finite number of sample frequencies. We cannot compute these functions outside the frequency band nor can we compute these functions between the frequency samples. Thus, claims of optimal gain, noise, and stability hold only at the sample frequencies. If $|S_1|$ is nearly 1 on the sample frequencies, there may be instability in the physical amplifier not observable on frequency samples. Chapter 10 offers a discussion of this sampling problem as a topic of research.

6.3 Single-Frequency Matching Examples

This section presents the amplifier performance with matching circuits of increasing complexity. To see the objective functions, the stability functions are suppressed and we plot only the gain and noise:

$$\gamma(\mathbf{x}) := \begin{bmatrix} +\|G_T(S_G(\mathbf{x}), S_L(\mathbf{x}))\|_{-\infty} \\ \|F(S_G(\mathbf{x}))\|_{\infty} \end{bmatrix}.$$

Call the collection $\gamma(X)$ of these vectors the *Gain-Noise image* of the amplifier:

$$\gamma(X) := \{\gamma(\mathbf{x}) : \mathbf{x} \in X\},$$

where X is the subset of $\mathbf{R}^M$ of admissible element values. Figure 6.6 illustrates a typical Gain-Noise image in the Gain-Noise plane. The horizontal axis

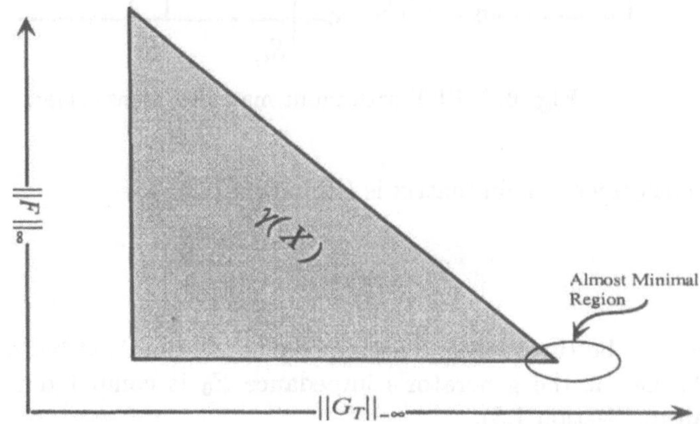

Fig. 6.6. Generic amplifier's Gain-Noise image $\gamma(X)$ in the Gain-Noise plane.

measures increasing gain by moving to the right. The vertical axis measures decreasing noise by moving from top to bottom. On this plot, the "lower right" is the multiobjective target region. The roughly triangular shape is common across a variety of amplifiers and argues that either a minimum element exists, or that the minimal elements are confined to a small region at the lower right tip $\gamma(X)$. For brevity, refer to this tip as the *almost minimal region.*

For single-frequency matching, the amplifier is a model of a Field-Effect Transistor (FET) biased for minimum noise figure as in Section 1.13. The amplifier's scattering matrix at 4 GHz is [106, Example 11.5]:

$$S = \begin{bmatrix} 0.6\angle-60^\circ & 0.05\angle 26^\circ \\ 1.9\angle 81^\circ & 0.5\angle-60^\circ \end{bmatrix}.$$

The noise parameters as in Chapter 2 are: Z_0=50 ohms, F_{min}=1.6 dB, $S_{\text{opt}} = 0.62\angle 100^\circ$, R_N=20 ohms. The Goal Attainment Method was run with the weight vector $\mathbf{w}$ giving equal weight to gain and noise: $w_G = 1$, $w_F = 1$. The design goal γ_u will always have the least constraints on the stability: $S_{1,u}$, $S_{2,u} = 1$. We design for minimum noise: $F_u = 1$. The design gain G_u will be typically set to 10 dB.

6.3.1 Input Transformer Matching

Figure 6.7 shows an input transformer is used to match the amplifier. There is no output matching circuit [67]. Using the chain formalism of Chapter 1,

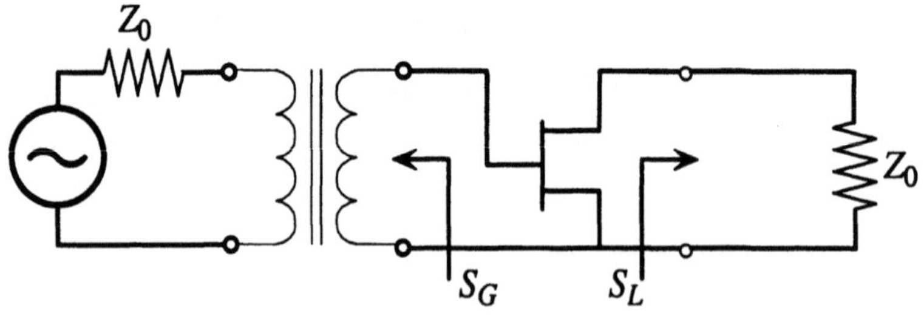

Fig. 6.7. FET with input matching transformer.

the transformer's chain matrix is (Equation 1.2):

$$T_{\text{transformer}} = \begin{bmatrix} n^{-1} & 0 \\ 0 & n \end{bmatrix},$$

where n is the turns ratio. The impedance Z_G obtained by looking into the transformer at the generator's impedance Z_0 is computed using the chain formalism (Section 1.3):

$$Z_G = \mathcal{G}(T_{\text{transformer}}; Z_0) = n^{-2} Z_0.$$

With $Z_0 = 50$ ohms, we can vary Z_G from 1 to 1000 ohms by varying n. The input and output reflectances are computed from Chapter 1:

$$S_G = \frac{Z_G - Z_0}{Z_G + Z_0} \quad \text{and} \quad S_L = 0.$$

For brevity, the amplifier's Gain-Noise function is written as

$$\gamma(Z_G) = \begin{bmatrix} G_T(Z_G) \\ F(Z_G) \end{bmatrix}.$$

At a single frequency, the norms $\| \circ \|_{-\infty}$ and $\| \circ \|_{\infty}$ can be suppressed. Figure 6.8 presents the amplifier's Gain-Noise image in the Gain-Noise plane. As Z_G varies from 1 to 1000 ohms, $\gamma(Z_G)$ sweeps out a smooth curve. Because

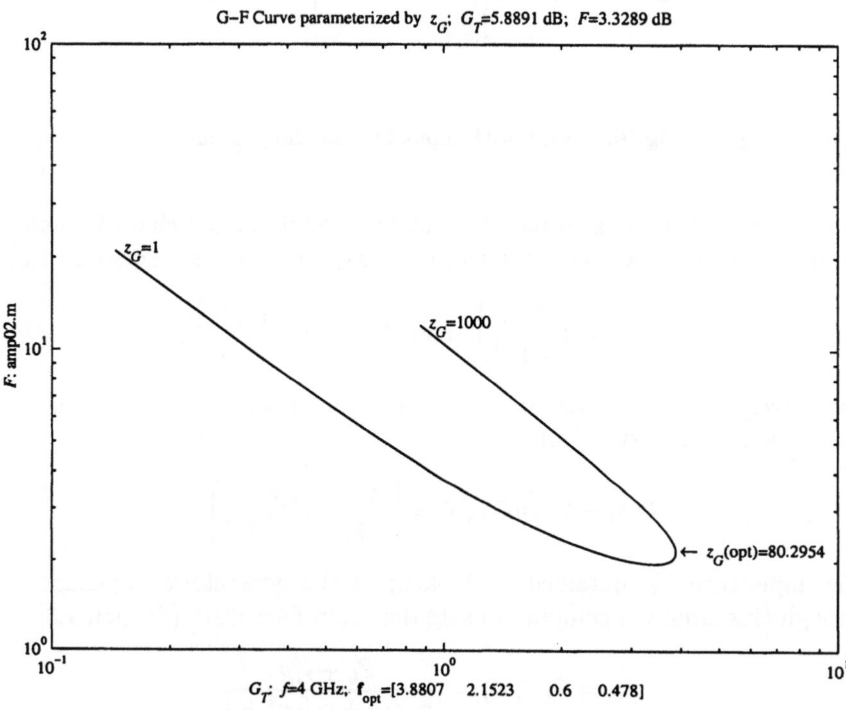

Fig. 6.8. Gain-Noise as a function of Z_G.

the Gain-Noise image is a smooth curve, a minimum element does not exist. Observe also that the Pareto set is confined to a relatively small portion of the curve. The Goal Attainment Method obtained a numerical Pareto optimum noted in the plot. The corresponding gain (dB) and noise figure (dB) are reported in the title of the plot. Not reported on the plot is that no loss of stability occurred over the variation of Z_G. The x-axis label reports S_1 and S_2 in the last two entries of the vector. The first two entries are the gain and noise in absolute value and are also listed in the title in dB.

6.3.2 Input LC Matching

Figure 6.9 shows the input matching circuit is a series inductor and a shunt capacitor. The input matching circuits are generated by sweeping over inductor $L = l_1$ and capacitor $C = c_1$ values. There is no output matching circuit. Thus,

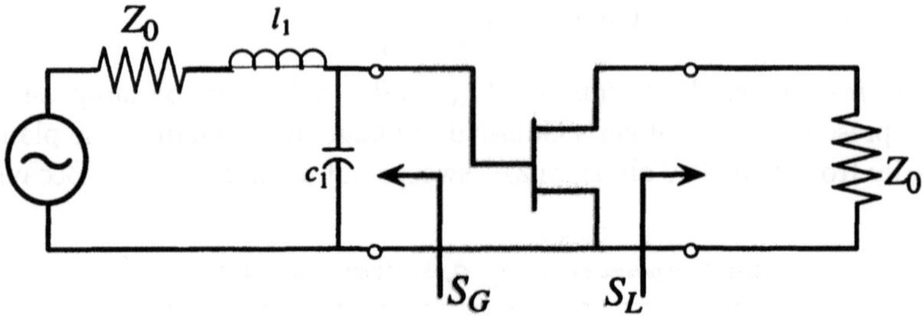

Fig. 6.9. FET with input LC matching circuit.

we are simply matching using an input LC circuit. In the chain formalism, the shunt capacitor and series inductor have chain matrices (Equation 1.3):

$$T_C(p) = \begin{bmatrix} 1 & 0 \\ pc_1 & 1 \end{bmatrix} \quad \text{and} \quad T_L(p) = \begin{bmatrix} 1 & pl_1 \\ 0 & 1 \end{bmatrix},$$

respectively. The chain matrix of the shunt capacitor and series inductor is their product (Equation 1.4):

$$T(p) = T_C(p)T_L(p) = \begin{bmatrix} 1 & l_1 p \\ c_1 p & c_1 l_1 p^2 + 1 \end{bmatrix}.$$

The impedance Z_G obtained by looking at the generator's impedance Z_0 through this circuit is computed using the chain formalism (Section 1.3):

$$Z_G = \mathcal{G}(T; Z_0) = \frac{Z_0 + l_1 p}{c_1 Z_0 p + c_1 l_1 p^2 + 1}.$$

The input and output reflectances are computed from Chapter 1:

$$S_G = \frac{Z_G - Z_0}{Z_G + Z_0} \quad \text{and} \quad S_L = 0.$$

As always, the normalizing impedance is $Z_0 = 50$ ohms. For brevity, the amplifier's Gain-Noise function is written as

$$\gamma(l_1, c_1) = \begin{bmatrix} G_T(l_1, c_1) \\ F(l_1, c_1) \end{bmatrix}.$$

Figures 6.10 and 6.11 separately plot G_T and F as functions of l_1 and c_1. Both surfaces show large flat regions that slow an optimizer with steep valleys and ridges that hide optimal points.

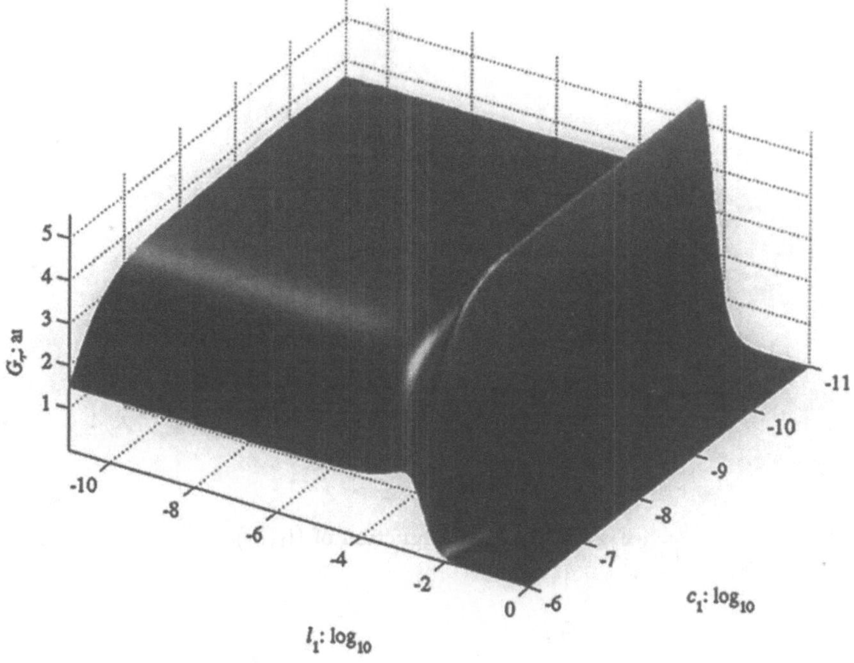

Fig. 6.10. G_T as a function of (l_1, c_1).

To check stability, Figure 6.12 plots $|S_2|$ and we see it never exceeds 0.55. Likewise, $|S_1|$ is always less than 0.6. Consequently, we can optimize $\gamma(l_1, c_1)$ without the stability functions.

Figure 6.13 presents a scatter plot of the transducer power gain G_T and the noise figure F. By generating random (l_1, c_1)'s and plotting the resulting $\gamma(l_1, c_1)$'s a sketch of the Gain-Noise image $\gamma(X)$ starts to emerge. Extreme values were retained to visualize the structure of $\gamma(X)$. Those (l_1, c_1) that have a "good" $\gamma(l_1, c_1)$—high gain and low noise—provide a "good" starting point for the Goal Attainment Method. The Goal Attainment Method was started near zero: $l_1 = 10^{-7}$ pH, $c_1 = 10^{-7}$ pF. The resulting numerical optimum is marked with the arrow labeled "optimum" and reported in the title in dB. The optimal values are reported on the plot: $c_{1,\mathrm{opt}} = 0.00$ pF, $l_{1,\mathrm{opt}} = 1.52$ pH. With the shunt capacitor essentially zero or an open circuit, a near-optimal input LC matching circuit consists only of a series inductor to deliver transducer power gain $G_T = 7$ dB and noise figure $F = 2.2$ dB.

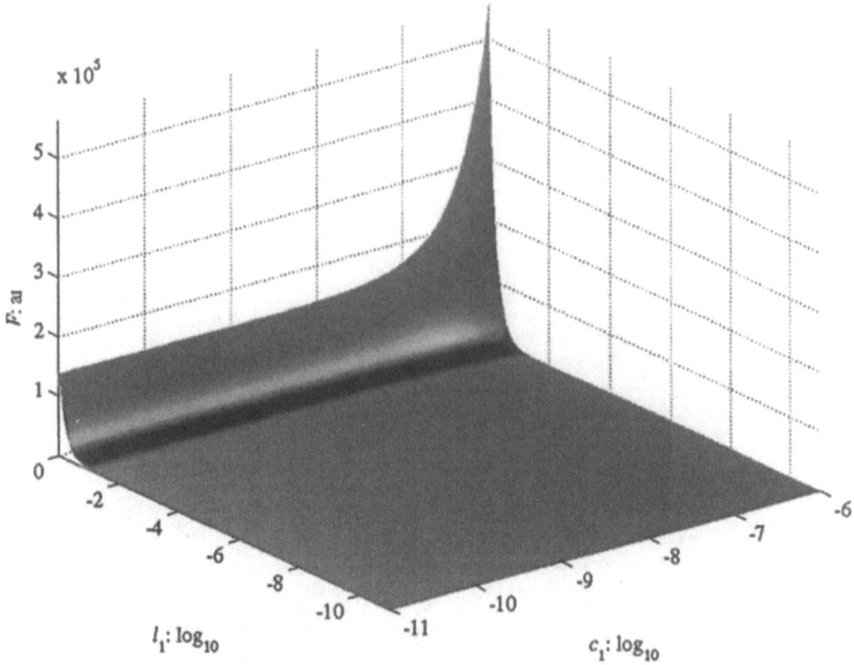

Fig. 6.11. F as a function of (l_1, c_1).

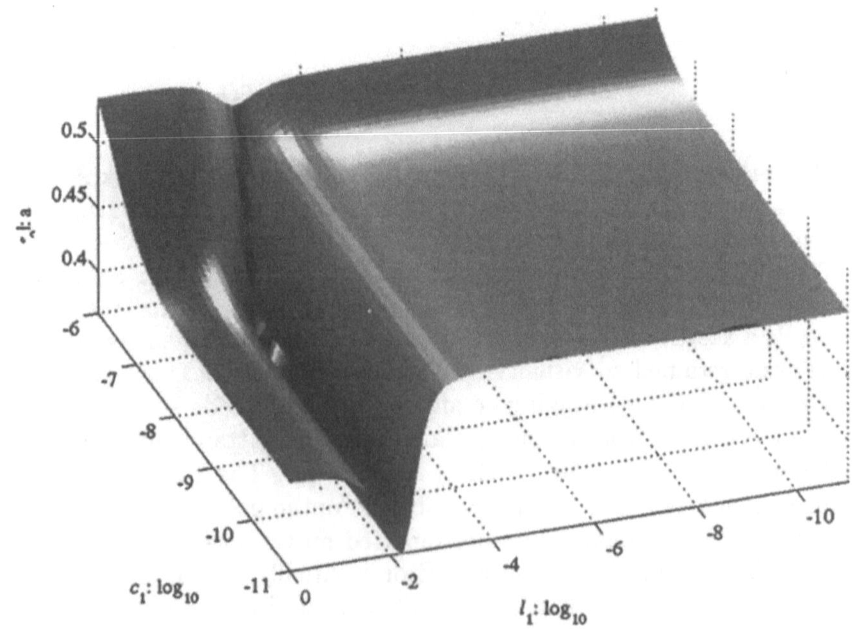

Fig. 6.12. $|S_2|$ as a function of (l_1, c_1).

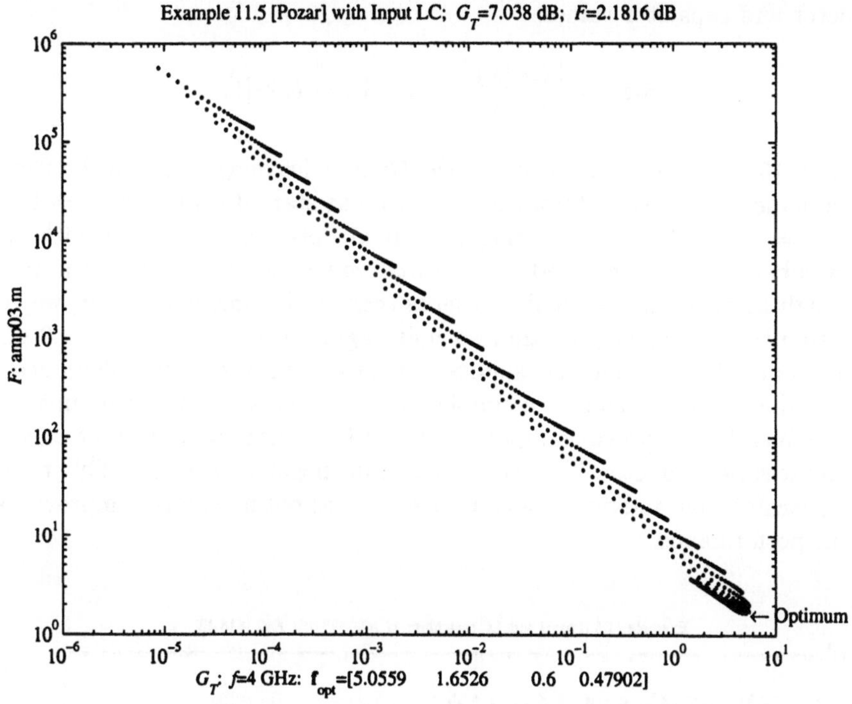

Fig. 6.13. Gain-Noise as a function of (l_1, c_1).

6.3.3 Input and Output LC Matching

Figure 6.14 extends the previous circuit by adding an output matching circuit. The same chain matrix computations compute S_G and S_L. As in the previous circuit, there was no loss of stability. Consequently, we suppress these

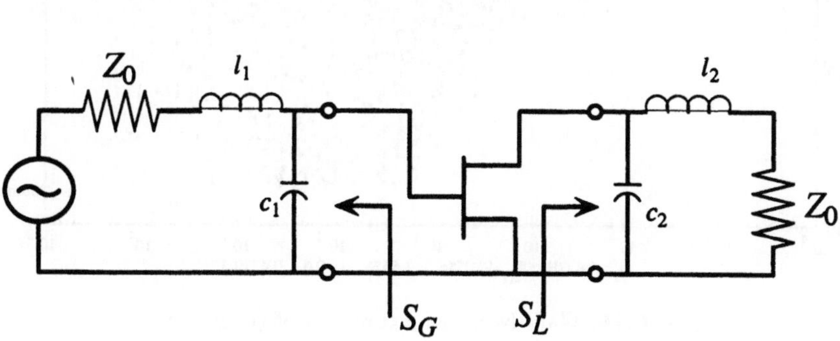

Fig. 6.14. FET with input and output matching circuits.

reflectances and write the amplifier's Gain-Noise function as a function of the inductor and capacitor values:

$$\gamma(\mathbf{x}) = \begin{bmatrix} G_T(\mathbf{x}) \\ F(\mathbf{x}) \end{bmatrix} \quad \mathbf{x} = [l_1\ c_1\ l_2\ c_2]^T.$$

Figure 6.15 presents a scatter plot of the Gain-Noise image $\gamma(X)$ in the Gain-Noise plane. The Goal Attainment Method was started near zero: $l_1 = 10^{-7}$ pH, $c_1 = 10^{-7}$ pF. The resulting numerical optimum is marked with the arrow labeled "optimum" and the optimal values are reported on the plot. Although there are more circuit elements than in the previous circuit, we get the same transducer power gain and noise figure: $G_T = 7$ dB, $F = 2.2$ dB. Up to roundoff, the shunt capacitors are zero or open circuits. The output series inductor is also zero or a simple wire. Consequently, the matching circuits reduce to the previous input matching LC circuit consisting of only of the input series inductor. This is an important negative example. This result demonstrates that adding extra components does not necessarily improve the circuit performance.

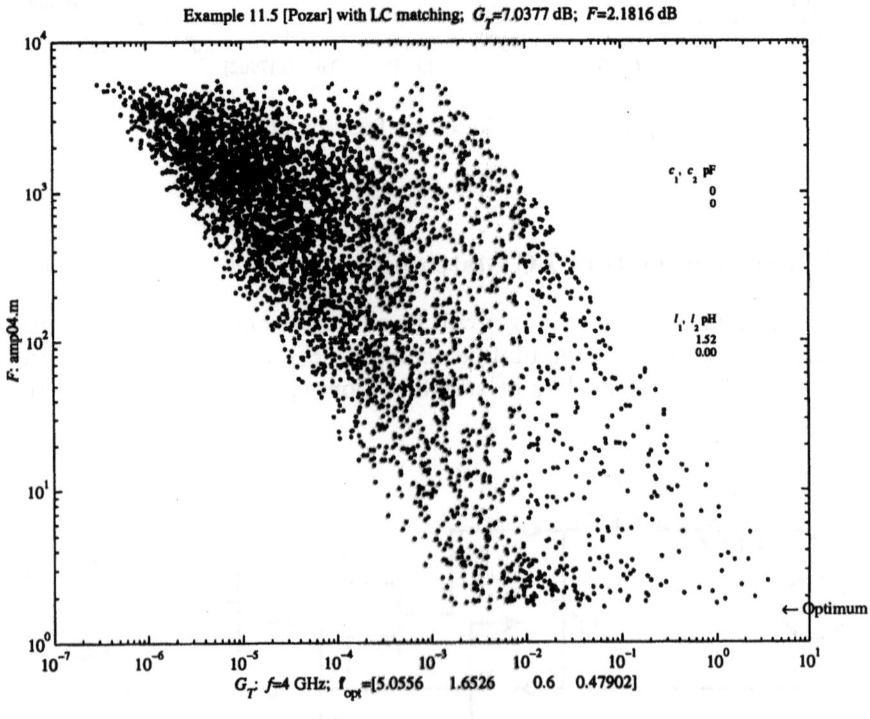

Fig. 6.15. Gain-Noise as a function of $(l_1\ c_1\ l_2\ c_2)$.

6.3.4 Stub Matching

Figure 6.16 shows a typical matching circuit constructed from series lines and shunt stubs [106, Example 11.5]. The lines and stubs are constructed from

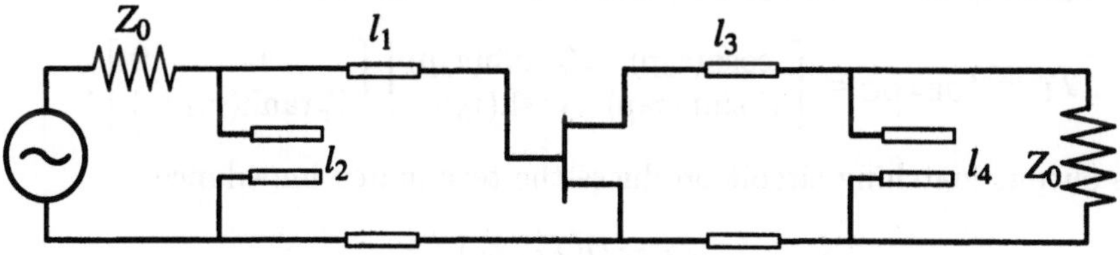

Fig. 6.16. FET and series/shunt matching circuits.

transmission lines. The transmission line is discussed in Chapter 4 and reviewed here. A uniform, lossless transmission line of characteristic impedance Z_c and *commensurate* length l is called a *unit element* (UE) and has *chain matrix* [6, Eq. 8.1]

$$T_{\mathrm{UE}}(p) = \begin{bmatrix} \cosh(\tau p) & Z_c \sinh(\tau p) \\ Y_c \sinh(\tau p) & \cosh(\tau p) \end{bmatrix},$$

where τ is the one-way delay $\tau = l/c$ determined by the speed of propagation c. If the UE is terminated in a load z_L, the terminated impedance of the resulting *stub* is (Equation 1.5):

$$z_T(p) = \mathcal{G}(T_{\mathrm{UE}}(p), z_L) = \frac{\cosh(\tau p) z_L + Z_c \sinh(\tau p)}{Y_c \sinh(\tau p) z_L + \cosh(\tau p)}.$$

As a special case, an open-circuit stub has admittance $y_{\mathrm{oc}}(p) = Y_c \tanh(\tau p)$. Consequently, the chain matrix of the open-circuit stub is (Section 1.2)

$$T_{\mathrm{OC}} = \begin{bmatrix} 1 & 0 \\ Y_c \tanh(\tau_2 p) & 1 \end{bmatrix}.$$

In Figure 6.16, the input matching circuit is a line with length l_1 and a stub of length l_2. Both line and stub take $Z_c = 50$ ohms. Their one-way travel times are τ_1 and τ_2. The chain matrix for the input matching circuit, looking out from Port 1 of the amplifier toward the generator, is the product of the UE and the open-circuit stub:

$$T_G = T_{\mathrm{UE}} T_{\mathrm{OC}} = \begin{bmatrix} \cosh(\tau_1 p) & Z_c \sinh(\tau_1 p) \\ Y_c \sinh(\tau_1 p) & \cosh(\tau_1 p) \end{bmatrix} \begin{bmatrix} 1 & 0 \\ Y_c \tanh(\tau_2 p) & 1 \end{bmatrix}.$$

This input matching circuit gives the terminated impedance

$$Z_G = \mathcal{G}(T_G, Z_0)$$

as a function of l_1 and l_2. The input reflectance

$$S_G = \frac{Z_G - Z_0}{Z_G + Z_0}$$

is also a function of l_1 and l_2. Likewise, the output matching circuit is a line with length l_3 and a stub of length l_4 and has chain matrix

$$T_L = T_{\mathrm{UE}} T_{\mathrm{OC}} = \begin{bmatrix} \cosh(\tau_3 p) & Z_c \sinh(\tau_3 p) \\ Y_c \sinh(\tau_3 p) & \cosh(\tau_3 p) \end{bmatrix} \begin{bmatrix} 1 & 0 \\ Y_c \tanh(\tau_4 p) & 1 \end{bmatrix}.$$

This output matching circuit produces the terminated impedance

$$Z_L = \mathcal{G}(T_L, Z_0)$$

as a function of l_3 and l_4. The output reflectance

$$S_L = \frac{Z_L - Z_0}{Z_L + Z_0}$$

is a function of l_3 and l_4. Thus, the amplifier's Gain-Noise function is

$$\gamma(\mathbf{x}) = \begin{bmatrix} G_T(\mathbf{x}) \\ F(\mathbf{x}) \end{bmatrix} \qquad \mathbf{x} = [l_1\ l_1\ l_3\ l_4]^T.$$

The amplifier designer searches over the admissible line lengths to optimize the amplifier's performance.

Figure 6.17 presents a scatter plot of G_T and F in the Gain-Noise plane. By plotting random $\gamma(\mathbf{x})$'s, a sketch of the Gain-Noise image $\gamma(X)$ is obtained. Good $\mathbf{x}$'s start the Goal Attainment Method, where "good" means that $\gamma(\mathbf{x})$ is in the lower right of the Gain-Noise plane. The resulting numerical optimum is marked with the arrow labeled "optimum" and the optimal line lengths are reported on the plot. Table 6.1 compares this optimum design result with Pozar's design [106, page 631].

Table 6.1. Stub matching: lengths in wavelength λ, gain and noise in dB, $Z_c = 50$ ohms.

Variable	Description	Pozar	Simulation
l_1	Input line length	0.226	0.231
l_2	Input shunt stub	0.144	0.121
l_3	Output line length	0.25	0.231
l_4	Output shunt stub	0.136	0.146
G_T	Transducer power gain	8.36	8.4
F	Noise figure	2.0	2.1

The slight variations in parameters and the continuity of the amplifier functions explain slight variations in performance. Because neither design is

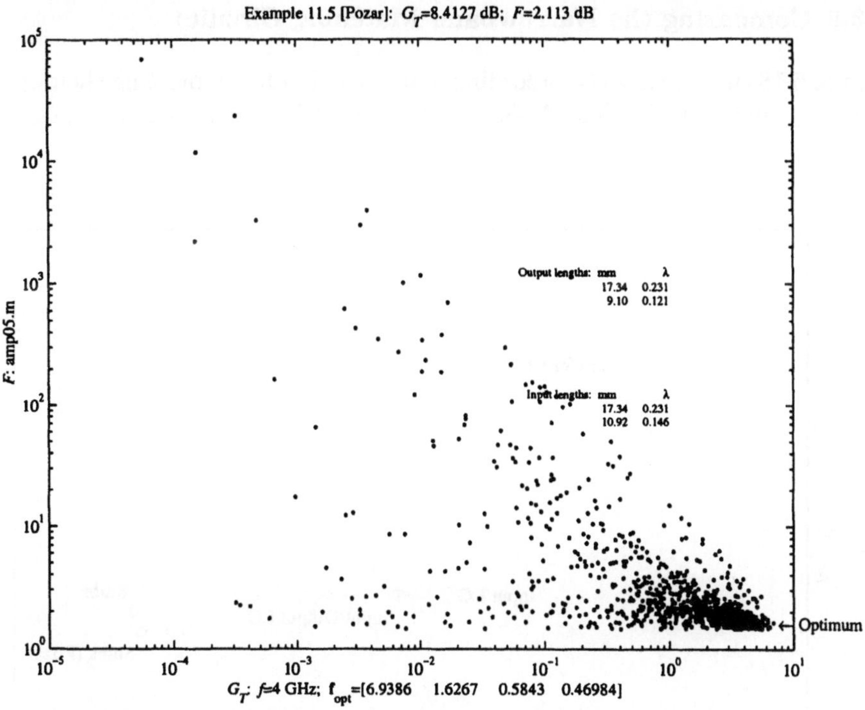

Fig. 6.17. Gain-Noise trade-offs using line and stub matching circuits.

better than the other, and both designs are in the almost minimal region (illustrated in Figure 6.6), one would be tempted to argue that both designs are Pareto optima or nearly so. This implies the minimum point does not exist but that the almost minimal region has a boundary that contains the minimal elements. However, it is one thing to argue that a given point looks Pareto, it is another thing to demonstrate a given point is Pareto.

Question 3. Numerically determine whether a given design, such as Pozar's design or the simulation, is a Pareto optimum.

Assuming we could answer that question, the next step is to compute *all* the Pareto optima and map out the minimal elements in the spirit of Das and Dennis [33]. Yet all this optimization is focused on a single circuit topology. The amplifier designer needs to know the optimal matching performance over many circuit topologies.

6.3.5 Comparing the Narrowband Matching Circuits

Figure 6.18 summarizes the preceding matching circuits by plotting their optimum G_T and F in the Gain-Noise plane. The big improvements are obtained

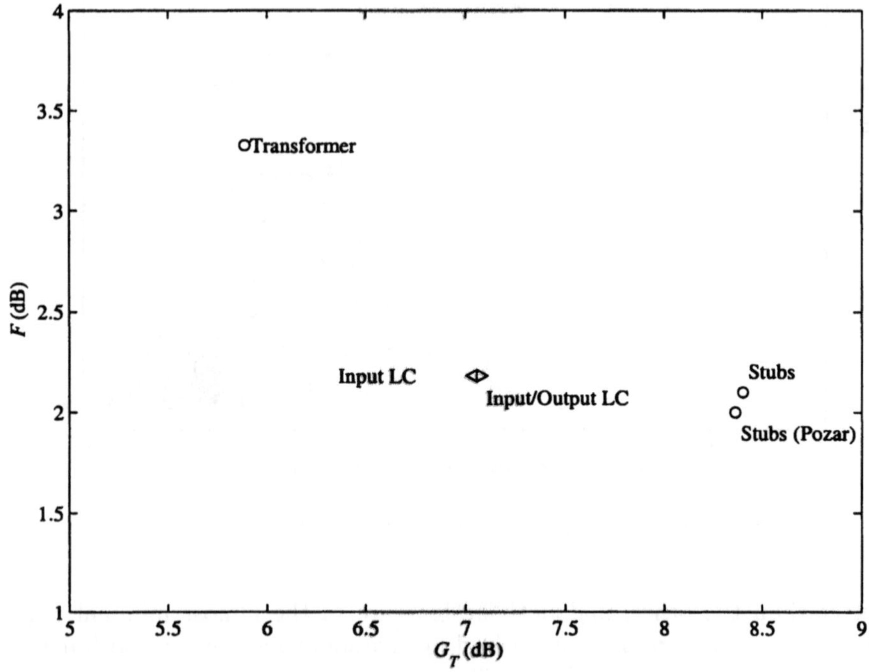

Fig. 6.18. Single-frequency matching performances.

by finding a different circuit. For example, adding more components to the LC matching circuits does not improve performance. The plot also shows what the amplifier designer is up against. Do we keep throwing more and more circuits at the problem? When do we stop the search for matching circuits? This plot leads us to ask the following:

What are the best possible matching performances?

Knowing the ultimate performance limits lets the amplifier designer benchmark the suboptimal performance of real-world matching circuits. The problem of computing these performance limits is the topic of subsequent chapters. Our more pressing task is to examine the matching performances of wideband amplifiers.

6.4 Wideband Matching Examples

Two real-world, wideband, low-noise, high-gain communication amplifiers are matched in this section. The first amplifier is the NE32484A, whose scattering and noise measurements are courtesy of the AnSoft Corporation. The second amplifier is NEC Corporation's NE321000, using the measurements from the NEC website [98]. Matching will be done by the shunt stub and line circuits as in Figure 6.16. Upper bounds on the line length l are used to avoid the poles in tanh and the periodicity of cosh and sinh. With the one-way delay $\tau = l/c$, the cosh's periodicity is

$$\cosh(\tau p) = \cosh(j\tau\omega) = \cos(\tau\omega).$$

By requiring $0 \leq \tau\omega < \pi/2$ across the frequency band $[\omega_{\min}, \omega_{\max}]$, the vanishing of the cosine is avoided. Thus, the bounds on the line length l are

$$0 \leq l < \frac{c}{4f_{\max}} =: l_{\max}, \tag{6.1}$$

where $2\pi f_{\max} = \omega_{\max}$.

6.4.1 NE32484A: 0.1–18 GHz

Figure 6.19 displays the scattering functions for the NE32484A amplifier operating from 0.1–18 GHz. The scattering matrix is the complex-valued matrix

$$S(j\omega) = \begin{bmatrix} S_{11}(j\omega) & S_{12}(j\omega) \\ S_{21}(j\omega) & S_{22}(j\omega) \end{bmatrix} \quad (\omega = 2\pi f).$$

Figure 6.19 plots the $S_{kl}(j\omega)$'s as curves in the complex plane. The real and imaginary axes are labeled "$\Re$" and "$\Im$," respectively. The label for each curve starts at the low-frequency position. There is a strong resemblance between these measured scattering functions and the simulated scattering functions of the model transistor amplifier in Section 1.13. For reference only, the NE32484A is biased using a drain-source voltage $V_{ds} = 2$V and drain-source current $I_{ds} = 10$mA. Figure 1.17 illustrates a classic biasing circuit.

The design goal we selected puts the least constraints on stability and asks for a low-noise design with a goal of 10 dB gain:

$$\gamma_u = \begin{bmatrix} -G_u \\ F_u \\ S_{1,u} \\ S_{2,u} \end{bmatrix} = \begin{bmatrix} -10^{10/10} \\ \min\{F_{\min}\} \\ 1 \\ 1 \end{bmatrix}.$$

As in the single-frequency examples, the amplifier's Gain-Noise image $\gamma(X)$ is sketched by randomly generating line lengths $\mathbf{x} = [l_1\ l_2\ l_3\ l_4]^T$, using these lengths to generate a matching circuit, and plotting the amplifier's Gain-Noise function

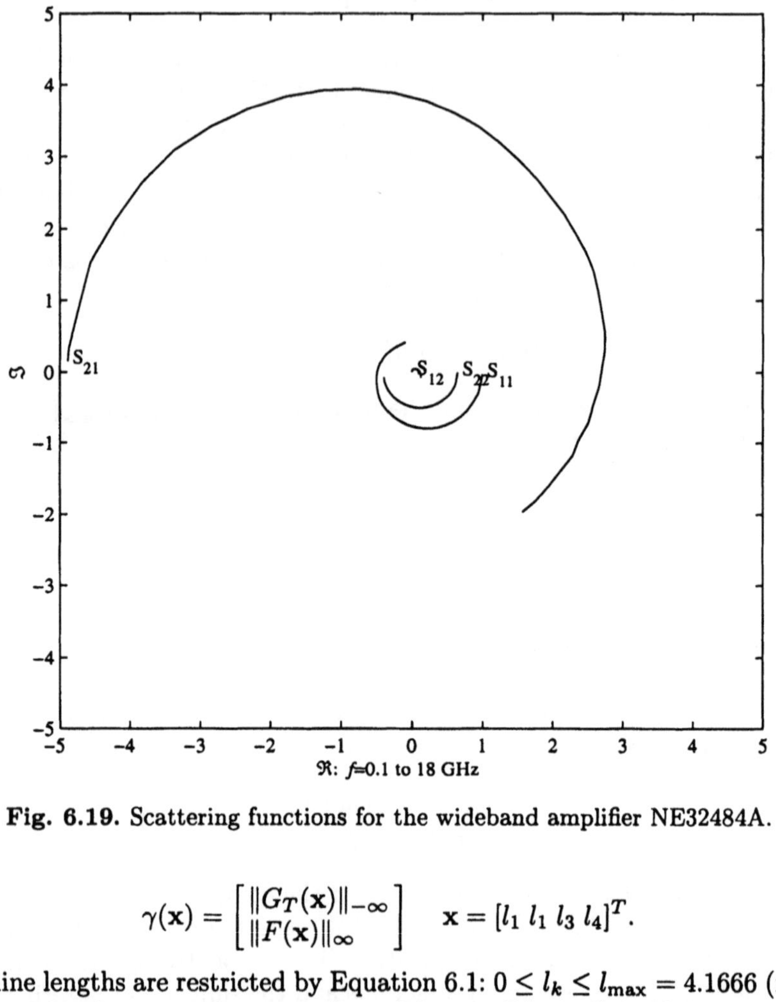

Fig. 6.19. Scattering functions for the wideband amplifier NE32484A.

$$\gamma(\mathbf{x}) = \begin{bmatrix} \|G_T(\mathbf{x})\|_{-\infty} \\ \|F(\mathbf{x})\|_{\infty} \end{bmatrix} \quad \mathbf{x} = [l_1 \; l_1 \; l_3 \; l_4]^T.$$

The line lengths are restricted by Equation 6.1: $0 \leq l_k \leq l_{\max} = 4.1666$ (mm). Figure 6.20 plots a random sampling of the amplifier's Gain-Noise image

$$\gamma(X) := \{\gamma(\mathbf{x}) : 0 \leq \mathbf{x} \leq l_{\max}\}.$$

The almost minimum region is even more "pointy" than in the single-frequency examples. A circuit with the greatest gain is the starting point for the Goal Attainment Method. The lengths reported on the figure are in millimeters and wavelength λ at 10 GHz. The resulting numerical optimum is marked with the arrow. The minimum noise level is obtainable with a gain of approximately 9 dB. We observe that the input stability is nearly violated: $\|S_1\|_{\infty} = 0.99899$. This fragile stability is characteristic of our wideband amplifiers in this circuit topology.

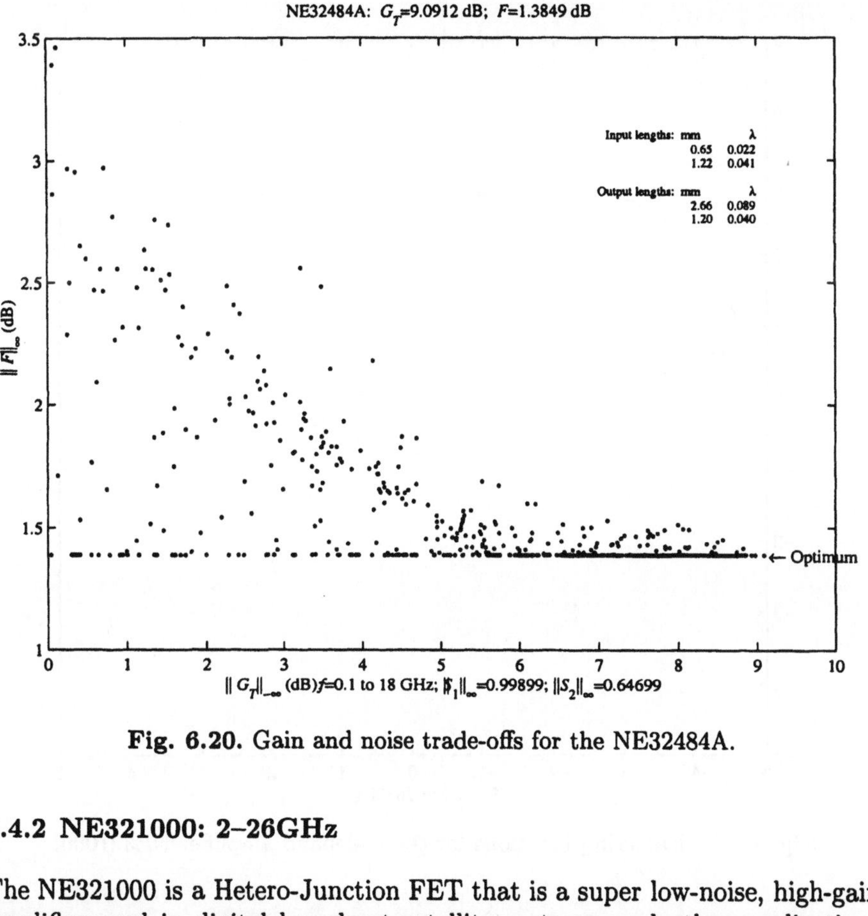

Fig. 6.20. Gain and noise trade-offs for the NE32484A.

6.4.2 NE321000: 2–26GHz

The NE321000 is a Hetero-Junction FET that is a super low-noise, high-gain amplifier used in digital broadcast satellite systems, and other applications [98]. Figure 6.21 displays the scattering functions for the NE321000 amplifier operating from 2–26 GHz. Referring to the classic biasing circuit in Figure 1.17, we report for reference only that the NE321000 is biased using a drain-source voltage $V_{ds} = 2$V and drain-source current $I_{ds} = 10$mA. Finally, we observe the close resemblance to the the simulated transistor amplifier in Section 1.13.

Matching is done using the shunt stub and line circuits of Figure 6.16. Upper bounds on the line lengths $\mathbf{x} = [l_1\ l_2\ l_3\ l_4]^T$ are set by Equation 6.1. The amplifier's Gain-Noise image $\gamma(X)$ is sketched by randomly selecting the line lengths and plotting the resulting Gain-Noise vectors:

$$\gamma(\mathbf{x}) = \begin{bmatrix} \|G_T(\mathbf{x})\|_{-\infty} \\ \|F(\mathbf{x})\|_\infty \end{bmatrix} \qquad \mathbf{x} = [l_1\ l_1\ l_3\ l_4]^T.$$

Figure 6.22 presents a random sampling of the Gain-Noise image

$$\gamma(X) = \{\gamma(\mathbf{x}) : 0 \le \mathbf{x} \le l_{\max}\}.$$

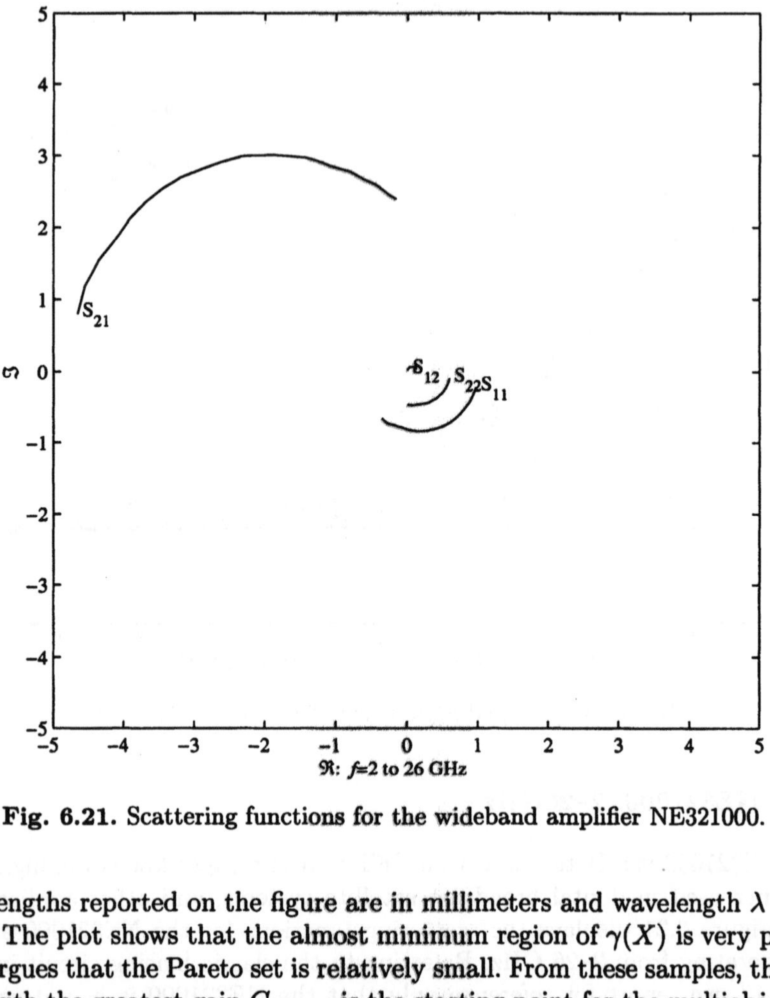

Fig. 6.21. Scattering functions for the wideband amplifier NE321000.

The lengths reported on the figure are in millimeters and wavelength λ at 10 GHz. The plot shows that the almost minimum region of $\gamma(X)$ is very pointy and argues that the Pareto set is relatively small. From these samples, the circuit with the greatest gain $G_{T,\max}$ is the starting point for the multiobjective optimization. The design goal was set with the least constraints on stability, and aims at a low-noise design with relatively high gain:

$$\gamma_u = \begin{bmatrix} -G_u \\ F_u \\ S_{1,u} \\ S_{2,u} \end{bmatrix} = \begin{bmatrix} -10 \times G_{T,\max} \\ \min\{F_{\min}\} \\ 1 \\ 1 \end{bmatrix}.$$

The resulting numerical optimum is marked with the arrow. At this near-optimal solution, $G_T = 10.1$ dB and $F = 1.8$ dB. However, just as in the previous section, input stability is nearly violated with $\|S_1\|_\infty = 0.998$.

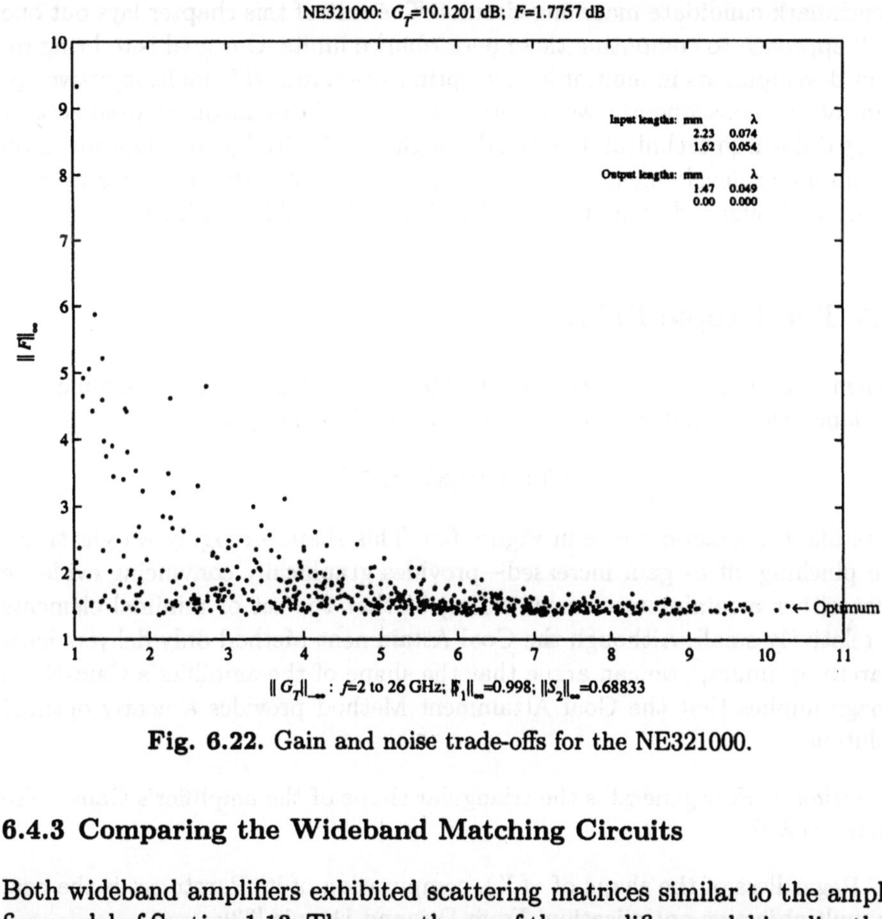

Fig. 6.22. Gain and noise trade-offs for the NE321000.

6.4.3 Comparing the Wideband Matching Circuits

Both wideband amplifiers exhibited scattering matrices similar to the amplifier model of Section 1.13. The smoothness of the amplifier model implies that "filling in" the sample points (i.e., by spline functions) is one approach that overcomes problems of sparsely sampled data. That is, with only a few sample points and lots of circuit parameters, the amplifier designer could report amazing matching performance—on the sample points. However, amplifier performance off the sample points could be disastrous. This "overfitting phenomena" is embedded in the literature and can trap the unwary. Section 10.6 details this overfitting phenomena and related research topics.

Both wideband amplifiers also exhibited a Gain-Noise image similar to the single-frequency matching examples of Section 6.3. Additional complications for wideband matching is that the stability cannot be ignored and that the stability margin is very small. More precisely, stability is barely holding on the sample points, which makes numerical wideband stability suspect.

As in the single-frequency matching examples, the amplifier designer faces the same questions: are matching circuits to be thrown at the amplifier hoping to get a lucky hit of better gain, less noise, and more stability? Or can we compute absolute performance limits—best possible gain, least noise, best possible stability? These performance limits allow the amplifier designer to benchmark candidate matching circuits. The rest of this chapter lays out one H^∞ approach to computing these performance limits. Our goal is to bring recent developments in multiobjective optimization into H^∞ multiobjective optimization. Consequently, we maintain a high-level discussion to avoid getting bogged down in technical details. Although H^∞ Multiobjective Optimization is still under development, recent results explain the shape of the Gain-Noise image and offer a design approach for absolutely stable amplifiers.

6.5 The Utopic Point

When applying the Goal Attainment Method to these selected amplifiers, it became evident that all the amplifier's Gain-Noise images

$$\gamma(X) = \{\gamma(\mathbf{x}) : \mathbf{x} \in X\}$$

look like the generic image in Figure 6.6. This shape—roughly a right triangle pinching off as gain increased—provides graphically convincing evidence that either a minimum element exists, or that the set of minimal elements is relatively small. Although the Goal Attainment Method only delivers local Pareto optimum,[2] we can argue that the shape of the amplifier's Gain-Noise image implies that the Goal Attainment Method provides a nearly optimal solution.

Question 4. How general is the triangular shape of the amplifier's Gain-Noise image $\gamma(X)$?

Regardless of the shape of $\gamma(X)$, computation of its Pareto set is the goal in multiobjective optimization. From Das and Dennis [33]:

> Very often in engineering applications, the desired result helpful in facilitating design is a whole collection of Pareto optimal points, representative of the entire spectrum of efficient solutions. Thus, ideally, the desired solution is the entire Pareto optimal set.

The collection of the Pareto optimal points is also called the *Pareto optimal front*. Pareto optimal front computations are an active topic in electrical engineering [80]. Das and Dennis introduce the *Normal-Bound Intersection* (NBI) method for computing the Pareto set [33]. The NBI method relies on the individual global minimizers:

$$\mathbf{x}_n = \operatorname{argmin}\{\gamma_n(\mathbf{x}) : \mathbf{x} \in X\} \quad (n = 1, \ldots, N).$$

These minimizers determine the *utopic point*

[2] [33] A point $\mathbf{x} \in X$ is a local Pareto optimum provided there is a neighborhood B of $\mathbf{x}$ in $\mathbf{R}^N$ such that any $\mathbf{y} \in B$ with $\gamma(\mathbf{y}) \leq \gamma(\mathbf{x})$ implies $\gamma(\mathbf{y}) = \gamma(\mathbf{x})$; or [75] for each $\mathbf{y} \in B$, there exists an index $m = 1, \ldots, M$ such that $\gamma_m(\mathbf{y}) \geq \gamma_m(\mathbf{x})$.

$$\gamma_\diamond := \begin{bmatrix} \gamma_1(\mathbf{x}_1) \\ \gamma_2(\mathbf{x}_2) \\ \vdots \\ \gamma_N(\mathbf{x}_N) \end{bmatrix},$$

which is a pseudo-minimum of $\gamma(X)$. Figure 6.23 shows that the utopic point is within "line-of-sight" of the Pareto points by shooting along the weight vector $\mathbf{w}$ over all positive weights: $\mathbf{w} \geq 0$. The claim is by setting $\gamma_\diamond = \gamma_u$ and varying

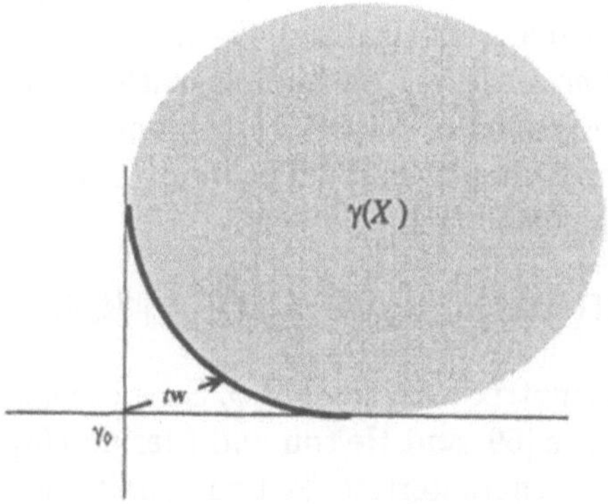

Fig. 6.23. The utopic point $\gamma_\diamond$.

the weight vector through all $\mathbf{w} \geq 0$, the Goal Attainment Method produces a superset of the Pareto set. In practice, we can only sample this superset. Das and Dennis [33] point out that this sampling may not be uniformly distributed. Their key claim is that the NBI method produces a uniform sampling of the Pareto set. This claim immediately causes the amplifier designer to ask how the NBI method compares to other multiobjective methods. For example, suppose we repeatedly applied the Goal Attainment Method to build up a superset of the Pareto set.

Question 5. How does the NBI method compare to the Goal Attainment Method to compute the Pareto set of the Amplifier Matching Problem?

The following sections mix this multiobjective optimization with the H^∞ theory and apply it to the Amplifier Matching Problem. For example, the amplifier designer is limited by the time and effort it takes to assess each matching circuit and contend with its numerical difficulties [22]. No matter how good an individual matching circuit, there remains the fundamental question: *What good can the amplifier's performance be?* That is, what is the largest gain (AMP-1), smallest noise (AMP-2), and smallest stability (AMP-3) that can be attained by any matching circuit. What is really being asked for is the utopic point. The H^∞ theory can estimate the utopic point. The utopic point provides the amplifier designer a benchmark of the suboptimal matching circuits and motivates the study of the individual amplifier functions undertaken in Chapter 7. We begin by generalizing multiobjective optimization to the H^∞ setting.

6.6 H^∞ Multiobjective Optimization

This section reviews recent developments in H^∞ multiobjective optimization. The discussion is strictly mathematical and distinct from the previous sections that focused on amplifier matching by fiddling with circuit parameters. Our goal is a multiobjective version of Nehari's Theorem.

Recall from Chapter 3 that Nehari's Theorem computes the distance between $\phi \in C(\mathbf{T})$ and the disk algebra:

$$\inf\{\|\phi - h\|_\infty : h \in \mathcal{A}(\mathbf{D})\} = \|\mathcal{H}_\phi\|,$$

where $\mathcal{H}_\phi$ is the Hankel matrix with symbol ϕ. Suppose h_{opt} is a best approximation. Helton and Howe [69] and Helton and Merino [74] observed that the *error function* $\phi - h_{\text{opt}}$ is characterized by two conditions:

- Flatness: $|\phi - h_{\text{opt}}| = \text{constant}$.
- Winding: $0 < \text{wind}[\phi - h_{\text{opt}}]$.

As illustrated in Figure 6.24, the *winding number* counts the net rotations that a continuous curve $\phi : \mathbf{T} \to \mathbf{C}$ makes around zero [74]. Helton's deep

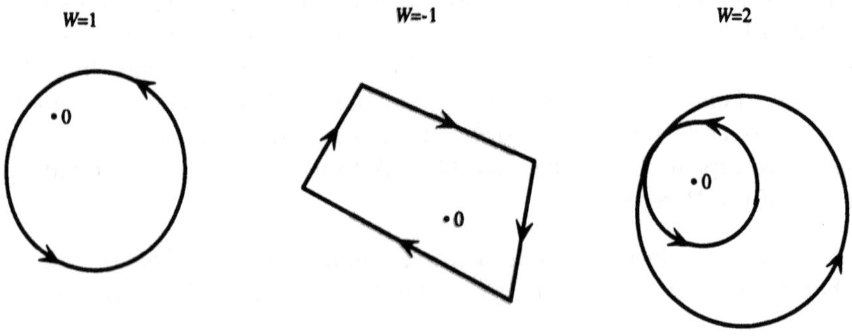

Fig. 6.24. Winding numbers.

contribution was to see how the Flatness and Winding conditions generalize to nonlinear and nondifferentiable functionals [69], [74]. Helton introduced a *performance function* $\Gamma : \mathbf{T} \times \mathbf{C} \to \mathbf{R}_+$ that is continuous and nonnegative. This performance function induces the *objective function* $\gamma : L^\infty(\mathbf{T}) \to \mathbf{R}_+$ defined as

$$\gamma(h) := \text{ess.sup}\{\Gamma(z, h(z)) : z \in \mathbf{T}\}.$$

Consider minimizing γ restricted to the disk algebra:

$$\inf\{\gamma(h) : h \in \mathcal{A}(\mathbf{D})\}.$$

For all but the most trivial performance functions, γ is not differentiable. Nevertheless, γ can be analyzed by the differential of the performance function Γ. Let ∂ denote the *complex differential* of Γ with respect to $h = u + jv$ [74, Eq. 9.2]:

$$\partial\Gamma(z,h) = \frac{\partial\Gamma}{\partial h}(z,h) = \frac{1}{2}\left\{\frac{\partial\Gamma}{\partial u}(z,h) - j\frac{\partial\Gamma}{\partial v}(z,h)\right\}.$$

With a slight abuse of notation, consider $\Gamma : L^\infty(\mathbf{T}) \to L^\infty(\mathbf{T})$ as the mapping

$$h \to \Gamma(z, h(z)).$$

The differential linearizes Γ as the mapping on L^∞ as [74, Eq. 11.3]:

$$\Gamma(z, h(z) + \Delta h(z)) = \Gamma(z, h(z)) + 2\Re[\partial\Gamma(z, h(z))\Delta h(z)] + \mathcal{O}[\Delta h(z)^2].$$

Although γ is not differentiable, it can be analyzed by applying *Kolmogorov Theory* to $\partial\Gamma$ [20, pages 6–11]. The Descent Lemma is a typical Kolmogorov result.

Lemma 6.6.1 (Descent) *Let $\mathcal{H}$ be a closed linear subspace of $C(\mathbf{T}, \mathbf{C})$. Let $\Gamma \in C^2(\mathbf{T} \times \mathbf{C}, \mathbf{R})$. Define $\gamma : \mathcal{H} \to \mathbf{R}$ by*

$$\gamma(h) := \sup\{\Gamma(z, h(z)) : z \in \mathbf{T}\}.$$

Denote the associated critical set *of h by*

$$\text{crit}[\gamma(h)] := \{z \in \mathbf{T} : \gamma(h) = \Gamma(z, h(z))\}.$$

Assume $\mathcal{H}$ is boundedly compact. If $h \in \mathcal{H}$ is not a local minimum, there is a nonzero $\Delta h \in \mathcal{H}$ such that

$$0 \geq \Re[\partial\Gamma(z, h(z))\Delta h(z)]$$

for all $z \in$ crit$[\gamma(h)]$. If the inequality is strict, Δh is a direction of descent: for all $t \in (0, t_0)$ there holds $\gamma(h) > \gamma(h + t\Delta h)$.

The Descent Lemma lets us linearize nondifferentiable γ with the differential $\partial\Gamma$. The Descent Lemma says that if $\mathcal{H}$ has sufficiently general interpolating properties, extreme conditions must hold at a minimizer so that no Δh can be found that is a direction of descent.

For example, suppose we are minimizing over the disk algebra $\mathcal{A}(\mathbf{D})$. Suppose the critical set is not all of the unit circle:

$$\text{crit}[\gamma(h)] \neq \mathbf{T},$$

Mergelyan's Theorem [110, page 423] produces $\Delta h \in \mathcal{A}(\mathbf{D})$ such that

$$\Delta h(z) = -\overline{\partial \Gamma(z, h(z))} \qquad (z \in \text{crit}[\gamma(h)]).$$

Consequently, Δh is a direction of descent. By the Descent Lemma, $h \in \mathcal{A}(\mathbf{D})$ cannot be a local minimum. Consequently, it is necessary that the critical set be all of $\mathbf{T}$ at a local minimum. That is, the *Flatness condition* must hold at a local minimum. Likewise, when the winding number of $\partial\Gamma(z, h(z))$ is nonpositive, its phase admits an analytic extention in the disk algebra of that gives a direction of descent [74, Theorem 10.3.1]. Consequently, Δh is a direction of descent and, by the Descent Lemma, $h \in \mathcal{A}(\mathbf{D})$ cannot be a local minimum. Thus, it is necessary that $\text{wind}[\partial\Gamma(z, h(z))] > 0$ or the *Winding condition* hold at a local minimum. Helton and Merino demonstrated that the Flatness and Winding Condition are also sufficient to characterize local minima.

Theorem 6.6.1 (Disk Algebra Minimizers) [68], [69], [74, Theorem 9.3.1] *Let $\Gamma : \mathbf{T} \times \mathbf{C} \to \mathbf{R}_+$ be C^3 and define the mapping $\gamma : \mathcal{A}(\mathbf{D}) \to \mathbf{R}_+$ by*

$$\gamma(h) := \sup\{\Gamma(z, h(z)) : z \in \mathbf{T}\}.$$

Assume $h_{\text{loc}} \in \mathcal{D}(\mathbf{T})$ and $z \to \partial\Gamma(z, h_{\text{loc}}(z))$ never vanishes. Then $h_{\text{loc}} \in \mathcal{A}(\mathbf{D})$ is a strict local minimizer of γ if and only if

- *Flatness:* $\Gamma(z, h_{\text{loc}}(z)) = \text{constant}$ *on* $z \in \mathbf{T}$.
- *Winding:* $0 < \text{wind}[\partial\Gamma(z, h_{\text{loc}}(z))]$.

This variational approach of the Descent Lemma illustrates the basic geometric idea exploited by Ball, Foias, and Helton [15]:

Nehari's Theorem may be invoked on manifolds in L^∞ that contain H^∞ in each tangent space.

Typically, Kolmogorov theory is applied to arbitrary sets in Banach spaces using tangent cones and contingent cones—the *Nonsmooth Analysis* of Clarke [32]. Although we stated the Descent Lemma for linear subspaces of $C(\mathbf{T}, \mathbf{C})$, we ask how far Nehari's Theorem can be pushed using Clarke's Nonsmooth Analysis:

Question 6. Can Nehari's Theorem be invoked on subsets of L^∞ that contain H^∞ in each contingent cone?

Theorem 6.6.1 characterizes the local minimizers of γ over the disk algebra. Under connectedness conditions on the sublevel sets of Γ, Helton and Merino found that *local solutions are unique* [74, Theorem 9.4.1]. Consequently, *any local minimizer must be the unique global minimizer.*

Theorem 6.6.2 [68], [69], [74, Theorem 9.4.1] *Let* $\Gamma : \mathbf{T} \times \mathbf{C} \to \mathbf{R}_+$ *be* C^1 *and define the mapping* $\gamma : \mathcal{A}(\mathbf{D}) \to \mathbf{R}_+$ *by*

$$\gamma(h) := \sup\{\Gamma(z, h(z)) : z \in \mathbf{T}\}.$$

Assume that the slices of the sublevel sets:

$$S(z, c) := \{h \in \mathbf{C} : \Gamma(z, h) \leq c\}$$

satisfy the Standard Assumption*:*

SA-1 $S(z,c)$ *is connected.*
SA-2 $S(z,c)$ *is simply connected.*
SA-3 The mapping $z \to S(z,c)$ *is uniformly bounded.*
SA-4 The mapping $z \to S(z,c)$ *has area uniformly bounded away from zero.*
SA-5 $\partial\Gamma(z,h)/\partial h$ *does not vanish on the boundary of* $S(z,c)$.

Then any local minimizer $h_{\mathrm{loc}} \in \mathcal{A}(\mathbf{D})$ *of* γ *is unique.*

It is informative to compare this general unicity result with the unicity obtained in Corollary 3.10.5. In Corollary 3.10.5, the sublevel sets are disks; these disks are contained in the unit disk; and these disks have nonvanishing radii. Consequently, the Standard Assumption holds and the unicity of Corollary 3.10.5 follows as a special case of Theorem 6.6.2. We will invoke unicity in the following chapters. This completes our brief survey of univariate H^∞ optimization.

Helton also generalized Nehari's Theorem to *multivariate optimization.* The continuous performance function $\Gamma : \mathbf{T} \times \mathbf{C}^L \to \mathbf{R}_+$ induces the objective function $\gamma : L^\infty(\mathbf{D}, \mathbf{C}^L) \to \mathbf{R}_+$ as

$$\gamma(\mathbf{h}) := \text{ess.sup}\{\Gamma(z, \mathbf{h}(z)) : z \in \mathbf{T}\}, \qquad \mathbf{h} = \begin{bmatrix} h_1 \\ h_2 \\ \vdots \\ h_L \end{bmatrix}.$$

Consider minimizing γ restricted to the $\mathbf{C}^L$-valued disk algebra:

$$\inf\{\gamma(\mathbf{h}) : \mathbf{h} \in \mathcal{A}(\mathbf{D}, \mathbf{C}^L)\}.$$

Although a minimax approach can be made [74, Chapter 17], the differential of Γ provides a straightforward approach [75]. With

$$\partial\Gamma := \left[\frac{\partial\Gamma}{\partial h_1} \; \frac{\partial\Gamma}{\partial h_2} \; \cdots \; \frac{\partial\Gamma}{\partial h_L}\right],$$

the mapping $\Gamma : L^\infty(\mathbf{T}, \mathbf{C}^L) \to L^\infty(\mathbf{T})$ linearizes as

$$\Gamma(z, \mathbf{h}(z) + \Delta\mathbf{h}(z)) = \Gamma(z, \mathbf{h}(z)) + 2\Re[\partial\Gamma(z, \mathbf{h}(z))\Delta\mathbf{h}(z)] + \mathcal{O}[\Delta\mathbf{h}^2].$$

At a local optima $\mathbf{h}_{\text{loc}}$, the Flatness condition still holds:

$$\Gamma(z, \mathbf{h}_{\text{loc}}(z)) = \text{constant},$$

but the Winding condition generalizes to its vector analog. Roughly speaking, the phase of the differential at a local optimum should come from an analytic function. More formally, there is $\mathbf{g} \in H^1(\mathbf{D}, \mathbf{C}^L)$ and measurable $u : \mathbf{T} \to \mathbf{R}_+$ such that [75, Theorem 2.1]:

$$u(z)\partial\Gamma(j\omega, \mathbf{h}_{\text{loc}}(z)) = z\mathbf{g}(z)^T \quad \text{a.e.}$$

That is, it is possible to align the amplitude of the differential to get an analytic function.

This "alignment" condition generalizes to H^∞ *multiobjective optimization.* Let $\Gamma_n : \mathbf{T} \times \mathbf{C}^L \to \mathbf{R}_+$ denote the nth performance function for $n = 1, \ldots, N$. Define the objective functions $\gamma_n : \mathcal{A}(\mathbf{D}, \mathbf{C}^L) \to \mathbf{R}_+$ as

$$\gamma_n(\mathbf{h}) = \sup\{\Gamma_n(z, \mathbf{h}(z)) : z \in \mathbf{T}\}.$$

The multiobjective function $\gamma : \mathcal{A}(\mathbf{D}, \mathbf{C}^L) \to \mathbf{R}_+^N$ is

$$\gamma(\mathbf{h}) = \begin{bmatrix} \gamma_1(\mathbf{h}) \\ \gamma_2(\mathbf{h}) \\ \vdots \\ \gamma_N(\mathbf{h}) \end{bmatrix}$$

with differential

$$\partial\Gamma(z, \mathbf{h}) := \begin{bmatrix} \frac{\partial\Gamma_1}{\partial h_1} & \frac{\partial\Gamma_1}{\partial h_2} & \cdots & \frac{\partial\Gamma_1}{\partial h_L} \\ \frac{\partial\Gamma_2}{\partial h_1} & \frac{\partial\Gamma_2}{\partial h_2} & \cdots & \frac{\partial\Gamma_2}{\partial h_L} \\ \vdots & & & \vdots \\ \frac{\partial\Gamma_N}{\partial h_1} & \frac{\partial\Gamma_N}{\partial h_2} & \cdots & \frac{\partial\Gamma_N}{\partial h_L} \end{bmatrix}.$$

Under sufficient smoothness, convexity, and rank conditions, the Flatness and Alignment conditions still hold at local optima.

Theorem 6.6.3 (Helton and Vityaev) [75, Theorem 2.1] *Suppose* $\mathbf{h}_{\text{loc}} \in H^\infty(\mathbf{D}, \mathbf{C}^L)$ *is a local Pareto optimum for* γ*: there is an* $\epsilon > 0$ *such that for each* $\mathbf{h} \in H^\infty(\mathbf{D}, \mathbf{C}^L)$ *with* $\|\mathbf{h} - \mathbf{h}_{\text{loc}}\|_\infty < \epsilon$*, there is an index* n *for which*

$$\gamma_n(\mathbf{h}) \geq \gamma_n(\mathbf{h}_{\text{loc}}).$$

(a) Assume more variables (L) than objectives (N): $L \geq N$.

(b) Assume rational performance functions:

$$\Gamma_n(z, \mathbf{h}) = \left| \frac{P_n(z, \mathbf{h})}{Q_n(z, \mathbf{h})} \right|^2, \quad (n = 1, \ldots, N),$$

where P_n and Q_n are holomorphic polynomials in $\mathbf{h}$ with rational coefficients in $z \in \mathbf{T}$ that have no poles on $\mathbf{T}$.

(c) Assume smoothness at the minimizer: $\mathbf{h}_{\text{loc}} \in H^\infty(\mathbf{D}, \mathbf{C}^L) \cap C^1(\mathbf{T})$.

(d) Assume smoothness of the differential: $\partial\Gamma(z, \mathbf{h}_{\text{loc}}(z)) \in C^1(\mathbf{T})$.

(e) Assume $\mathbf{h}_{\text{loc}}$ is a strict Pareto point: For each $n' = 1, \ldots, N$, there exists $\Delta\mathbf{h}_{n'} \in H^\infty(\mathbf{D}, \mathbf{C}^L) \cap C^1(\mathbf{T})$, a constant $C > 0$, and a $t_0 > 0$ such that for all $t \in (0, t_0)$ and all $n = 1, \ldots, N$, $n \neq n'$ there holds

$$\gamma_n(\mathbf{h}_{\text{loc}} + t\Delta\mathbf{h}_{n'}) - \gamma_n(\mathbf{h}_{\text{loc}}) < -Ct.$$

(f) Assume full rank: $\text{rank}[\partial\Gamma(z, \mathbf{h}_{\text{loc}}(z))] = N$.

Then

- *Flatness: For $n = 1, \ldots, N$, and for $z \in \mathbf{T}$*

$$\Gamma_n(z, \mathbf{h}_{\text{loc}}(z)) = \text{constant}.$$

- *Alignment: There is $\mathbf{g} \in H^2(\mathbf{D}, \mathbf{C}^L)$ and a nonzero $\mathbf{u} \in C^1(\mathbf{T}, \mathbf{R}_+^N)$*

$$\mathbf{u}(z)^T \partial\Gamma(z, \mathbf{h}_{\text{loc}}(z)) = z\mathbf{g}(z)^T \quad \text{a.e.}$$

The strict Pareto says all the objective functions matter. Graphically, the image of the strict Pareto points—ignoring the Ct condition—are illustrated in Figure 6.25. The open circles indicate that these image points cannot correspond to strict Pareto points. That is, if $\mathbf{h}_{\text{loc}}$ is a global minimum for γ_n:

$$\gamma_n(\mathbf{h}_{\text{loc}}) \leq \gamma_n(H^\infty),$$

then $\mathbf{h}_{\text{loc}}$ cannot be a strict Pareto point.

Theorem 6.6.3 is a direct descendent classical H^∞ theory and is at the current limits of H^∞ research. Theorem 6.6.3 answers the question of identifying local minimizers. The Flatness condition provides a straightforward test—simply plot $\Gamma_n(z, \mathbf{h}_{\text{loc}}(z))$ on $z \in \mathbf{T}$ and test for flatness. The Alignment condition is more challenging, especially using sampled data rather than rational functions, and promises to be a challenging research topic [74], [76]. Theorem 6.6.3 applies to the problem of computing all the Pareto points in an H^∞ multiobjective problem. Likewise, automating Theorem 6.6.3 in a Goal Attainment Method or the NBI Method await implementation. Our application of Theorem 6.6.3 focuses on the information it gives about a strict Pareto point, our observations regarding the shape of the amplifier's Gain-Noise image, and the consequences for the design of stable amplifiers.

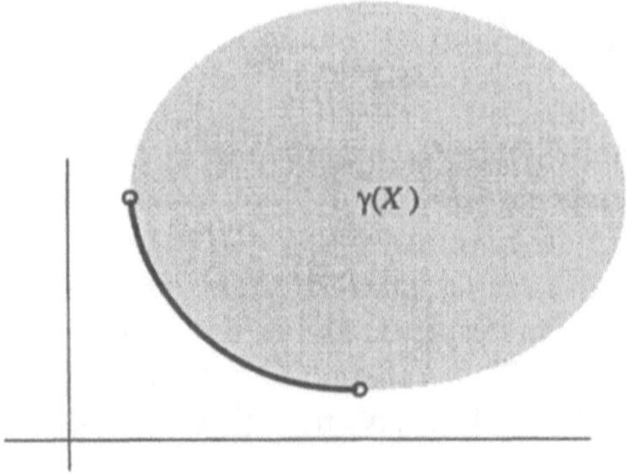

Fig. 6.25. Images of strict Pareto points.

6.7 Utopic Points and Stable Amplifier Design

The amplifier designer is limited by the time and effort it takes to assess each matching circuit and contend with its numerical difficulties [22]. And for any given design, the amplifier designer always has nagging doubts. Could new parameters be found that give better performance? Could another matching circuit topology give better performance? Both questions point to the fundamental question:

- What are the performance limits of the amplifier?

If the performance limits were known, the amplifier designer would like to know all possible design solutions. That is,

- What are all the Pareto points?

With respect to a specific circuit topology, the amplifier's multiobjective function

$$\gamma(\mathbf{x}) = \begin{bmatrix} -\|G_T(S_G(\mathbf{x}), S_L(\mathbf{x}))\|_{-\infty} \\ \|F(S_G(\mathbf{x}))\|_{\infty} \\ \|S_1(S_L(\mathbf{x}))\|_{\infty} \\ \|S_2(S_G(\mathbf{x}))\|_{\infty} \end{bmatrix} \qquad (\mathbf{x} \in X)$$

is bounded by the utopic point $\gamma_{\diamond,X}$

$$\gamma_{\diamond,X} \leq \gamma(X)$$

so that $\gamma_{\diamond,X}$ is pseudo-minimum element of $\gamma(X)$ and supports the computations of the Pareto points (Section 6.5).

To bound the amplifier's performance over all the matching circuits, observe that $S_G(\mathbf{x})$ and $S_L(\mathbf{x})$ belong to the orbits of $S_{G,0}$ and $S_{L,0}$ as discussed in Chapter 5. If these reflectances are zero and the matching circuits are lumped and lossless, the orbits are dense in $\Re\overline{B}\mathcal{A}_1^\infty(\mathbf{C}_+)$ by Darlington's Theorem of Chapter 5. The amplifier function generalizes to the disk algebra as

$$\gamma(S_G,S_L) = \begin{bmatrix} -\|G_T(S_G,S_L)\|_{-\infty} \\ \|F(S_G)\|_\infty \\ \|S_1(S_L)\|_\infty \\ \|S_2(S_G)\|_\infty \end{bmatrix}$$

with the corresponding minimization problem:

$$\inf\{\gamma(S_G,S_L) : S_G, S_L \in \Re\overline{B}\mathcal{A}_1(\mathbf{C}_+)\}.$$

How does Theorem 6.6.3 apply to this optimization problem? Theorem 6.6.3 has $\mathcal{A}_1(\mathbf{C}_+,\mathbf{C}^L)$ as the domain of a general γ. Motivated by the Goal Attainment Method, we can incorporate the unit ball constraints as follows:

$$\gamma(S_G,S_L) = \begin{bmatrix} -\|G_T(S_G,S_L)\|_{-\infty} \\ \|F(S_G)\|_\infty \\ \|S_1(S_L)\|_\infty \\ \|S_2(S_G)\|_\infty \\ \|S_G\|_\infty \\ \|S_L\|_\infty \end{bmatrix}, \quad \mathbf{w} = \begin{bmatrix} w_G \\ w_F \\ 0 \\ 0 \\ 0 \\ 0 \end{bmatrix}, \quad \gamma_u = \begin{bmatrix} -G_{T,u} \\ F_u \\ S_{1,u} \\ S_{2,u} \\ 1 \\ 1 \end{bmatrix},$$

to get $\mathcal{A}_1(\mathbf{C}_+,\mathbf{C}^2)$ as the domain of the amplifier's γ. The real constraint may also be incorporated but may be ignored provided the gain, noise, and stability functions are also real. Likewise, the range of the gain objective is not positive but can be adjusted in several ways: The positivity constraint can be generalized; we can use $\|G_T^{-1}\|_\infty$; we can add a large constant and floor; etc. However, the substantial impediments to applying Theorem 6.6.3 to the amplifier are its rank conditions: (a) that the number of objectives must not exceed the number of variables, and (f) $\mathrm{rank}[\partial\Gamma] \geq L$. Conversations with Helton indicate that the rank conditions are fundamental and cannot be relaxed.

We make several assumptions to apply Theorem 6.6.3. First, assume a local Pareto point $(S_{G,\mathrm{loc}}, S_{L,\mathrm{loc}})$ does exist in the open unit ball of the disk algebra. The localness allows us to drop the unit ball constraint. Second, assume the amplifier is unconditionally stable as discussed in Chapter 2. That is, we can drop the stability constraints and apply Theorem 6.6.3 to

$$\gamma(S_G,S_L) = \begin{bmatrix} -\|G_T(S_G,S_L)\|_{-\infty} \\ \|F(S_G)\|_\infty \end{bmatrix}$$

on a neighborhood of $(S_{G,\mathrm{loc}}, S_{L,\mathrm{loc}})$. The performance function

$$\Gamma(S_G,S_L) = \begin{bmatrix} -G_T(S_G,S_L) \\ F(S_G) \end{bmatrix}$$

has differential

$$\partial \Gamma = \begin{bmatrix} -\frac{\partial G_T}{\partial S_G} & -\frac{\partial G_T}{\partial S_L} \\ \frac{\partial F}{\partial S_G} & 0 \end{bmatrix}.$$

Under reasonable assumptions, we can verify that $\det[\partial \Gamma] \neq 0$ on the unit ball.

Lemma 6.7.1 *Assume $S_{\text{opt}} \in BL^{\infty}(j\mathbf{R})$ but does not belong to $H^{\infty}(\mathbf{C}_+)$. Then for all $S_G \in \overline{B}H^{\infty}(\mathbf{C}_+)$:*

$$\frac{\partial F}{\partial S_G}(S_G) \neq 0.$$

Proof: From Section 2.3,

$$F(S_G) = F_{\min} + \frac{4R_N}{Z_0} \frac{|S_G - S_{\text{opt}}|^2}{(1 - |S_G|^2)|1 + S_{\text{opt}}|^2}$$

so that

$$\frac{\partial F}{\partial S_G} = \frac{4R_N}{Z_0} \frac{(\overline{S_G} - \overline{S_{\text{opt}}})(1 - \overline{S_G} S_{\text{opt}})}{(1 - |S_G|^2)^2}.$$

Then $\partial F / \partial S_G = 0$ if and only if either $S_G = S_{\text{opt}}$ or $1 = \overline{S_G} S_{\text{opt}}$. The first possibility is excluded by the assumption that S_{opt} is not in $H^{\infty}(\mathbf{C}_+)$. The second possibility is excluded by the assumption that $\|S_{\text{opt}}\|_{\infty} < 1$. ///

The differential $\partial F = 0$ correctly identifies S_{opt} as the global minimum of the noise figure. However, S_{opt} is the reflectance obtained by tuning for minimum noise at each frequency—there is no guarantee that S_{opt} is analytic.

Lemma 6.7.2 *Assume the amplifier is strictly unconditionally stable. That is, for any S_G, $S_L \in \overline{B}H^{\infty}(j\mathbf{R})$,*

$$\|S_2(S_G)\|_{\infty}, \|S_1(S_L)\|_{\infty} < 1.$$

Assume the amplifier is nonconstant or that $S_2(S_G) \neq$ constant for any input reflectance. Then for all S_G, $S_L \in \overline{B}H^{\infty}(\mathbf{C}_+)$:

$$\frac{\partial G_T}{\partial S_L}(S_G, S_L) \neq 0.$$

Proof: Transducer power gain takes several forms [53, Eq. 3.2.2]:

$$G_T(S_G, S_L) = |S_{21}|^2 \frac{1 - |S_G|^2}{|1 - S_G S_{11}|^2} \frac{1 - |S_L|^2}{|1 - S_2(S_G) S_L|^2}.$$

Strict stability implies $G_T(S_G, S_L)$ is well defined with differential

$$\frac{\partial G_T}{\partial S_L} = |S_{21}|^2 \frac{1-|S_G|^2}{|1-S_G S_{11}|^2} \frac{(1-\overline{S_L S_L})(S_2-\overline{S_L})}{(|1-S_2(S_G)S_L|^2)^4}.$$

Because $|1 - S_2 S_L| \neq 0$, $\partial G_T/\partial S_L = 0$ if and only if $\overline{S_L} = S_2(S_G)$. This is precisely the condition for the optimal matching at the output [53, Eq. 3.6.2]. Now observe that both S_2 and S_L are H^∞ functions. For equality to hold with the conjugate in place, both functions must be real and constant. This is excluded by the nonconstant assumption. ///

Under the assumptions of both lemmas, $\det[\partial\Gamma] \neq 0$ and we have (a) and (f) of Theorem 6.6.3. Assume sufficient smoothness so that (c) and (d) hold. To have (b) hold, approximate the amplifier functions with rational functions. Now assume (e), that $(S_{G,\text{loc}}, S_{L,\text{loc}})$ is strict Pareto. Apply Theorem 6.6.3 to $(S_{G,\text{loc}}, S_{L,\text{loc}})$. The flatness condition is

$$G_T(S_{G,\text{loc}}, S_{L,\text{loc}}) = \text{constant},$$

$$F(S_{G,\text{loc}}) = \text{constant}.$$

The alignment condition is

$$\mathbf{u}(j\omega)^T \partial\Gamma(j\omega, \mathbf{h}_{\text{loc}}(j\omega)) = \frac{j\omega - 1}{j\omega + 1}\mathbf{g}(j\omega)^T,$$

where $\mathbf{g} = [g_1\ g_2]^T \in H^2(\mathbf{C}_+, \mathbf{C}^2)$ and $\mathbf{u} = [u_1\ u_2]^T \in C_1^1(j\mathbf{R}, \mathbf{R}_+^2)$ is nonzero. For the noise figure, flatness

$$F(S_{G,\text{loc}}) = \text{constant}$$

and alignment

$$u_1(j\omega)\frac{\partial F}{\partial S_G}(S_{G,\text{loc}}) = \frac{j\omega - 1}{j\omega + 1} g_2(j\omega) \quad \text{a.e.}$$

imply that $S_{G,\text{loc}}$ is a *local H^∞ minimizer of F*. Reaching back to the uniqueness results of Chapter 3, we assert that $S_{G,\text{loc}}$ is a global minimizer of F. There are a several ways see this. A sophisticated approach applies Theorem 6.6.2 under sufficient smoothness conditions of F. Indeed, Section 7.1 will show that F is a particularly simple function. A second approach reaches back to Corollary 3.10.5 but requires that we know that the *sublevel sets of F are disks.* This is common engineering knowledge and explored in excruciating detail in Chapter 7. By either approach, sufficient smoothness guarantees that $S_{G,\text{loc}}$ is actually a *global minimizer of F*:

$$\|F(S_{G,\text{loc}})\|_\infty = F_{\overline{B}H^\infty} := \inf\{\|F(S_G)\|_\infty : S_G \in \overline{B}H^\infty(\mathbf{C}_+)\}.$$

According to the comments following Theorem 6.6.3, a global minimizer cannot be a strict Pareto point. Consequently, we have the following conjecture:

Question 7 (Stable Amplifier Conjecture). Is the stable amplifier's utopic point actually the minimum element of its Gain-Noise image?

That is, the optimal performance is delivered by the images of the minimum element—there is no other point in the amplifier's Gain-Noise image with better gain and less noise. Figure 6.26 illustrates the Stable Amplifier Conjecture. The shape of the Gain-Noise image tells us how to compute the

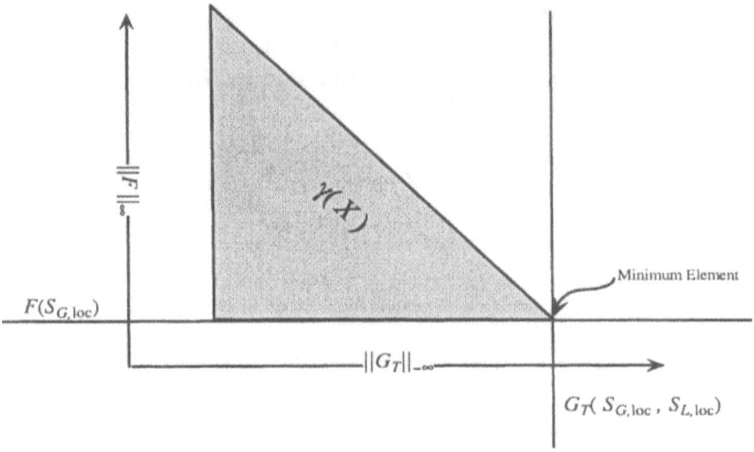

Fig. 6.26. Geometry of the Stable Amplifier Conjecture.

minimum element. The minimum element is computed by first minimizing the noise figure F to get $S_{G,\mathrm{loc}}$. Holding $S_{G,\mathrm{loc}}$ fixed, now maximize the gain

$$\min\{-\|G(S_{G,\mathrm{loc}}, S_L)\|_{-\infty} : S_L \in \overline{B}H^\infty(\mathbf{C}_+)\}$$

to get $S_{L,\mathrm{loc}}$. The uniqueness arguments of the noise minimizer should apply to the gain minimizer. Consequently, the $(S_{G,\mathrm{loc}}, S_{L,\mathrm{loc}})$ will be the global minimizer. Thus, the stable amplifier conjecture gives an algorithm for computing the best possible performance of the amplifier. In addition, the figure partly answers Question 4 regarding the shape of the Gain-Noise images and the location of the Pareto points in the almost minimum region.

This completes the discussion of the stable amplifier. We return to the more general conditionally stable amplifier. The "formal" utopic point for the amplifier function on the disk algebra is

$$\gamma_{\circ,\overline{B}A_1} := \begin{bmatrix} \inf\{-\|G_T(S_G, S_L)\|_{-\infty} : S_G, S_L \in \Re\overline{B}A_1(\mathbf{C}_+)\} \\ \inf\{\|F(S_G)\|_\infty : S_G \in \Re\overline{B}A_1(\mathbf{C}_+)\} \\ \inf\{\|S_1(S_L)\|_\infty : S_L \in \Re\overline{B}A_1(\mathbf{C}_+)\} \\ \inf\{\|S_2(S_G)\|_\infty : S_G \in \Re\overline{B}A_1(\mathbf{C}_+)\} \end{bmatrix}.$$

Extending the domain of γ to H^∞ gives the corresponding minimization problem

$$\inf\{\gamma(S_G, S_L) : S_G, S_L \in \overline{B}H^\infty(\mathbf{C}_+)\}$$

and the "formal" utopic point

$$\gamma_{\diamond,\overline{B}H^\infty} := \begin{bmatrix} \inf\{-\|G_T(S_G, S_L)\|_{-\infty} : S_G, S_L \in \overline{B}H_1^\infty(\mathbf{C}_+)\} \\ \inf\{\|F(S_G)\|_\infty : S_G \in \overline{B}H^\infty(\mathbf{C}_+)\} \\ \inf\{\|S_1(S_L)\|_\infty : S_L \in \overline{B}H^\infty(\mathbf{C}_+)\} \\ \inf\{\|S_2(S_G)\|_\infty : S_G \in \overline{B}H^\infty(\mathbf{C}_+)\} \end{bmatrix}$$

that bounds amplifier performance as

$$\gamma_{\diamond,\overline{B}H^\infty} \leq \gamma_{\diamond,\overline{B}\mathcal{A}_1} \leq \gamma_{\diamond,X}.$$

The quotes on "formal" signify some qualifications on the domain of γ. In the next chapter, we will see that gain and stability are intimately linked in our wideband amplifiers (Section 6.4). Because these amplifiers are only conditionally stable, it is possible to find matching circuits that yield infinite gain at some frequencies. To discard these unstable amplifiers requires that the domain of the gain function be restricted. Under these restrictions, we develop sufficient conditions so that equality holds:

$$\gamma_{\diamond,\overline{B}H^\infty} = \gamma_{\diamond,\overline{B}\mathcal{A}_1} = \lim_{\dim(X)\to\infty} \gamma_{\diamond,X}$$

for the stability and noise components. For the gain, we go after utopic points under the amplifier designer's stability constraints. These computations inform the amplifier designer if the desired constraints are feasible (i.e., there may be no circuit meeting the constraints). If the design constraints are feasible, the amplifier designer can benchmark any particular matching circuit. For example, if a matching circuit gets within 0.5 dB of the best possible gain-stability trade-off, as we will see in Chapter 8, the amplifier designer is spared a lengthy search over other competing circuits.

7

H^∞ Multidisk Methods

Suppose that our amplifier designer is given the amplifier of Section 6.3. At 4 GHz, the amplifier has scattering matrix

$$S = \begin{bmatrix} 0.6\angle -60^\circ & 0.05\angle 26^\circ \\ 1.9\angle 81^\circ & 0.5\angle -60^\circ \end{bmatrix},$$

with noise parameters $Z_0 = 50$ ohms; $F_{\min} = 1.6$ dB; $S_{\text{opt}} = 0.62\angle 100^\circ$; $R_N = 20$ ohms. Figure 7.1 shows the amplifier, matching circuits, input reflectance S_G and output reflectance S_L. Suppose the amplifier designer wants matching

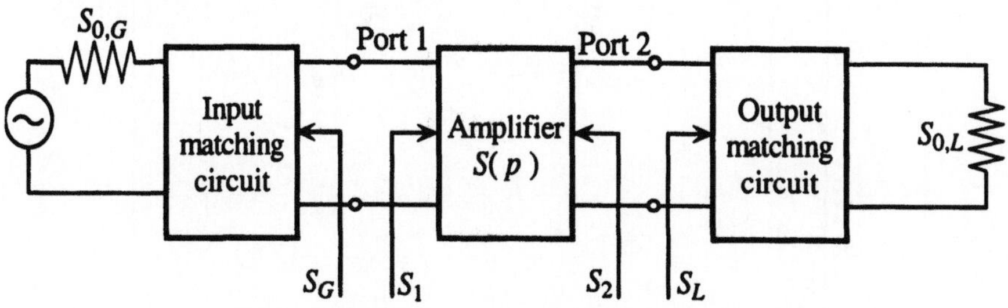

Fig. 7.1. Amplifier and matching circuits.

circuits to deliver a transducer power gain G_T in excess of 8.6 dB but suppress the noise figure F below 2 dB. The amplifier designer is asking

Can matching circuits be found with reflectances S_G and S_L such that $F(S_G) \le 2$ dB and $G_T(S_G, S_L) \ge 8.6$ dB?

The multidisk method simply plots the noise and gain disks and checks if they intersect. The transducer power gain factors as (Section 2.1):

$$G_T(S_G, S_L) = G_0 + G_G(S_G, S_L) + G_L(S_L) \quad [\text{dB}].$$

The amplifier gain is $G_0 = 5.58$ dB; the output gain G_L has a maximum value of 1.25 dB, so the input gain G_G must exceed 1.8 dB [106, page 630]. For ease of presentation, make the *unilateral approximation* (Section 2.1):

$$G_G(S_G, S_L) \approx G_G(S_G) = \frac{1 - |S_G|^2}{|1 - S_G S_{11}|^2}.$$

The unilateral approximation decouples the input and output reflectance so $G_T(S_G, S_L)$ is maximized by separately maximizing $G_G(S_G)$ and $G_L(S_L)$. In terms of the input reflectance, the design problem is to find an S_G in the unit disk with

- $F(S_G) \leq 2$ dB;
- $G_G(S_G) \geq 1.8$ dB.

Figure 7.2 shows that these gain and noise constraints correspond to gain and noise disks. The multidisk method simply checks if the gain and noise disks

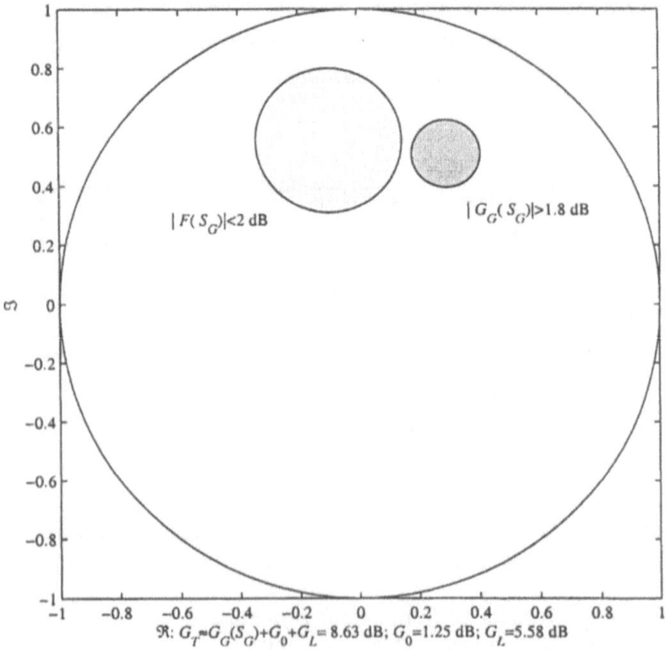

Fig. 7.2. Noise and input gain disks.

intersect. Our amplifier designer can glance at Figure 7.2 to see that the gain and noise disks do not intersect. That is, no input reflectance S_G can meet the design constraints. Consequently, no input matching circuit can meet the design constraints either. Our amplifier designer is forced to accept either more noise or less gain to get a realizable design. Once the noise and input gain disks overlap, any S_G in their intersection yields a line-stub matching circuit [106, Example 11.5]. However, this matching is good at a single frequency.

The wideband problem extends these disks as functions of frequency. That is, the amplifier designer specifies noise, gain, and stability constraints over a frequency band. At each frequency, these gain, noise, and stability constraints lead to the corresponding gain, noise, and stability regions, typically disks or the exterior of disks. Plotting these regions as functions of frequency gives rise to the gain, noise, and stability *tubes*.

An L^∞ *feasible design* requires that these tubes have a nonempty intersection over all frequencies. Physically, this nonempty intersection means that matching circuits can be found that meet the gain, noise, and stability constraints—on a frequency-by-frequency basis—rather than over all frequencies. Mathematically, this nonempty intersection means that there are input and output reflectances S_G, $S_L \in \overline{B}L^\infty(j\mathbf{R})$ that interpolate the constraint tube.

An L^∞ feasible design does not guarantee that these reflectances came from single input and output matching circuits. If these reflectances did come from terminated matching circuits, S_G and S_L must be *analytic*:

$$S_G, S_L \in \overline{B}H^\infty(\mathbb{C}_+).$$

An analytic and L^∞ feasible solution is an H^∞ *feasible design*: the gain, noise, and stability tubes have nonempty intersection with the closed unit ball of H^∞. This chapter develops an H^∞ Multidisk theory to handle the feasibility of these tubes as follows:

Noise: Section 7.1 discusses the noise figure and associated noise tube. The noise tube is the sublevel set of the noise figure

$$[F \le F_u] := \{S_G \in BL^\infty(j\mathbf{R}) : F(S_G; j\omega) \le F_u(j\omega) \text{ a.e.}\},$$

where $F_u(j\omega)$ is the frequency-dependent noise bound specified by the amplifier designer. The noise figure gives the strongest and simplest result because the noise figure F is really a function on $BL^\infty(j\mathbf{R})$. The H^∞ theory determines the best possible wideband noise bound the amplifier designer can demand. Specifically, we determine if F_u is an H^∞ feasible design:

$$\emptyset \ne \overline{B}H^\infty(\mathbb{C}_+) \cap [F \le F_u]?$$

If the intersection is nonempty, a passive reflectance S_G does exist that meets the constraint $F(S_G; j\omega) \le F_u(j\omega)$.

Stability: Section 7.2 develops the stability tubes. From Section 2.2, the input and output stability functions are

$$S_1(S_L) = S_{11} + S_{12}S_L(1 - S_{22}S_L)^{-1}S_{21},$$

$$S_2(S_G) = S_{22} + S_{12}S_G(1 - S_{11}S_G)^{-1}S_{12}.$$

Because the amplifier is active, these reflectances need not be passive. Our amplifier designer wants not only stability

$$\|S_1(S_L)\|_\infty \leq,$$

$$\|S_2(S_G)\|_\infty \leq 1$$

but stability as a function of frequency

$$|S_1(S_L; j\omega)| \leq S_{1,u}(j\omega) \leq 1,$$

$$|S_2(S_G; j\omega)| \leq S_{2,u}(j\omega) \leq 1.$$

The stability tubes are the corresponding sublevel sets

$$[|S_1| \leq S_{1,u}] := \{S_L \in L^\infty(j\mathbf{R}) : |S_1(S_L; j\omega)| \leq S_{1,u}(j\omega) \text{ a.e.}\},$$

$$[|S_2| \leq S_{2,u}] := \{S_G \in L^\infty(j\mathbf{R}) : |S_2(S_G; j\omega)| \leq S_{2,u}(j\omega) \text{ a.e.}\}.$$

The complication is that the stability tubes have cross sections that are either the interior of the stability circle, which is a disk, or the exterior to the stability circle, or a half-plane. Moreover, these cross sections may or may not intersect the unit disk. The H^∞ Multidisk Method handles stability by first checking if the stability constraints are L^∞ feasible:

$$\emptyset \neq \overline{B}L^\infty(j\mathbf{R}) \cap [|S_1| \leq S_{1,u}]?$$

$$\emptyset \neq \overline{B}L^\infty(j\mathbf{R}) \cap [|S_2| \leq S_{2,u}]?$$

If the stability constraints are L^∞ feasible, a tube with a *circular cross section* (CXS) can be threaded through the intersection. This CXS tube can be tested for intersection with $H^\infty(\mathbf{C}_+)$. A nonempty intersection means there are passive reflectances S_G or S_L that meet the stability constraints.

Gain: Section 7.3 develops a similar approach for the gain. The amplifier designer wants a transducer power gain

$$G_T(S_G, S_L; j\omega) \geq G_{T,u}(j\omega).$$

This is handled by factoring G_T into its input and output gains:

$$G_G(S_G, S_L; j\omega) \geq G_{G,u}(j\omega),$$

$$G_L(S_L; j\omega) \geq G_{L,u}(j\omega).$$

The output gain is the easiest to analyze. The output gain tube is the superlevel set:

$$[G_L \geq G_{L,u}] := \{S_L \in L^\infty(j\mathbf{R}) : G_L(S_L; j\omega) \geq G_{L,u}(j\omega) \text{ a.e.}\}.$$

We first determine if $G_{L,u}$ is L^∞ feasible;

$$\emptyset \neq \overline{B}L^\infty(j\mathbf{R}) \cap [G_L \geq G_{L,u}]?$$

The complication is that without a stability constraint, the gain analysis is not very practical. Thus, the amplifier designer determines if the gain and stability constraints are L^∞ stable and feasible:

$$\emptyset \neq \overline{B}L^\infty(j\mathbf{R}) \cap [|S_1| \leq S_{1,u}] \cap [G_L \geq G_{L,u}]?$$

If the output gain is L^∞ stable and feasible, the next step is to ask if the gain and stability constraints are H^∞ stable and feasible:

$$\emptyset \neq \overline{B}H^\infty(\mathbf{C}) \cap [|S_1| \leq S_{1,u}] \cap [G_L \geq G_{L,u}]?$$

As in the stability analysis, we thread the L^∞ stable and feasible region with a tube of circular cross section and determine if this CXS tube intersects H^∞. More complicated is the input gain. The amplifier designer asks for input gain

$$G_G(S_G, S_L; j\omega) \geq G_{G,u}(j\omega)$$

while guaranteeing stability:

$$(S_G, S_L) \in [|S_2| \leq S_{2,u}] \times [|S_1| \leq S_{1,u}].$$

Thus, the L^∞ stable and feasible test for the input gain is

$$\emptyset \neq \overline{B}L^\infty(j\mathbf{R}, \mathbf{C}^2) \cap [G_G \geq G_{G,u}] \cap \{[|S_2| \leq S_{2,u}] \times [|S_1| \leq S_{1,u}]\}?$$

We offer new results showing how the input gain links to the stability and develops an H^∞ stability and feasibility test. However, even finding a good visualization of this set for the amplifier designer is a challenging graphical problem in $j\mathbf{R} \times \mathbf{C} \times \mathbf{C}$.

The H^∞ Multidisk Method: Section 7.4 pulls all the constraints together. The L^∞ test is to find center and radius functions

$$C_X := \begin{bmatrix} S_{X,G} \\ S_{X,L} \end{bmatrix}, \quad R_X = \begin{bmatrix} R_{X,G} & 0 \\ 0 & R_{X,L} \end{bmatrix},$$

such that the disk

$$\overline{D}(C_X, R_X) := \{\phi \in L^\infty(j\mathbf{R}, \mathbf{C}^2) : (\phi - C_X)(\phi - C_X)^H \leq R_X^2\}$$

is contained in the intersection of the amplifier's disks:

$$\begin{aligned} \overline{D}(C_X, R_X) \subseteq \overline{B}L^\infty(j\mathbf{R}, \mathbf{C}^2) &\cap [G_G \geq G_{G,u}] \\ &\cap \{[F \leq F_u] \times [|S_1| \leq S_{1,u}]\} \\ &\cap \{[|S_2| \leq S_{2,u}] \times [G_L \geq G_{L,u}]\}. \end{aligned}$$

This CXS disk is tested for intersection with H^∞:

$$\emptyset \neq H^\infty(\mathbb{C}_+, \mathbb{C}^2) \cap \overline{D}(C_X, R_X)?$$

Because there is no cross coupling,

$$\overline{D}(C_X, R_X) = \overline{D}(C_{X,G}, R_{X,G}) \times \overline{D}(C_{X,L}, R_{X,L}),$$

so intersection with H^∞ can be tested on each individual disk. Finally, we have not exploited the full disk structure with the scaling matrix P_X:

$$\overline{D}(C_X, R_X, P_X) := \left\{\phi \in L^\infty(j\mathbf{R}, \mathbb{C}^2) : (\phi - C_X)P^2(\phi - C_X)^H \leq R_X^2\right\}.$$

Thus, research on these coupled disks, the visualization of such disks, the interaction between R_X and P_X, and fitting the "biggest" disk in the amplifier's multidisks are all open topics. Specific questions on these topics are made explicit in this chapter. More general research topics are discussed in Chapter 10.

7.1 Noise Tubes

The wideband noise disks are determined from the amplifier designer's operational constraints. Because these disks are indexed by frequency, plotting each disk as a function of frequency generates the *noise tube*. Referring to Figure 7.1, the input matching circuit determines the noise figure F. The amplifier designer selects a class of lossless input matching circuits. Let $\mathcal{U}_G \subseteq U^+(2)$ denote the corresponding class of scattering matrices. The amplifier designer wants to minimize the noise figure F over $\mathcal{U}_G$. Each $\mathcal{S}_G \in \mathcal{U}_G$ determines the noise figure F at each frequency ω as

$$F(S_G; j\omega) = F_{\min}(j\omega) + \frac{4R_N(j\omega)}{Z_0} \frac{|S_G(j\omega) - S_{\text{opt}}(j\omega)|^2}{(1 - |S_G(j\omega)|^2)|1 + S_{\text{opt}}(j\omega)|^2}, \quad (7.1)$$

where the generator's reflectance S_G is

$$S_G := \mathcal{F}_2(\mathcal{S}_G, S_{0,G}) = \mathcal{S}_{G,22} + \mathcal{S}_{G,21}S_{0,G}(1 - \mathcal{S}_{G,11}S_{0,G})^{-1}\mathcal{S}_{G,12}.$$

Before starting the analysis of the noise figure, two observations are helpful. First, the noise figure minimizer is $S_G = S_{\text{opt}}$. If S_{opt} belongs to the unit ball of H^∞, the noise figure minimization problem is solved. However, $S_{\text{opt}}(j\omega)$ is the reflectance that minimizes the noise figure at each individual frequency. There is no guarantee that the resulting function is analytic. Thus, we assume that S_{opt} is only L^∞. Second, $\|S_G\|_\infty > 1$ produces a negative noise figure. To avoid this complication, the domain of F is always restricted to some subset that forces S_G in the unit ball.

By carefully restricting the domain of the noise figure, its sublevel sets are shown to be disks contained in $BL^\infty(j\mathbf{R})$. These disks prove that the noise figure is weak* lower semicontinuous. This continuity and the weak* compactness $\overline{B}L^\infty(j\mathbf{R})$ produce minimizers for the noise figure. Moreover, these disks let us apply Nehari's Theorem (Chapter 3) to determine the H^∞ feasibility of a noise figure design.
Assume the following:

NF-1 Z_0 is a positive constant.
NF-2 $F_{\min} \in L^\infty(j\mathbf{R})$ and $F_{\min} \geq 1$.
NF-3 $R_N \in L^\infty(j\mathbf{R})$ is even, positive, and $R_N^{-1} \in L^\infty(j\mathbf{R})$.
NF-4 $S_{\text{opt}} \in B\Re L^\infty(j\mathbf{R})$.

These assumptions, coupled with $S_G \in BL^\infty(j\mathbf{R})$, make $F(S_G)$ a well-defined function in $L^\infty(j\mathbf{R})$. We will also take considerable freedom with the domains of F. Let $[\mathcal{U}_G, S_{0,G}]$ denote the open subset of $\mathcal{U}_G$ consisting of those scattering matrices $\mathcal{S}_G$ that satisfy $\|\mathcal{F}_2(\mathcal{S}_G, S_{0,G})\|_\infty < 1$. The noise figure F is a well-defined mapping of $[\mathcal{U}_G, S_{0,G}]$ into $L^\infty(j\mathbf{R})$. Denote this mapping as

$$F : [\mathcal{U}_G, S_{0,G}] \to L^\infty(j\mathbf{R}).$$

If $\mathcal{U}_G = U^+(2,\infty)$ and $S_{0,G} = 0$, Darlington's Theorem(Chapter 5) gives that the S_G's are dense in $\overline{B}\Re\mathcal{A}_1(\mathbf{C}_+)$. Converting the domain of the noise figure to these 1-ports gives the noise figure as the mapping

$$F : B\Re\mathcal{A}_1(\mathbf{C}_+) \to L^\infty(j\mathbf{R}).$$

Minimizers over this domain have the physical interpretation that there exists a lumped, lossless 2-port of finite degree that realizes the minimum with arbitrary precision. We can extend the domain so as to apply Nehari's Theorem to the mapping

$$F : BH^\infty(\mathbf{C}_+) \to L^\infty(j\mathbf{R}).$$

Finally, we can extend the noise figure's domain as

$$F : BL^\infty(j\mathbf{R}) \to L^\infty(j\mathbf{R}).$$

This is the most general domain for the noise figure. The amplifier designer wants the noise figure F to be bounded by user-specified $F_u \in L^\infty(j\mathbf{R})$.

The Noise Figure Problem. Determine if there exists an S_G in $\overline{B}H^\infty(\mathbf{C}_+)$ such that $F_u(j\omega) \geq F(S_G; j\omega)$ a.e.

If a solution exists, then the amplifier designer knows that specified noise figure F_u is H^∞ realizable. The next refinement asks whether there exists an input reflectance $S_G \in B\Re\mathcal{A}_1(\mathbf{C}_+)$ such that $F_u(j\omega) \geq F(S_G; j\omega)$ If a solution exists, then the amplifier designer knows that there is a lumped, lossless 2-port that solves the Noise Figure Problem to arbitrary precision.

Although $F_u \geq F(S_G)$ is a disk [106, pages 628–632], the technical complication is that these disks are not necessarily in $BL^\infty(j\mathbf{R})$. If $F_u = F_{\min}$, then the noise disk has radius zero—only S_{opt} belongs to the noise disk. Consequently, this noise disk belongs to $BL^\infty(j\mathbf{R})$ if and only if S_{opt} is strictly passive. This explains the open unit ball assumption in NF-4. Strict passivity also lets us increase F_u from $F_{\min}$ while keeping the noise disk in the open ball. Therefore, the trick is to put an upper bound on F_u that keeps the noise disk strictly inside the unit disk. This is the rationale for Equation 7.2.

Theorem 7.1.1 *Assume NF-1, NF-2, NF-3, and NF-4 hold. Then the noise figure F is well defined and a continuous mapping*

$$F : BL^\infty(j\mathbf{R}) \to L^\infty(j\mathbf{R}).$$

Select $\Delta s \in [0,1)$ that constrains the amplifier designer's $F_u \in L^\infty(j\mathbf{R})$ as

$$\frac{(\Delta s - |S_{\text{opt}}|)^2}{1 - \Delta s^2} \times \frac{4R_N/Z_0}{|1 + S_{\text{opt}}|^2} \geq F_u - F_{\min} \geq 0. \tag{7.2}$$

Define the noise figure parameter *[106, Eq. 11.59]*

$$N_u = \frac{F_u - F_{\min}}{4R_N/Z_0}|1 + S_{\text{opt}}|^2,$$

the noise figure center and radius functions *[106, Eq. 11.61]*

$$C_{F_u} = \frac{S_{\text{opt}}}{N_u + 1}; \quad R_{F_u} = \frac{\sqrt{N_u(N_u + 1 - |S_{\text{opt}}|^2)}}{N_u + 1},$$

and the noise figure disk

$$\overline{D}(C_{F_u}, R_{F_u}) := \{\phi \in L^\infty(j\mathbf{R}) : |\phi(j\omega) - C_{F_u}(j\omega)| \leq R_{F_u}(j\omega) \text{ a.e.}\}.$$

Then the following are equivalent:

(a) F_u is H^∞ realizable; there is an $S_G \in \overline{B}H^\infty(\mathbf{C}_+)$ such that $F_u(j\omega) \geq F(S_G; j\omega)$.
(b) The noise disk $\overline{D}(C_{F_u}, R_{F_u})$ has nonempty intersection with the closed unit ball of $H^\infty(\mathbf{C}_+)$: $\emptyset \neq \overline{B}H^\infty(\mathbf{C}_+) \cap \overline{D}(C_{F_u}, R_{F_u})$.
(c) $\mathcal{H}^*_{C_{F_u}} \mathcal{H}_{C_{F_u}} \leq \mathcal{T}_{R^2_{F_u}}$.

Proof: The assumptions NF-1, NF-2, NF-3, and NF-4 show that F is well defined as a mapping on the open unit ball $BL^\infty(j\mathbf{R})$. To see continuity, observe that F is the composition of the following continuous functions:

- By NF-2, the shift $\phi \mapsto \phi + F_{\min}$ is continuous on $L^\infty(j\mathbf{R})$.
- By NF-1, NF-3, and NF-4 the scaling

$$\phi \mapsto \frac{4R_N}{Z_0}\frac{\phi}{|1 + S_{\text{opt}}|^2}$$

is continuous on $L^\infty(j\mathbf{R})$.

- If $\phi \in BL^\infty(j\mathbf{R})$, the inversion $\phi \mapsto (1-|\phi|^2)^{-1}$ is continuous on $L^\infty(j\mathbf{R})$.
- By NF-4, $\phi \mapsto |\phi - S_{\text{opt}}|^2$ is continuous on $L^\infty(j\mathbf{R})$.

Thus, F is continuous on $BL^\infty(j\mathbf{R})$. The next part of the proof computes a bound on the sublevel sets of F to guarantee strict inclusion in the open unit ball. The positivity constraints in NF-3 show that N_u is well defined and $N_u \in L^\infty(j\mathbf{R})$, likewise for the noise figure center and radius functions. NF-3 gives that R_N is strictly positive. Thus,

$$\frac{4R_N/Z_0}{|1+S_{\text{opt}}|^2} \geq Z_0^{-1}\|R_N\|_{-\infty} > 0.$$

As Δs increases to 1, this strictly positive lower bound shows that Equation 7.2 is eventually true. Equation 7.2 is equivalent to following inequality on the noise figure parameter:

$$N_u \leq \frac{(\Delta s - |S_{\text{opt}}|)^2}{1-\Delta s^2}.$$

Straightforward but tedious algebra shows that this is equivalent to

$$|C_{F_u}| + R_{F_u} \leq \Delta s < 1 \iff \overline{D}(C_{F_u}, R_{F_u}) \subseteq \Delta s \times \overline{B}L^\infty(j\mathbf{R}). \tag{7.3}$$

Thus, Equation 7.2 forces the noise disk into $BL^\infty(j\mathbf{R})$. Because the noise disks belong to the unit ball and the domain of F is $BL^\infty(j\mathbf{R})$, the sublevel sets are the noise disks [106, Eq. 11.60]:

$$F_u \geq F(S_G) \iff S_G \in \overline{D}(C_{F_u}, R_{F_u}).$$

Thus, F_u is H^∞ realizable if and only if

$$F_u \geq F(S_G \in \overline{B}H^\infty(\mathbf{C}_+)) \iff \overline{B}H^\infty(\mathbf{C}_+) \cap \overline{D}(C_{F_u}, R_{F_u}) \neq \emptyset.$$

This demonstrates that (a) and (b) are equivalent. Because the noise disk is contained in $BL^\infty(j\mathbf{R})$, it follows that

$$\overline{B}H^\infty(\mathbf{C}_+) \cap \overline{D}(C_{F_u}, R_{F_u}) = H^\infty(\mathbf{C}_+) \cap \overline{D}(C_{F_u}, R_{F_u}).$$

Nehari's Theorem applies to the last intersection and gives the equivalence of (b) and (c) using Lemma 3.7.1 or [65, page 1132], [66, Theorem 4.2]. The even assumption in NF-3 lets us omit the conjugation operator $\check{R}_{F_u}$ [66, page 41]. The adjoint "*" can be placed on either Hankel operator. ///

Theorem 7.1.1 takes Δs as a constant. For more design freedom, Δs can be a measurable function. An important special case occurs when F_u is constant:

$$F_u \geq F(S_G; j\omega) \iff F_u \geq \|F(S_G)\|_\infty.$$

Figure 7.3 illustrates a noise tube with constant F_u for the NE32484A wideband amplifier discussed in Section 6.4. For each frequency, the collection of

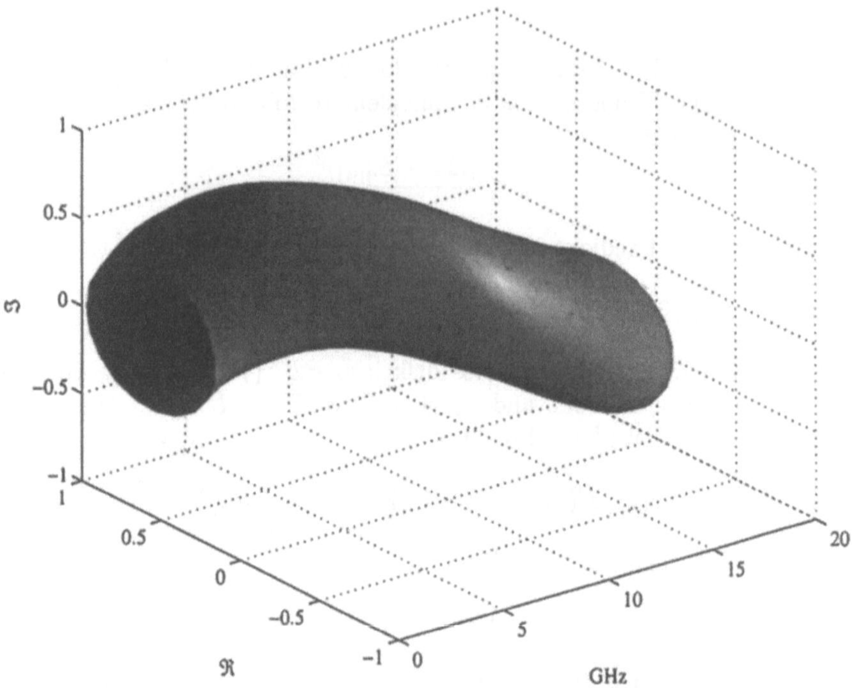

Fig. 7.3. Noise tube of the wideband amplifier NE32484A.

reflectances $S_G \in \mathbf{D}$ that satisfy

$$F(S_G; j\omega) \leq F_u$$

constitute a noise disk as illustrated in Figure 7.2:

$$\overline{D}(C_{F_u}(j\omega), R_{F_u}(j\omega)) \subset \mathbf{D}.$$

Plotting these noise disks as a function of frequency generates the noise tube. Figure 7.3 shows this noise tube with "GHz" labeling the frequency axis and "ℜ" and "ℑ" labeling the real and imaginary axis, respectively.

If $S_G \in BH^\infty(\mathbf{C}_+)$ satisfies $F(S_G; j\omega) \leq F_u$, the analytic curve

$$\omega \mapsto S_G(j\omega)$$

"threads" the noise tube. Figure 7.4 looks at the noise tube from the high frequencies. We see that the noise tube is "pinching off" at the higher frequencies. As F_u is decreased, the noise tube will pinch off more and more

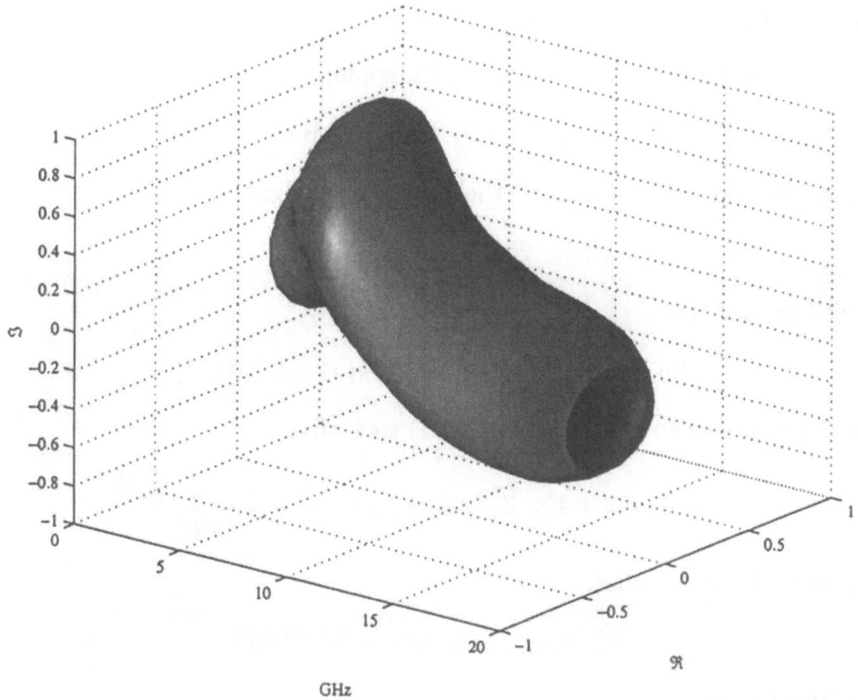

Fig. 7.4. View of the high-frequency "pinching" of the noise tube of the wideband amplifier NE32484A.

until no analytic curve can thread the noise tube. This leads us to ask the following question: *What is the smallest noise figure that is H^∞ realizable?* Because F_u is constant,

$$F(S_G, j\omega) \leq F_u \iff \|F(S_G)\|_\infty \leq F_u,$$

so we are really asking if we can compute

$$\inf\{\|F(S_G)\|_\infty : S_G \in \overline{B}H^\infty(\mathbf{C}_+)\}.$$

This infimum is computed by Nehari's Theorem. With the best possible noise figure computed, we can ask if the infimum is really a minimum. That is, does a minimizer S_G exist? And if a minimizer exists, is it unique? And if a unique minimizer S_G exists, can we use S_G to extract an input matching circuit? The following results address these questions. However, extracting a matching circuit is rather difficult. Chapter 10 discusses several research on this circuit extraction problem.

Theorem 7.1.2 *Assume NF-1, NF-2, NF-3, and NF-4. Then*

$$\inf\{\|F(S_G)\|_\infty : S_G \in BH^\infty(\mathbf{C}_+)\}$$

admits minimizers.

Proof:. Theorem 7.1.1 shows that the mapping

$$\|F\|_\infty : BL^\infty(j\mathbf{R}) \to \mathbf{R}_+$$

taking the input reflectances to their worst noise figure

$$S_G \mapsto \|F(S_G)\|_\infty$$

is well defined. Let

$$F_{BH^\infty} := \inf\{\|F(S_G)\|_\infty : S_G \in BH^\infty(\mathbf{C}_+)\}.$$

NF-3 permits us to select a $\Delta s_+ \in [0,1)$ such that

$$F_+ := F_{BH^\infty} + 1 \le Z_0^{-1}\|R_N\|_{-\infty} \times \frac{(\Delta s_+ - \|S_{\mathrm{opt}}\|_\infty)^2}{1 - \Delta s_+^2}.$$

Equation 7.3 shows that

$$\overline{D}(C_{F_+}, R_{F_+}) \subseteq \Delta s_+ \times \overline{B}L^\infty(j\mathbf{R}).$$

By construction,

$$F_{BH^\infty} := \inf\{\|F(S_G)\|_\infty : S_G \in BH^\infty(\mathbf{C}_+) \cap \overline{D}(C_{F_+}, R_{F_+})\}.$$

Observe that $\|F\|_\infty$ restricted to $\overline{D}(C_{F_+}, R_{F_+})$ has sublevel sets that are disks in $\Delta s_+ \times \overline{B}L^\infty(j\mathbf{R})$. These disk are weak* compact by Lemma 3.7.1. That is,

$$\|F\|_\infty : \overline{D}(C_{F_+}, R_{F_+}) \to \mathbf{R}_+$$

is lower semicontinuous in the weak* topology of $L^\infty(j\mathbf{R})$. Because $H^\infty(\mathbf{C}_+)$ is weak* closed,

$$\|F\|_\infty : \overline{D}(C_{F_+}, R_{F_+}) \cap \overline{B}H^\infty(\mathbf{C}_+) \to \mathbf{R}_+$$

is also weak* lower semicontinuous on its weak* compact domain. The existence of minimizers follow by the Weierstrass Theorem. ///

With the existence of minimizers settled, the next problem is to estimate

$$F_{BH^\infty} := \min\{\|F(S_G)\|_\infty : S_G \in BH^\infty(\mathbb{C}_+)\}.$$

Theorem 7.1.1(c) characterizes solutions as

$$\mathcal{T}_{R^2_{F_u}} - \mathcal{H}^*_{C_{F_u}} \mathcal{H}_{C_{F_u}} \geq 0. \tag{7.4}$$

By estimating the Fourier coefficients using a Fast Fourier Transform (FFT) and truncating the Toeplitz and Hankel operators, we obtain a matrix approximation to Equation 7.4. When the smallest eigenvalue of this matrix approximation changes sign, we have an approximation to F_{BH^∞}. Section 10.7 offers a detailed description of this algorithm.

Figure 7.5 plots such an approximation. The details of the computation are described in Section 10.7. What Figure 7.5 reports is that

$$F_{BH^\infty} \approx 1 \quad [\text{dB}].$$

A smaller F_u violates $F_u \geq F_{\min}$. Moreover, the 1 dB minimum accords with the "pinching" of the noise tube observed in Figure 7.4.

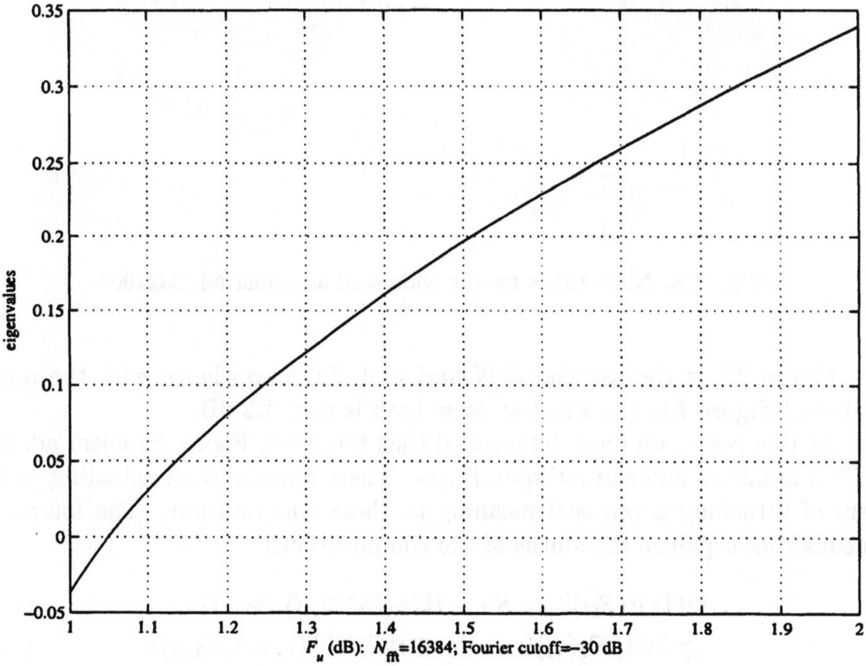

Fig. 7.5. Nehari noise curve for the wideband amplifier NE32484A.

It is worthwhile to compare this amplifier to the more recent NEC321000 discussed in Section 6.4. Figure 7.6 plots noise tubes with $F_u = 1, 2, 3, 4$ dB. The generous disks at 2, 3 and 4 dB indicates that an H^∞ function can "thread" these tubes.

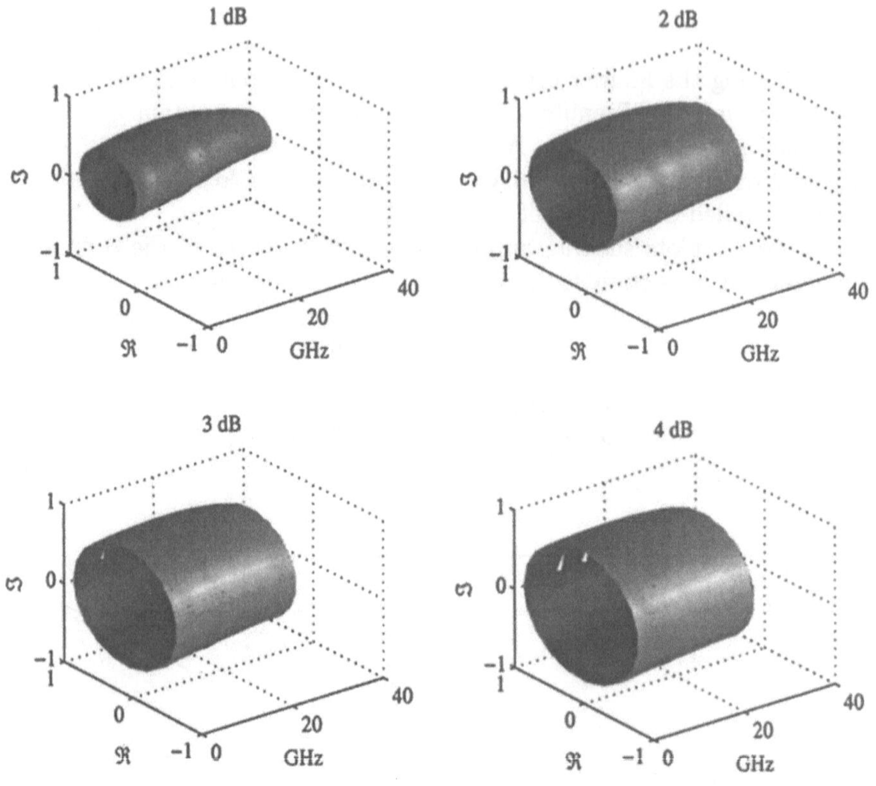

Fig. 7.6. Noise tubes for the wideband amplifier NEC321000.

Figure 7.7 is the associated Nehari plot. In concordance with the noise tubes of Figure 7.6, the smallest noise level is near 1.2 dB.

At this point, we have determined that the Noise Figure Problem admits H^∞ minimizers and can estimate F_{BH^∞}. There remains the fascinating problem of attaching a physical meaning to these computations. The following inequalities organize the status of the computations:

$$\inf\{\|F(S_G)\|_\infty : S_G \in B\mathcal{F}_2(U^+(2,d), S_{0,G})\} \tag{7.5}$$

$$\geq \inf\{\|F(S_G)\|_\infty : S_G \in B\mathcal{F}_2(U^+(2,\infty), S_{0,G})\} \tag{7.6}$$

$$\geq \inf\{\|F(S_G)\|_\infty : S_G \in \Re BA_1(\mathbf{C}_+)\}|_{S_{0,G}=0} \tag{7.7}$$

$$\overset{\text{Theorem 7.1.2}}{\geq \min} \{\|F(S_G)\|_\infty : S_G \in BH^\infty(\mathbf{C}_+)\}. \tag{7.8}$$

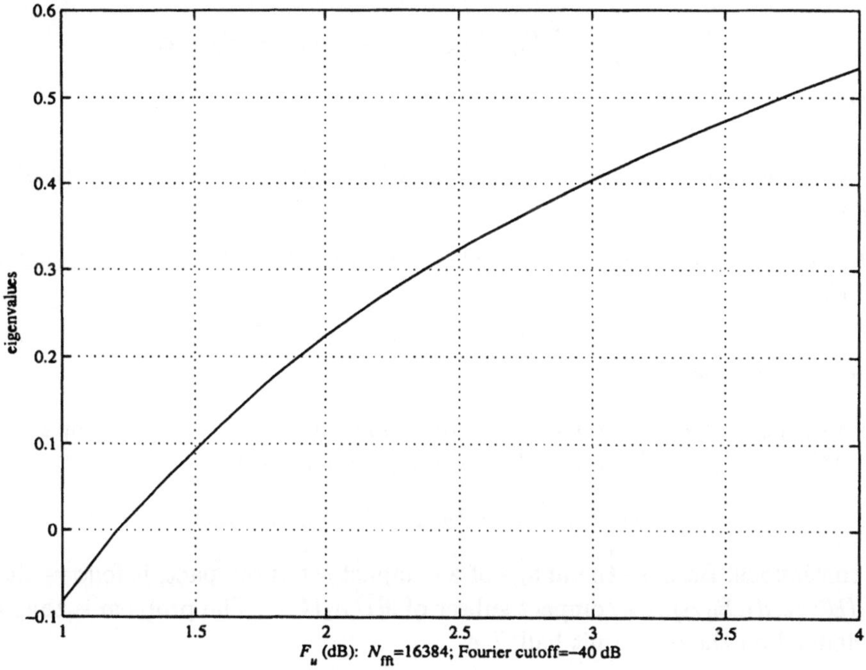

Fig. 7.7. Nehari noise curve for the wideband amplifier NEC321000.

The first line minimizes the noise figure over the lumped circuits of degree d. Because $U^+(2,\infty)$ contains all the lumped circuits, set inclusion forces the inequality between (7.5) and (7.6). If we make the assumption that the generator's reflectance is $S_{0,G} = 0$, Darlington's Theorem says that $B\mathcal{F}_2(U^+(2,\infty),0)$ is dense in $\Re B\mathcal{A}_1(\mathbb{C}_+)$. This set inclusion forces the inequality between (7.6) and (7.7). Finally, the inclusion of the disk algebra in $H^\infty(\mathbb{C}_+)$ gives the inequality between (7.7) and (7.8). The remainder of this section aims at turning these "inf" 's into "min" 's and the inequalities into equalities. Continuity of F eliminates the inequality between (7.6) and (7.7).

Theorem 7.1.3 *Assume NF-1, NF-2, NF-3, and NF-4 hold. Then*

$$\begin{aligned}&\inf\{\|F(S_G)\|_\infty : S_G \in B\mathcal{F}_2(U^+(2,\infty),0)\}\\&= \inf\{\|F(S_G)\|_\infty : S_G \in \Re B\mathcal{A}_1(\mathbb{C}_+)\}.\end{aligned}$$

Proof: Because F is defined only on the open unit ball, we must watch B and $\overline{B}$ carefully. Let F_U and $F_{\mathcal{A}}$ denote the left and right infima, respectively. We always have $F_U \geq F_{\mathcal{A}}$ by set inclusion. To prove equality, let $\epsilon > 0$ and select $S_G \in \Re B\mathcal{A}_1(\mathbb{C}_+)$ such that $\|F(S_G)\|_\infty \leq F_{\mathcal{A}} + \epsilon$. By Darlington's Theorem, $B\mathcal{F}_2(U^+(2,\infty),0)$ is dense in $\Re B\mathcal{A}_1(\mathbb{C}_+)$ so that there is a sequence

$$\mathcal{F}_2(U^+(2,\infty),0) \ni S_{G,n} \to S_G.$$

The continuity of F (Theorem 7.1.1) lets us take the limit to get

$$F_U \leq \lim_{n\to\infty} \|F(S_{G,n})\|_\infty = \|F(S_G)\|_\infty \leq F_{\mathcal{A}} + \epsilon.$$

///

Continuity of F also lets us pick up a "min" in (7.5) because $U^+(2,d)$ is compact.

Theorem 7.1.4 *Let $S_{0,G} \in BH^\infty(\mathbf{C}_+)$. Assume NF-1, NF-2, NF-3, and NF-4. Then*

$$\inf\{\|F(S_G)\|_\infty : S_G \in \mathcal{F}_2(U^+(2,d), S_{0,G})\}$$

admits minimizers.

Proof: There are two steps to this reasoning. First, $U^+(2,d)$ is compact in $H^\infty(\mathbf{C}_+, \mathbf{C}^{2\times 2})$ by Theorem 4.2.2. Second, the mapping

$$U^+(2,d) \ni S_G \mapsto \mathcal{F}_2(S_G, S_{0,G}) \in L^\infty(j\mathbf{R})$$

is continuous. Because the image of a compact set is compact, it follows that $\mathcal{F}_2(U^+(2,d), S_{0,G})$ is a compact subset of $\Re\overline{B}\mathcal{A}_1(\mathbf{C}_+)$. The problem is that F is defined on the open unit ball. Let

$$F_d := \inf\{\|F(S_G)\|_\infty : S_G \in \mathcal{F}_2(U^+(2,d), 0)\}.$$

Let $S_+ \in \mathcal{F}_2(U^+(2,d), 0)$ such that

$$F_d < \|F(S_+)\|_\infty =: F_+.$$

As in the proof of Theorem 7.1.2, construct the center C_{F_+} and radius functions R_{F_+}. Then NF-1, NF-2, NF-3, and NF-4 imply that

$$F_d = \inf\{\|F(S_G)\|_\infty : S_G \in \mathcal{F}_2(U^+(2,d), 0) \cap \overline{D}(C_{F_+}, R_{F_+})\}.$$

Because F is continuous and the intersection is compact, we get the minimum in (7.5). ///

Summarizing the progress so far, there holds

$$\begin{aligned}
&\overset{\text{Theorem 7.1.4}}{\min} \{\|F(S_G)\|_\infty : S_G \in B\mathcal{F}_2(U^+(2,d), S_{0,G})\} \\
&\geq \inf\{\|F(S_G)\|_\infty : S_G \in B\mathcal{F}_2(U^+(2,\infty), S_{0,G})\} \\
&\overset{\text{Darlington}}{=} \inf\{\|F(S_G)\|_\infty : S_G \in \Re B\mathcal{A}_1(\mathbf{C}_+)\}|_{S_{0,G}=0} \\
&\geq \overset{\text{Theorem 7.1.2}}{\min} \{\|F(S_G)\|_\infty : S_G \in BH^\infty(\mathbf{C}_+)\}.
\end{aligned}$$

The next goal is equality between (7.7) and (7.8). Because of the real constraints in the noise figure assumptions, F is real symmetric with convex sublevel sets. By variations of the arguments in Section 3.12, we claim

$$\begin{aligned}&\inf\{\|F(S_G)\|_\infty : S_G \in \Re B\mathcal{A}_1(\mathbb{C}_+)\}\\&= \inf\{\|F(S_G)\|_\infty : S_G \in B\mathcal{A}_1(\mathbb{C}_+)\}.\end{aligned}$$

Assume NF-1, NF-2, NF-3, and NF-4 hold. Theorem 7.1.2 produces a minimizer

$$F_{BH^\infty} = \min\{\|F(S_G)\|_\infty : S_G \in BH^\infty(\mathbb{C}_+)\} = \|F(S_{G,\min})\|_\infty$$

with noise figure center and noise functions that satisfy

$$S_{G,\min} \in \overline{D}(C_{\min}, R_{\min}) \cap H^\infty(\mathbb{C}_+).$$

If $Q_{\min}$ is a spectral factor of $R_{\min}$, the membership in the noise disk is equivalent to $|Q_{\min}^{-1} S_{G,\min} - Q_{\min}^{-1} C_{\min}| \le 1$. Because we are at an H^∞ minimum,

$$\|Q_{\min}^{-1} C_{\min} + H^\infty(\mathbb{C}_+)\|_\infty = 1.$$

With continuity of $Q_{\min}^{-1} C_{\min}$, Theorem 3.9.2 implies equality with the disk algebra:

$$1 = \|Q_{\min}^{-1} C_{\min} + H^\infty(\mathbb{C}_+)\|_\infty = \|Q_{\min}^{-1} C_{\min} + \mathcal{A}_1(\mathbb{C}_+)\|_\infty.$$

With Dini-continuity of $Q_{\min}^{-1} C_{\min}$, Corollary 3.10.5 implies equality with the disk algebra and the existence of a unique minimizer in the disk algebra.

If a minimizer exists for the norm in the disk algebra, the reasoning may be reversed to get a minimizer for the noise figure in the disk algebra. The interesting part is to use this argument when only a *minimizing sequence* exists in the disk algebra.

Theorem 7.1.5 *Assume*

NF-1c Z_0 *is a positive constant.*
NF-2c $F_{\min} \in 1 \dotplus C_0(j\mathbf{R})$ *and* $F_{\min} \ge 1$.
NF-3c $R_N \in 1 \dotplus C_0(j\mathbf{R})$ *is even, positive, and* $R_N^{-1} \in 1 \dotplus C_0(j\mathbf{R})$.
NF-4c $S_{\text{opt}} \in B\Re\{1 \dotplus C_0(j\mathbf{R})\}$.

Assume also that $F_{BH^\infty} > \|F_{\min}\|_\infty$. *Then*

$$\inf\{\|F(S_G)\|_\infty : S_G \in B\mathcal{A}_1(\mathbb{C}_+)\} = \inf\{\|F(S_G)\|_\infty : S_G \in BH^\infty(\mathbb{C}_+)\}.$$

Proof: There always holds

$$\begin{aligned}F_{BA} &:= \inf\{\|F(S_G)\|_\infty : S_G \in B\mathcal{A}_1(\mathbb{C}_+)\}\\&\ge \min\{\|F(S_G)\|_\infty : S_G \in BH^\infty(\mathbb{C}_+)\} =: F_{BH^\infty}.\end{aligned}$$

Suppose the inequality is strict. Then there is an $S_G \in BH^\infty(\mathbb{C}_+)$ such that

$$F_{BA} > \|F(S_G)\|_\infty. \tag{7.9}$$

Observe that the continuity of F implies $[F < F_{BA}]$ is open. Moreover,

$$[F < F_{BA}] \subset D(C_{\mathcal{A}}, R_{\mathcal{A}}),$$

where the noise figure center and radius functions

$$C_{\mathcal{A}} = \frac{S_{\text{opt}}}{N_{\mathcal{A}}+1}, \quad R_{\mathcal{A}} = \frac{\sqrt{N_{\mathcal{A}}(N_{\mathcal{A}}+1-|S_{\text{opt}}|^2)}}{N_{\mathcal{A}}+1},$$

are determined from the noise figure parameter

$$N_{\mathcal{A}} = \frac{F_{B\mathcal{A}} - F_{\min}}{4R_N/Z_0}|1+S_{\text{opt}}|^2.$$

Let $R_{\mathcal{A}}$ have spectral factorization $R_{\mathcal{A}} = |Q_{\mathcal{A}}|$. By Theorem 3.11.1,

$$\|Q_{\mathcal{A}}^{-1}C_{\mathcal{A}} - Q_{\mathcal{A}}^{-1}S_G\|_\infty < 1.$$

If we assume that $Q_{\mathcal{A}}^{-1}C_{\mathcal{A}} \in 1 \dotplus C_0(j\mathbf{R})$, Theorem 3.9.2 forces equality:

$$1 > \|Q_{\mathcal{A}}^{-1}C_{\mathcal{A}} - H^\infty(\mathbf{C}_+)\|_\infty = \|Q_{\mathcal{A}}^{-1}C_{\mathcal{A}} - \mathcal{A}_1(\mathbf{C}_+)\|_\infty.$$

The strict equality lets us select $S_{\mathcal{A}} \in \mathcal{A}_1(\mathbf{C}_+)$ that satisfies

$$1 - \epsilon_0 > \|Q_{\mathcal{A}}^{-1}C_{\mathcal{A}} - Q_{\mathcal{A}}^{-1}S_{\mathcal{A}}\|_\infty,$$

for some $1 > \epsilon_0 > 0$. This forces the pointwise result:

$$(1-\epsilon_0)R_{\mathcal{A}} \geq |C_{\mathcal{A}} - S_{\mathcal{A}}|.$$

Wit some effort, we will show that this disk equation implies

$$\|F(S_{\mathcal{A}})\|_\infty < F_{B\mathcal{A}}.$$

This contradiction implies that Equation 7.9 cannot be true or that the inequality $F_{B\mathcal{A}} \geq F_{BH^\infty}$ cannot be strict. To start this demonstration, first observe that our assumptions make $R_{\mathcal{A}}$ uniformly positive:

$$\begin{aligned}
R_{\mathcal{A}} &\geq \frac{N_{\mathcal{A}}}{1+N_{\mathcal{A}}} \\
&\geq \|1+N_{\mathcal{A}}\|_\infty^{-1} N_{\mathcal{A}} \\
&\geq \|1+N_{\mathcal{A}}\|_\infty^{-1} \left(F_{B\mathcal{A}} - F_{\min}\right)\frac{(1-\|S_{\text{opt}}\|_\infty)^2}{4\|R_N\|_\infty/Z_0} \\
&\geq \|1+N_{\mathcal{A}}\|_\infty^{-1}(F_{B\mathcal{A}} - \|F_{\min}\|_\infty)\frac{(1-\|S_{\text{opt}}\|_\infty)^2}{4\|R_N\|_\infty/Z_0}.
\end{aligned}$$

Second, define for $\epsilon \in [0, \epsilon_0]$,

$$N(\epsilon) = \frac{(1-\epsilon)F_{B\mathcal{A}} - F_{\min}}{4R_N/Z_0}|1+S_{\mathrm{opt}}|^2,$$

and

$$C(\epsilon) = \frac{S_{\mathrm{opt}}}{N_{\mathcal{A}}(\epsilon)+1}, \quad R(\epsilon) = \frac{\sqrt{N(\epsilon)(N(\epsilon)+1-|S_{\mathrm{opt}}|^2)}}{N(\epsilon)+1}.$$

In $L^\infty(j\mathbf{R})$, $N(\epsilon) \to N_{\mathcal{A}}$, $C(\epsilon) \to C_{\mathcal{A}}$, and $R(\epsilon) \to R_{\mathcal{A}}$ as $\epsilon \to 0$. Then

$$\begin{aligned}|S_{\mathcal{A}} - C(\epsilon)| &\le |S_{\mathcal{A}} - C_{\mathcal{A}}| + |C_{\mathcal{A}} - C(\epsilon)| \\ &\le (1-\epsilon_0)R_{\mathcal{A}} + |C_{\mathcal{A}} - C(\epsilon)| \\ &\le (1-\epsilon_0)R(\epsilon) + |R_{\mathcal{A}} - R(\epsilon)| + |C_{\mathcal{A}} - C(\epsilon)|.\end{aligned}$$

Because the last two terms are bounded as $\mathcal{O}[\epsilon]$:

$$|S_{\mathcal{A}} - C(\epsilon)| \le R(\epsilon) - \epsilon_0 R(\epsilon) + \mathcal{O}[\epsilon].$$

Because $R_{\mathcal{A}}$ is uniformly positive, and $R(\epsilon)$ converges to $R_{\mathcal{A}}$, the last two terms are uniformly negative for all $\epsilon > 0$ sufficiently small. This puts $S_{\mathcal{A}} \in \overline{D}(C(\epsilon), R(\epsilon))$ or that $F(S_{\mathcal{A}}) \le (1-\epsilon)F_{B\mathcal{A}}$. Finally, it remains to demonstrate continuity of $Q_{\mathcal{A}}^{-1}C_{\mathcal{A}}$. By the continuity assumptions NF-2c, NF-3c, and NF-4c, both $C_{\mathcal{A}}$ and $R_{\mathcal{A}}$ belong to $1 \dotplus C_0(j\mathbf{R})$ so it remains to demonstrate continuity of $Q_{\mathcal{A}}^{-1}$. Wrap $R_{\mathcal{A}}$ on the unit circle by the Cayley transform in Lemma 3.5.1: $r_{\mathcal{A}} = R_{\mathcal{A}} \circ \mathbf{c}^{-1} \in C(\mathbf{T})$. Positivity implies that $\log(r_{\mathcal{A}}) \in C(\mathbf{T})$ and defines the outer function [42, page 24]:

$$q_{\mathcal{A}}(z) := \exp\left(\frac{1}{2\pi}\int_0^{2\pi} \frac{e^{jt}+z}{e^{jt}-z}\log(r_{\mathcal{A}}(e^{jt}))dt\right) \in \mathcal{A}(\mathbf{D}).$$

Lemma 3.5.1 gives that $Q_{\mathcal{A}} = q_{\mathcal{A}} \circ \mathbf{c} \in \mathcal{A}_1(\mathbf{C}_+)$ is an outer function. Thus, $Q_{\mathcal{A}}^{-1} \in \mathcal{A}_1(\mathbf{C}_+)$. ///

Under the assumptions NF-1c, NF-2c, NF-3c, and NF-4c there holds

$$\begin{aligned} &\overset{\text{Theorem 7.1.4}}{\min} \{\|F(S_G)\|_\infty : S_G \in B\mathcal{F}_2(U^+(2,d), S_{0,G})\} \\ &\ge \inf\{\|F(S_G)\|_\infty : S_G \in B\mathcal{F}_2(U^+(2,\infty), S_{0,G})\} \\ &\overset{\text{Darlington}}{=} \inf\{\|F(S_G)\|_\infty : S_G \in \Re B\mathcal{A}_1(\mathbf{C}_+)\}|_{S_{0,G}=0} \\ &\overset{\text{Theorem 7.1.5}}{=} \overset{\text{Theorem 7.1.2}}{\min} \{\|F(S_G)\|_\infty : S_G \in BH^\infty(\mathbf{C}_+)\}. \end{aligned}$$

What this means is that there is a sequence of lumped, lossless 2-ports $\{S_{G,n}\} \in U^+(2,\infty)$ that get arbitrarily close to the computable H^∞ minimum of the noise figure:

$$\|F(\mathcal{F}_2(S_{G,n}, 0))\|_\infty \to F_{BH^\infty}.$$

Slightly stronger continuity assumptions should produce a disk algebra minimizer (getting $Q_{\min}^{-1}C_{\min}$ to be Dini-continuous).

Question 8. What modifications of NF-1, NF-2, NF-3, and NF-4 produce a noise-minimizing S_G in the disk algebra?

This completes our work on the noise figure bound. For context, recall from Section 6.6 that the utopic point

$$\gamma_{\diamond,\overline{B}\mathcal{A}_1} := \begin{bmatrix} \inf\{-\|G_T(S_G,S_L)\|_{-\infty} : S_G, S_L \in \Re\overline{B}\mathcal{A}_1(\mathbf{C}_+)\} \\ \inf\{\|F(S_G)\|_\infty : S_G \in \Re\overline{B}\mathcal{A}_1(\mathbf{C}_+)\} \\ \inf\{\|S_1(S_L)\|_\infty : S_L \in \Re\overline{B}\mathcal{A}_1(\mathbf{C}_+)\} \\ \inf\{\|S_2(S_G)\|_\infty : S_G \in \Re\overline{B}\mathcal{A}_1(\mathbf{C}_+)\} \end{bmatrix}$$

needs to be computed. This section "fills in" the second component of $\gamma_{\diamond,\overline{B}\mathcal{A}_1}$. The rest of this chapter fills in the remaining components, starting with stability.

Before turning to the stability bounds, we close this section with two questions of a more practical nature. First, it is one thing to claim that a lumped 2-port exists that can get arbitrarily close to the noise bound but quite another to actually produce a circuit. The question of finding a practical input matching circuit that gets close to the best possible noise figure, albeit computable, is an active topic in amplifier design [106], [53] and is discussed in Section 10.5.

The second question relates to Darlington's Theorem. The basic assumption in this noise-figure computation is that the generator's reflectance $S_{G,0} = 0$. How can we get tight bounds without this constraint? This essentially asks how the various matching orbits of $S_{G,0} \neq 0$ reside in the unit ball of H^∞ and is also a major theme of Chapter 10.

7.2 Stability Tubes

For the amplifier Figure 2.1 to be stable, the input and output reflectances must be passive. That is, the only useful matching circuits are those that force

$$S_1 = S_{11} + \frac{S_{12}S_{21}S_L}{1 - S_{22}S_L} \in \overline{B}H^\infty(\mathbf{C}_+),$$

$$S_2 = S_{22} + \frac{S_{12}S_{21}S_G}{1 - S_{11}S_G} \in \overline{B}H^\infty(\mathbf{C}_+).$$

A more general approach to stability design starts with the amplifier designer specifying frequency-dependent stability bounds:

$$|S_1(S_L; j\omega))| \le S_{1,u}(j\omega) \le 1,$$

$$|S_2(S_G; j\omega))| \le S_{2,u}(j\omega) \le 1.$$

The stability problem is to characterize those S_L, $S_G \in \overline{B}H^\infty(\mathbf{C}_+)$ that satisfy these stability constraints.

It is claimed that an amplifier may be unconditionally stable: for all input and output reflectances S_L, $S_G \in \overline{B}H^\infty(\mathbb{C}_+)$

$$\|S_1(S_L)\|_\infty, \|S_2(S_G)\|_\infty \leq 1.$$

The beauty of an unconditionally stable amplifier is that the stability constraints may be ignored. An amplifier is unconditionally stable if and only if [106, Eqs. 11.31, 11.32]:

US-1 $K = \frac{1-|S_{11}|^2-|S_{22}|^2+|\det[S]|^2}{2|S_{12}S_{21}|} > 1.$
US-2 $|\det[S]| < 1.$

In our experience, unconditionally stable wideband amplifiers are not common. For example, Figure 7.8 shows that the NE32484A amplifier is not unconditionally stable (excepting the frequency band near 14 GHz). Consequently, this section develops conditional stability designs.

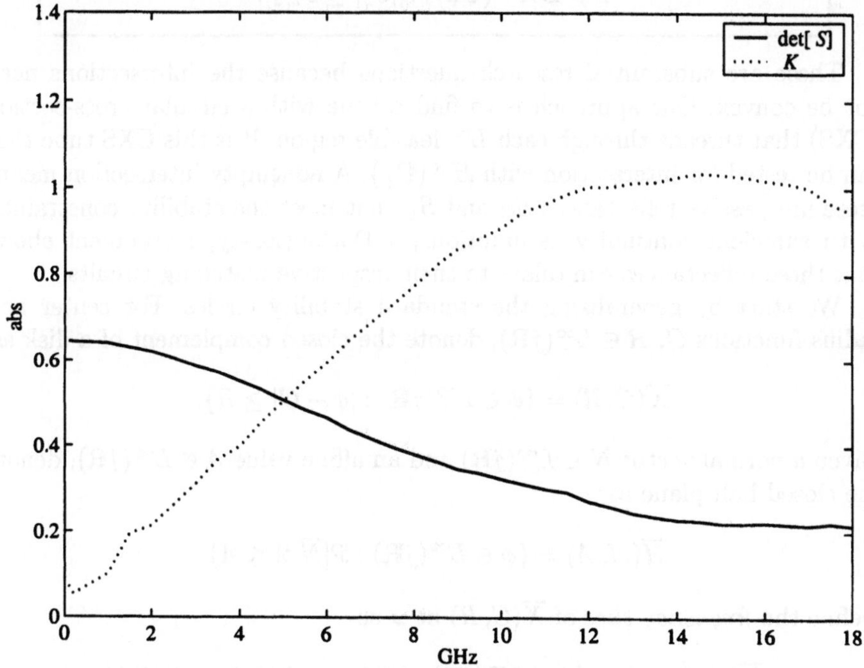

Fig. 7.8. The NE32484A is not unconditionally stable on 0.1–18 GHz.

Consider the input and output stability sublevel sets:

$$[|S_1| \leq S_{1,u}] := \{S_L \in L^\infty(j\mathbf{R}) : |S_1(S_L; j\omega))| \leq S_{1,u}(j\omega) \text{ a.e.}\},$$

$$[|S_2| \leq S_{2,u}] := \{S_G \in L^\infty(j\mathbf{R}) : |S_2(S_G; j\omega))| \leq S_{2,u}(j\omega) \text{ a.e.}\}.$$

The approach to stability breaks into two tests. The first test asks if the stability constraints are L^∞ *feasible*:

The L^∞ Feasibility Problem. Given $S_{1,u}$, $S_{2,u} \in L^\infty(j\mathbf{R})$, determine if

$$\emptyset \neq \overline{B}L^\infty(j\mathbf{R}) \cap [|S_1| \le S_{1,u}],$$
$$\emptyset \neq \overline{B}L^\infty(j\mathbf{R}) \cap [|S_2| \le S_{2,u}].$$

If the stability constraints are L^∞ feasible, the second test should question the H^∞ *feasibility* of the stability constraints:

The H^∞ Feasibility Problem. Given $S_{1,u}$, $S_{2,u} \in L^\infty(j\mathbf{R})$, determine if

$$\emptyset \neq \overline{B}H^\infty(\mathbf{C}_+) \cap [|S_1| \le S_{1,u}],$$
$$\emptyset \neq \overline{B}H^\infty(\mathbf{C}_+) \cap [|S_2| \le S_{2,u}].$$

These are substantial research questions because the intersections need not be convex. Our approach is to find a tube with a circular cross-section (CXS) that threads through each L^∞ feasible region. It is this CXS tube that can be tested for intersection with $H^\infty(\mathbf{C}_+)$. A nonempty intersection means there are passive reflectances S_G and S_L that meet the stability constraints. With sufficient continuity assumptions, a Darlington-type argument shows that these reflectances can dilate to their respective matching circuits.

We start by generalizing the standard stability circles. For center and radius functions C, $R \in L^\infty(j\mathbf{R})$, denote the closed complement of a disk as

$$\overline{X}(C,R) = \{\phi \in L^\infty(j\mathbf{R}) : |\phi - C| \ge R\}.$$

Given a normal vector $N \in L^\infty(j\mathbf{R})$ and an affine value $A \in L^\infty(j\mathbf{R})$, denote the closed half plane as

$$\overline{H}(N,A) = \{\phi \in L^\infty(j\mathbf{R}) : \Re[\overline{N}\phi] \le A\}.$$

Define the *frequency slice* of $\overline{X}(C,R)$ at ω as

$$\overline{X}(C,R;j\omega) = \{\phi \in L^\infty(j\mathbf{R}) : |\phi(j\omega) - C(j\omega)| \ge R(j\omega)\}$$

and likewise for other sets. At a single frequency, the frequency slice makes no sense. However, usage takes frequency slices over frequency bands of positive measure or in the space of continuous functions.

Port 1: Generalize the output stability center and radius functions [106, Eq. 11.28]

$$C_L = \frac{S_{1,u}^2 \overline{S_{22}} - \overline{\det[S]} S_{11}}{S_{1,u}^2 |S_{22}|^2 - |\det[S]|^2}, \quad R_L = S_{1,u} \left| \frac{S_{12} S_{21}}{S_{1,u}^2 |S_{22}|^2 - |\det[S]|^2} \right|.$$

The sublevel set $[|S_1| \leq S_{1,u}]$ has frequency slices that are either the interior of the stability disk, its exterior, or a half plane:

$$[S_1 \leq S_{1,u}; j\omega] = \begin{cases} \overline{D}(C_L, R_L; j\omega) & S_{1,u}(j\omega)|S_{22}(j\omega)| < |\det[S(j\omega)]| \\ \overline{X}(C_L, R_L; j\omega) & S_{1,u}(j\omega)|S_{22}(j\omega)| > |\det[S(j\omega)]| \\ \overline{H}(N_L, A_L; j\omega) & \text{otherwise} \end{cases}.$$

The half plane is determined by its normal and affine values:

$$N_L = -\overline{\det[S]} S_{11}, \quad A_L = (S_{1,u}^2 - |S_{11}|^2)/2 - \Re[S_{1,u}^2 S_{22}].$$

The appearance of the half plane shows that the stability regions do not "jump" from one part of the unit disk to another when the denominators change sign. As Figure 7.9 illustrates, the stability region stays in the same general location while the stability circle transits through a half plane.

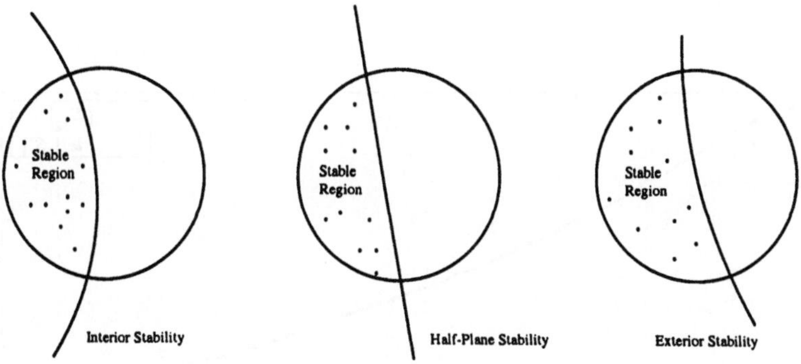

Fig. 7.9. Transition from interior to exterior stability.

Port 2: Generalize the input stability center and radius functions [106, Eq. 11.29]

$$C_G = \frac{S_{2,u}^2 \overline{S_{11}} - \overline{\det[S]} S_{22}}{S_{2,u}^2 |S_{11}|^2 - |\det[S]|^2}, \quad R_G = S_{2,u} \left| \frac{S_{12} S_{21}}{S_{2,u}^2 |S_{11}|^2 - |\det[S]|^2} \right|.$$

The sublevel set $[|S_2| \leq S_{2,u}]$ has frequency slices that are either the interior of the stability disk, its exterior, or a half plane:

$$[S_2 \le S_{2,u}; j\omega] = \begin{cases} \overline{D}(C_G, R_G; j\omega) & S_{2,u}(j\omega)|S_{11}(j\omega)| < |\det[S(j\omega)]| \\ \overline{X}(C_G, R_G; j\omega) & S_{2,u}(j\omega)|S_{11}(j\omega)| > |\det[S(j\omega)]| \\ \overline{H}(N_G, A_G; j\omega) & \text{otherwise} \end{cases},$$

where the half plane is determined by its normal and affine values:

$$N_G = -\overline{\det[S]}S_{22}, \quad A_G = (S_{2,u}^2 - |S_{22}|^2)/2 - \Re[S_{2,u}^2 S_{11}].$$

The following figures display the stability regions of the NE32484A amplifier (Section 6.4). We first assess the stability behavior of S_1 with $S_{1,u} = 1$. Figure 7.10 compares $|S_{22}|$ and $|\det[S]|$. The exterior stability holds for $f > 2$ GHz. Figure 7.11 plots the corresponding stability tube

$$[|S_1| = 1] = \{S_L \in L^\infty(j\mathbf{R}) : |S_1(S_L(j\omega))| = 1\}$$

and the unit tube

$$\{S_L \in L^\infty(j\mathbf{R}) : |S_L(j\omega)| = 1\}.$$

By Figure 7.10, the stable S_L's must thread the interior of the unit tube and the exterior of the stability tube. Next, we assess the stability behavior of S_2 with $S_{2,u} = 1$. Figure 7.12 shows exterior stability holding over all frequencies. Figure 7.13 shows the stability tube $[|S_2| = 1]$ and the unit tube. The stable S_G's must thread the interior of the unit tube and the exterior of the stability tube.

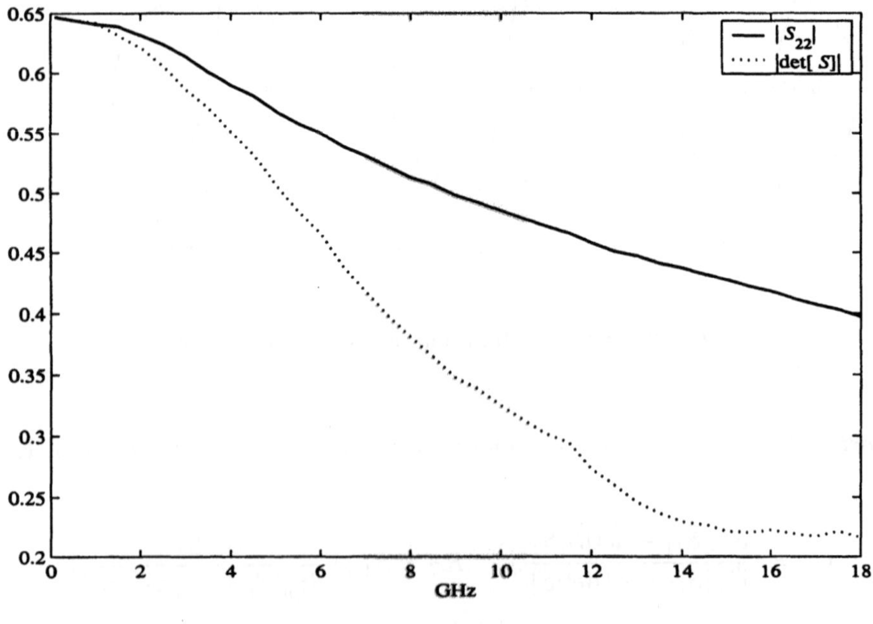

Fig. 7.10. Stability Test for S_1 of the NE32484A.

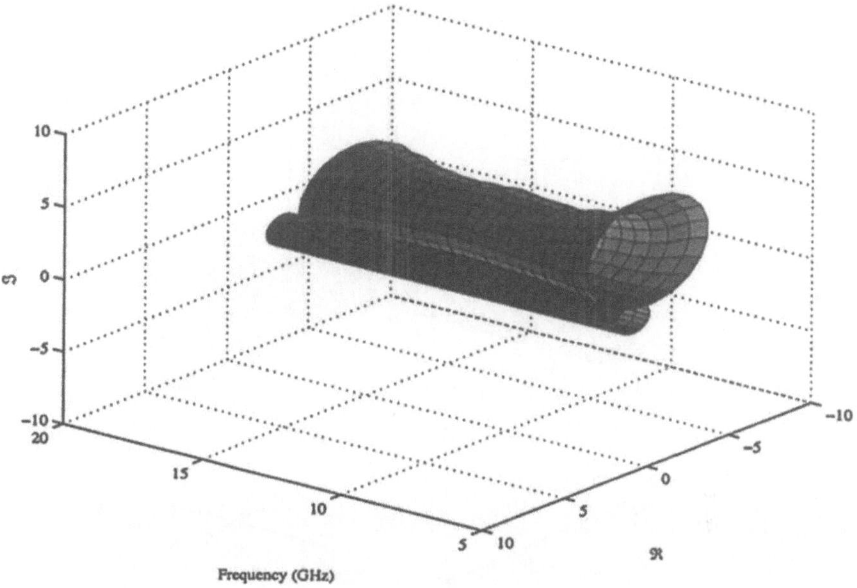

Fig. 7.11. NE32484A's S_1 stability tube and the unit tube.

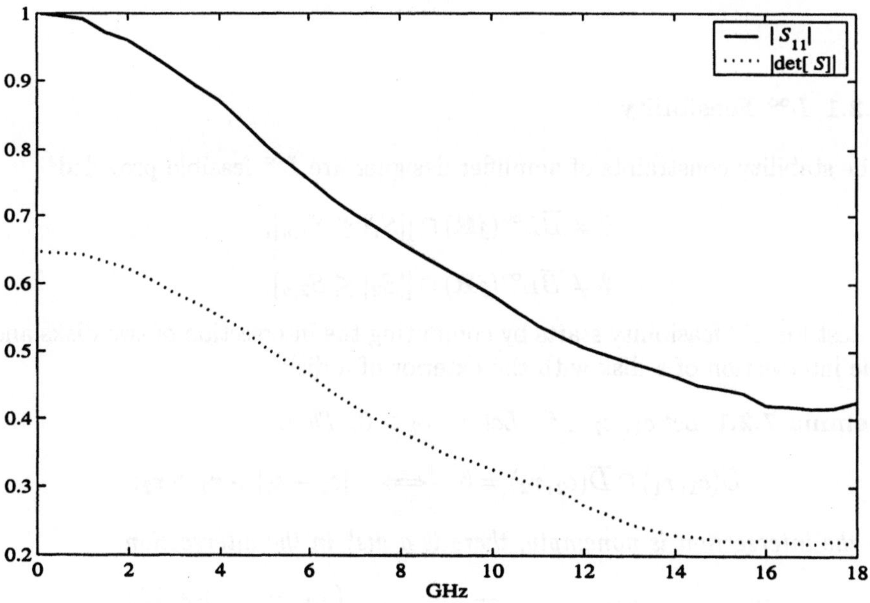

Fig. 7.12. Stability Test for S_2 of the NE32484A.

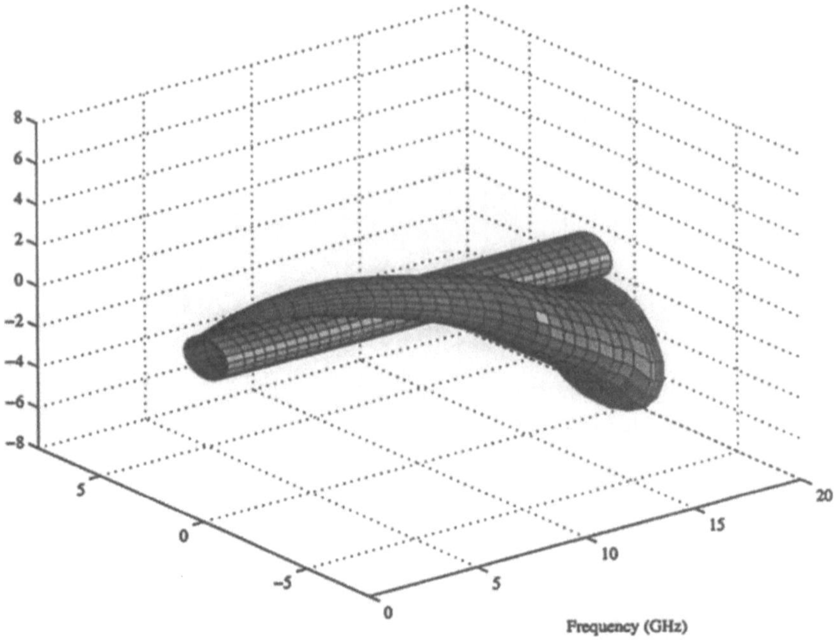

Fig. 7.13. NE32484A's S_2 stability tube and the unit tube.

7.2.1 L^∞ Feasibility

The stability constraints of amplifier designer are L^∞ feasible provided

$$\emptyset \neq \overline{B}L^\infty(j\mathbf{R}) \cap [|S_1| \leq S_{1,u}],$$

$$\emptyset \neq \overline{B}L^\infty(j\mathbf{R}) \cap [|S_2| \leq S_{2,u}].$$

A test for L^∞ feasibility starts by computing the intersection of two disks and the intersection of a disk with the exterior of a disk.

Lemma 7.2.1 *Let* c_1, $c_2 \in \mathbf{C}$. *Let* r_1, $r_2 > 0$. *Then*

$$\overline{D}(c_1, r_1) \cap \overline{D}(c_2, r_2) = \emptyset \quad \Longleftrightarrow \quad |c_1 - c_2| > r_1 + r_2.$$

If the intersection is nonempty, there is a disk in the intersection:

$$\overline{D}(c_1, r_1) \cap \overline{D}(c_2, r_2) \supseteq \overline{D}(c_X, r_X) \quad \begin{cases} c_X = \{e_1 + e_2\}/2 \\ r_X = |e_1 - e_2|/2 \end{cases},$$

where the "endpoints" e_1 *and* e_2 *are computed from the "convex coordinates" of the line segment between* c_1 *and* c_2

$$\begin{cases} e_1 = c_1 + t_1'(c_2 - c_1) \; t_1' := \min\{1 + t_2, t_1\} \quad t_1 = r_1|c_2 - c_1|^{-1} \\ e_2 = c_1 + t_2'(c_2 - c_1) \; t_2' := \max\{-t_1, 1 - t_2\} \; t_2 = r_2|c_2 - c_1|^{-1} \end{cases}.$$

Proof: The intersection test transcribes the convex coordinates of Figure 7.14. The convex coordinates of the line determined by c_1 and c_2 are

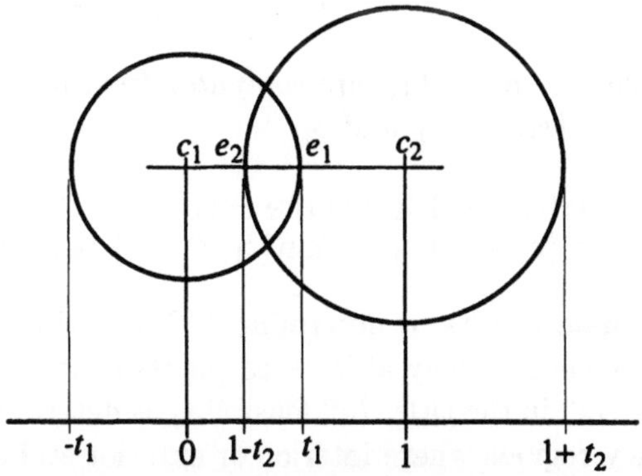

Fig. 7.14. Convex coordinates.

$$c(t) = c_1 + t(c_2 - t_1).$$

Figure 7.14 shows that the endpoints determine the center and radius of a disk inscribed in the intersection. The only technical complication is to make sure that the endpoints are properly truncated to avoid picking up extra area. Figure 7.14 shows that the bounds are as follows: $0 < t_1 \leq 1 + t_2$ and $-t_1 \leq 1 - t_2 < 1$. ///

Figure 7.15 presents the intersections between disk and the exterior of a disk.

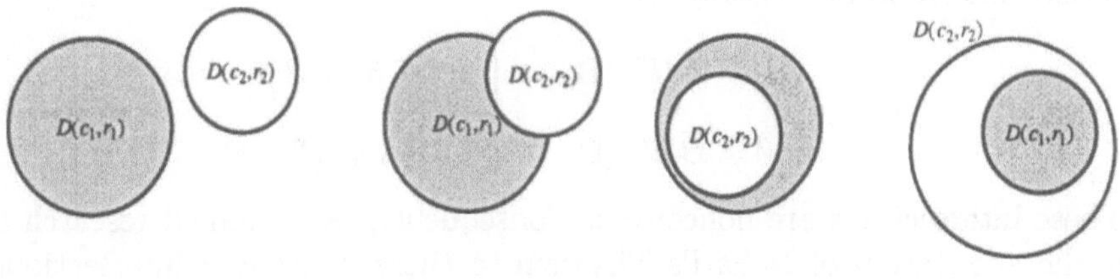

Fig. 7.15. $D(c_1, r_1) \cap X(c_2, r_2)$ configurations.

By adapting the convex coordinates of Figure 7.14 to these configurations, the following intersection test is obtained.

Lemma 7.2.2 *Let c_1, $c_2 \in \mathbf{C}$. Let r_1, $r_2 > 0$. Then*

$$\overline{D}(c_1, r_1) \cap \overline{X}(c_2, r_2) = \emptyset \iff |c_1 - c_2| < r_2 - r_1.$$

If the intersection is nonempty,

$$\overline{D}(c_1, r_1) \cap \overline{X}(c_2, r_2) \supseteq \overline{D}(c_X, r_X) \quad \begin{cases} c_X = \{e_1 + e_2\}/2 \\ r_X = |e_1 - e_2|/2 \end{cases},$$

where the "endpoints" e_1 and e_2 are computed from the "convex coordinates" of the line segment between c_1 and c_2

$$\begin{cases} e_1 = c_1 - t_1(c_2 - c_1) \; t_1 := r_1|c_2 - c_1|^{-1} \\ e_2 = c_1 + t_2'(c_2 - c_1) \; t_2' := \min\{t_1, 1 - t_2\} \; t_2 := r_2|c_2 - c_1|^{-1} \end{cases}.$$

Table 7.1 uses these results to determine L^∞ feasibility. The assumption is that the half plane occurs only at isolated points so it may be ignored; hence, the word "practical" in the title. L^∞ feasibility is determined by first computing the frequency regions where interior or exterior stability holds, and then applying the stability tests of Table 7.1.

Table 7.1. Practical tests of L^∞ feasibility.

Frequency Region	Interior	Exterior
Input port	$\|C_L\| \le R_L + 1$	$\|C_L\| \ge R_L - 1$
Output port	$\|C_G\| \le R_G + 1$	$\|C_G\| \ge R_G - 1$

7.2.2 H^∞ Feasibility

Once it is established that the stability constraints are L^∞ feasible, the next step is to test if there are any passive reflectances in the unit ball of H^∞ that satisfy the stability constraints:

$$\emptyset \ne \overline{B}H^\infty(\mathbf{C}_+) \cap [|S_1| \le S_{1,u_1}]?$$

$$\emptyset \ne \overline{B}H^\infty(\mathbf{C}_+) \cap [|S_2| \le S_{1,u_2}]?$$

These intersections are nonconvex. Consequently, an excellent research topic is the adaptation of Nehari's Theorem to these nonconvex intersections. In this regard, the Ph.D. thesis of David Schwartz is a good starting point [113].

We evade this substantial undertaking by using tubes that thread the L^∞ feasible regions determined by Table 7.1. That is, we use Lemmas 7.2.1 and 7.2.2 to compute tubes of circular cross section (CXS) that satisfy:

$$\overline{D}(C_{X,1}, R_{X,1}) \subseteq \overline{B}L^\infty(j\mathbf{R}) \cap [|S_1| \le S_{1,u_1}],$$

$$\overline{D}(C_{X,2}, R_{X,2}) \subseteq \overline{B}L^\infty(j\mathbb{R}) \cap [|S_2| \le S_{1,u_2}].$$

These CXS tubes are tested for intersection with H^∞:

$$\emptyset \ne \overline{D}(C_{X,1}, R_{X,1}) \cap H^\infty(\mathbb{C}_+)?$$

$$\emptyset \ne \overline{D}(C_{X,2}, R_{X,2}) \cap H^\infty(\mathbb{C}_+)?$$

Practically speaking, a nonempty intersection informs the amplifier designer that $S_{1,u}$ or $S_{2,u}$ can be realized by an input or output matching circuit. The quotes on "practically" means that we believe by slightly smoothing the CXS tube, the distance from the tube to the disk algebra is zero. Darlington's Theorem then informs us that there is a lumped matching circuit that gets arbitrarily close to the stability performance.

The tricky part about the CXS tubes is when the stability sublevel set switches from a disk to the exterior of a disk:

$$[|S_1| \le S_{1,u_1}] = \begin{cases} \overline{D}(C_L, R_L) & \text{interior} \\ \overline{X}(C_L, R_L) & \text{exterior} \end{cases},$$

$$[|S_2| \le S_{1,u_2}] = \begin{cases} \overline{D}(C_G, R_G) & \text{interior} \\ \overline{X}(C_G, R_G) & \text{exterior} \end{cases}.$$

Table 7.2 lists the center and radius functions of the CXS tube for the stability of S_1. The CXS tube is computed from Lemmas 7.2.1 and 7.2.2 by taking $c_1 = 0$, $r_1 = 1$, $c_2 = C_L$, and $r_2 = R_L$.

Table 7.2. CXS tube for S_1 that threads the L^∞ feasible region.

Interior Region	$[\lvert S_1\rvert \le S_{1,u}] = \overline{D}(C_L, R_L)$	
$C_{X,1} = \{t_1' + t_2'\}C_L/2$	$t_1' = \min\{1 + t_2, t_1\}$	$t_1 = \lvert C_L\rvert^{-1}$
$R_{X,1} = \lvert t_1' - t_2'\rvert\lvert C_L\rvert/2$	$t_2' = \max\{-t_1, 1 - t_2\}$	$t_2 = R_L\lvert C_L\rvert^{-1}$
Exterior Region	$[\lvert S_1\rvert \le S_{1,u}] = \overline{X}(C_L, R_L)$	
$C_{X,1} = \{-t_1 + t_2'\}C_L/2$		
$R_{X,1} = \lvert t_1 + t_2'\rvert\lvert C_L\rvert/2$	$t_2' = \min\{t_1, 1 - t_2\}$	

We close by examining the wideband S_1 stability of the NEC321000 transistor (Section 6.4). The stability constraint is taken to be the constant function:

$$S_{1,u}(j\omega) = \text{constant}.$$

By iterating Table 7.1, it was numerically found that L^∞ feasibility holds for $S_{1,u} \ge 0.87$. We arbitrarily select $S_{1,u} = 0.9$ and test for H^∞ feasibility using the CXS tube:

$$\emptyset \ne \overline{D}(C_{X,1}, R_{X,1}) \cap H^\infty(\mathbb{C}_+)?$$

Figure 7.16 plots the stability regions. For frequencies less than 8.1 GHz, the input stability sublevel set is the input circle and its interior:

$$[|S_1| \le S_{1,u}] = \overline{D}(C_L, R_L).$$

For frequencies greater than 8.1 GHz the sublevel set is the exterior of the disk:

$$[|S_1| \le S_{1,u}] = \overline{X}(C_L, R_L).$$

Because $S_{1,u} = 0.9$ is L^∞ feasible, there always holds

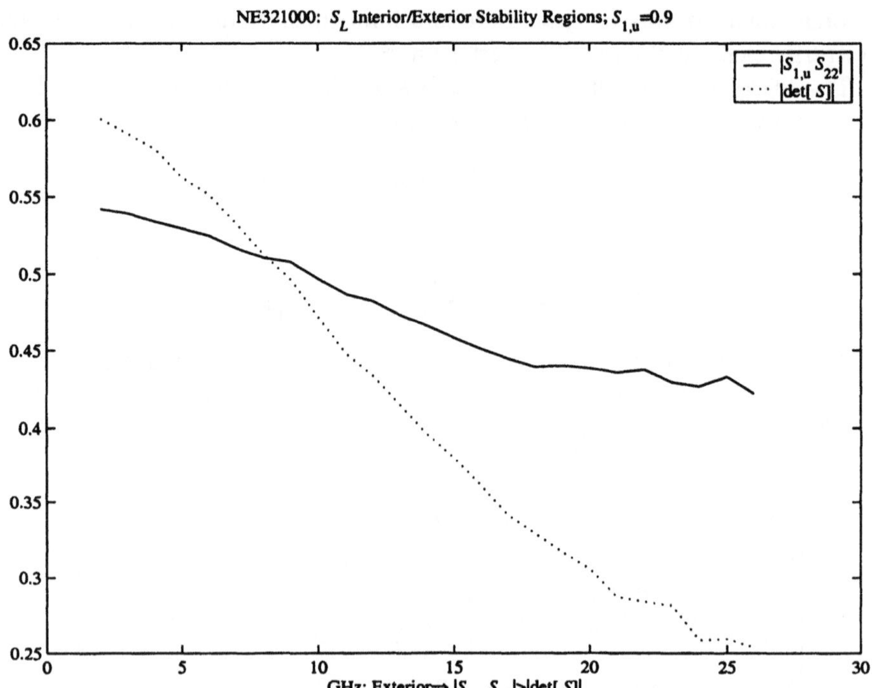

Fig. 7.16. Stability Regions for the NEC321000.

$$\overline{D}(C_{X,1}, R_{X,1}) \subseteq \overline{B}L^\infty(j\mathbf{R}) \cap [|S_1| \le S_{1,u}].$$

Figure 7.17 shows the frequency slices of the CXS tube during the transition from interior to exterior regions as frequency sweeps from 6 to 9 GHz. At each frequency, the plot shows the unit circle, the sublevel set boundary marked by portion of the large circle, and the CXS disk with center $C_{X,1}$ marked by the "+." The CXS disk is always in the stable region.

The simple implementation of Nehari's Theorem is

$$\emptyset \neq \overline{D}(C_X, R_X) \cap H^\infty(\mathbf{C}_+) \quad \Longleftrightarrow \quad \mathcal{H}_{C_X \circ \mathbf{c}^{-1}} \mathcal{H}^*_{C_X \circ \mathbf{c}^{-1}} \le T_{\tilde{R}^2{}_X \circ \mathbf{c}^{-1}},$$

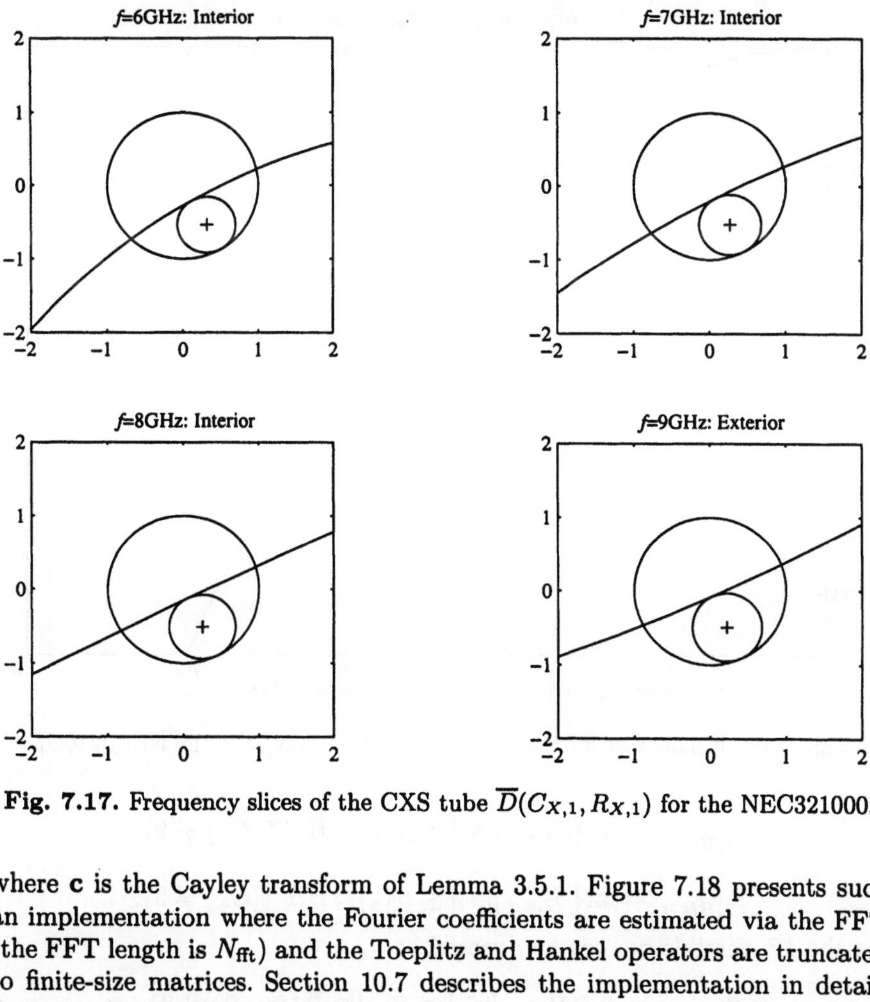

Fig. 7.17. Frequency slices of the CXS tube $\overline{D}(C_{X,1}, R_{X,1})$ for the NEC321000.

where $\mathbf{c}$ is the Cayley transform of Lemma 3.5.1. Figure 7.18 presents such an implementation where the Fourier coefficients are estimated via the FFT (the FFT length is N_{fft}) and the Toeplitz and Hankel operators are truncated to finite-size matrices. Section 10.7 describes the implementation in detail. Because the smallest eigenvalue is slightly positive, the amplifier designer estimates that for $S_{1,u} = 0.9$, there is a output reflectance S_L that attains this stability constraint in the CXS tube:

$$S_L \in \overline{D}(C_{X,1}, R_{X,1}) \cap H^\infty(\mathbf{C}_+) \subseteq [|S_1| \leq S_{1,u}] \cap \overline{B}H^\infty(\mathbf{C}_+).$$

Assuming continuity, a lumped input matching circuit can be found that realizes this stability performance. This completes the discussion of the stability bounds.

To use these bounds, recall the utopic point

$$\gamma_{\diamond,\overline{B}H^\infty} := \begin{bmatrix} \inf\{-\|G_T(S_G, S_L)\|_{-\infty} : S_G, S_L \in \overline{B}H^\infty(\mathbf{C}_+)\} \\ \inf\{\|F(S_G)\|_\infty : S_G \in \overline{B}H^\infty(\mathbf{C}_+)\} \\ \inf\{\|S_1(S_L)\|_\infty : S_L \in \overline{B}H^\infty(\mathbf{C}_+)\} \\ \inf\{\|S_2(S_G)\|_\infty : S_G \in \overline{B}H^\infty(\mathbf{C}_+)\} \end{bmatrix}$$

of Section 6.6. Let

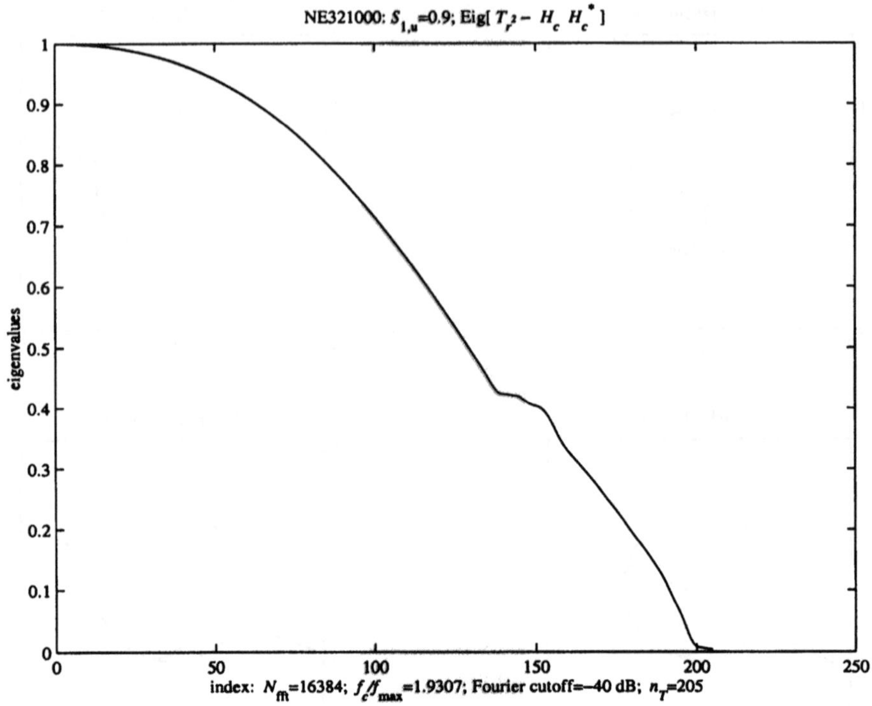

Fig. 7.18. Nehari test for the CXS tube $\overline{D}(C_{X,1}, R_{X,1})$ of the NEC321000.

$$S_{1,\overline{B}H^\infty} := \inf\{S_{1,u} : [|S_1| \leq S_{1,u}] \cap \overline{B}H^\infty(\mathbf{C}_+) \neq \emptyset\},$$

$$S_{2,\overline{B}H^\infty} := \inf\{S_{2,u} : [|S_2| \leq S_{2,u}] \cap \overline{B}H^\infty(\mathbf{C}_+) \neq \emptyset\}.$$

Let the L^∞ feasible bounds be denoted as

$$S_{1,\overline{B}L^\infty} := \inf\{S_{1,u} : [|S_1| \leq S_{1,u}] \cap \overline{B}L^\infty(j\mathbf{R}) \neq \emptyset\},$$

$$S_{2,\overline{B}L^\infty} := \inf\{S_{2,u} : [|S_2| \leq S_{2,u}] \cap \overline{B}L^\infty(j\mathbf{R}) \neq \emptyset\}.$$

These may be estimated by iterating Table 7.1. The problems of working with sampled data are raised in Section 10.6. Finally,

$$S_{1,X} := \inf\{S_{1,u} : \overline{D}(C_{X,1}, R_{X,1}) \cap H^\infty(\mathbf{C}_+) \neq \emptyset\},$$

$$S_{2,X} := \inf\{S_{2,u} : \overline{D}(C_{X,2}, R_{X,2}) \cap H^\infty(\mathbf{C}_+) \neq \emptyset\}.$$

Then the stability components of the utopic point $\gamma_{\circ,\overline{B}H^\infty}$ are bounded:

$$\begin{bmatrix} S_{1,\overline{B}L^\infty} \\ S_{2,\overline{B}L^\infty} \end{bmatrix} \leq \begin{bmatrix} S_{1,\overline{B}H^\infty} \\ S_{2,\overline{B}H^\infty} \end{bmatrix} \leq \begin{bmatrix} S_{1,X} \\ S_{2,X} \end{bmatrix}.$$

For the NEC321000 amplifier,

$$0.87 \leq S_{1,\overline{BL}\infty} \leq S_{1,\overline{BH}\infty} \leq S_{1,X} \leq 0.9.$$

This raises the possibility that these two easy bounds may be "good enough." That is, the use of the CXS tube to evade the nonconvex Nehari test may be justified.

Question 9. Are such tight stability bounds common in wideband amplifiers?

Finally, Das and Dennis remark that the utopic point $\gamma_{\diamond,\overline{BH}\infty}$ need not be exact. Consequently, using such approximations to the utopic point may be "good enough" for the Pareto problem of Section 6.6.

7.3 Gain Tubes

Before we start the gain analysis, it is worthwhile to orient ourselves with respect to the preceding analysis. Section 7.1 showed that the sublevel set $[F \leq F_u]$ of the noise fit in the unit disk and admitted a straight forward Nehari test. Section 7.2 showed that the stability sublevel sets $[|S_1| \leq S_{1,u}]$ and $[|S_2| \leq S_{2,u}]$ did not necessary fit in the unit disk and could toggle between interior and exterior stability. By threading the intersection of the stability sublevel sets and the unit disk with tubes of circular cross section (CXS), a Nehari test could be applied for stability.

The gain functions are still more complicated. For the input gain, S_G and S_L are coupled. For the input and output gain, gain without stability makes no sense. This section explores the gain functions under the stability constraint. We develop feasibility tests for the output gain. For the input gain, some feasibility tests are possible but the coupling forces us into a CXS solution again. Nevertheless, the analytic gain and stability bounds are novel. Section 7.4 uses these results where all the gain, noise, and stability constraints are invoked for an amplifier design.

The transducer power gain G_T is a function of the reflectances S_G and S_L determined from the input and output matching circuits [106, Eq. 11.19]:

$$G_T(S_G, S_L) = |S_{21}|^2 + \frac{1-|S_G|^2}{|1-S_1(S_L)S_G|^2} + \frac{1-|S_L|^2}{|1-S_{22}S_L|^2} \quad \text{[dB]},$$

where the input reflectance is

$$S_1(S_L) = S_{11} + S_{12}S_L(1-S_{22}S_L)^{-1}S_{21},$$

and the gains are typically known as

- Amplifier gain: $G_0 := |S_{21}|^2$.
- Input gain: $G_G(S_G, S_L) := \frac{1-|S_G|^2}{|1-S_1(S_L)S_G|^2}$.
- Output gain: $G_L(S_L) := \frac{1-|S_L|^2}{|1-S_{22}S_L|^2}$.

The Amplifier Matching Problem seeks S_G and S_L to maximize the gain, minimize the noise, and guarantee stability. In this section, we ignore the noise and focus on the gain. We seek matching circuits that exceed a specified gain $G_{T,u}(j\omega)$ at each frequency. The analysis is complicated because stability is intertwined with gain and the multivariate nature of the input gain.

The first issue to settle is whether any reflectances—stable or not—attain the desired gain performance. Denote the domain of the gain as

$$\mathcal{D}_0 := \left\{ \begin{bmatrix} S_G \\ S_L \end{bmatrix} \in L^\infty(j\mathbf{R}, \mathbf{C}^2_\infty) : (1 - S_1 S_G)^{-1}, (1 - S_1 S_G)^{-1} \in L^\infty(j\mathbf{R}) \right\},$$

where $\mathbf{C}^2_\infty$ denote $\mathbf{C}^2$ equipped with the sup-norm. If the amplifier is also an L^∞ function, the problem of L^∞ feasibility is how the superlevel sets of $G_T : \mathcal{D}_0 \to \mathbf{R}$ relate to the unit ball.

> L^∞ **Feasible Gain.** Given $G_{T,u} \in L^\infty(j\mathbf{R})$, determine if
>
> $$\emptyset \neq \overline{B}L^\infty(j\mathbf{R}, \mathbf{C}^2_\infty) \cap [G_T \geq G_{T,u}].$$

If the amplifier designer determines that $G_{T,u}$ is L^∞ feasible, the next question to as is if there are any circuits that realize this gain.

> H^∞ **Feasible Gain.** Given $G_{T,u} \in L^\infty(j\mathbf{R})$, determine if
>
> $$\emptyset \neq \overline{B}H^\infty(\mathbf{C}_+, \mathbf{C}^2_\infty) \cap [G_T \geq G_{T,u}].$$

What prevents a straight forward application of Nehari's Theorem is that the input gain is a function of both S_G and S_L. An important exception occurs when the amplifier is unilateral.

7.3.1 Unilateral Amplifiers

An amplifier is unilateral provided $S_{12} = 0$ and decouples the load from the input gain [106, Eq. 11.20a]:

$$G_G(S_G, S_L) = \frac{1 - |S_G|^2}{|1 - S_{11}S_G|^2}.$$

In a unilateral amplifier, G_T is maximized by separately maximizing the input $G_G(S_G)$ and output gain $G_L(S_L)$. This suggests replacing the jointly coupled and hard problem of maximizing G_T with its unilateral approximation:

$$G_{TU}(S_G, S_L) := |S_{21}|^2 + \frac{1 - |S_G|^2}{|1 - S_{11}S_G|^2} + \frac{1 - |S_L|^2}{|1 - S_{22}S_L|^2} \quad \text{[dB]}.$$

How good is the unilateral assumption for a wideband amplifier? The error ratio is [53, Eq. 3.5.1]:

$$\frac{G_T}{G_{TU}} = \frac{1}{|1-X|^2}, \quad X := \frac{S_{12}S_{21}S_G S_L}{(1-S_{11}S_G)(1-S_{22}S_L)}.$$

If S_{11} and S_{22} are passive, the error ratio is bounded by [53, Eq. 3.5.2], [106, Eq. 11.46]:

$$\frac{1}{(1+U)^2} \leq \frac{G_T}{G_{TU}} \leq \frac{1}{(1-U)^2},$$

where U denotes the *unilateral figure of merit*

$$U = \frac{|S_{12}S_{21}S_{11}S_{22}|}{(1-|S_{11}|^2)(1-|S_{22}|^2)}.$$

From Pozar [106, page 623]: "Usually, an error of a few tenths of a dB or less justifies the unilateral assumption." *So how well does the unilateral assumption hold for the amplifiers we are considering?* For the NE32484A amplifier, Figure 6.19 shows that S_{11} and S_{22} are passive and that S_{12} is relatively small. Figure 7.19 displays the bounds of G_T/G_{TU} for the NE32484A amplifier.

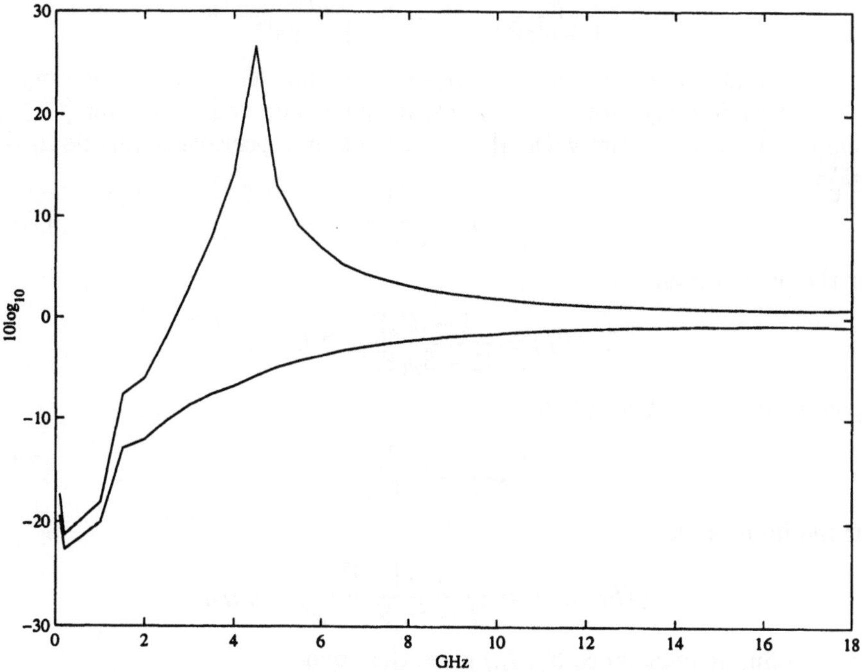

Fig. 7.19. Bounds for the unilateral figure of merit of the NE32484A.

In light of Pozar's remarks, the gap between the bounds demonstrates that the unilateral assumption is not good. Similar results hold for the NEC321000 amplifier. We can either continue with a unilateral assumption and maximize G_{TU} or contend with the more complete analysis. Regardless of the approach, the gain function needs to be connected to the stability bounds.

7.3.2 Generic Gain and Stability Bounds

Introduce the generic gain function $G : \mathcal{D} \subset \mathbf{C} \to \mathbf{R}$

$$G(\phi) = \frac{1 - |\phi|^2}{|1 - s\phi|^2}$$

with domain $\mathcal{D} = \{\phi \in \mathbf{C} : 1 \neq s\phi\}$. The superlevel sets

$$[G \geq g] := \{\phi \in \mathbf{C} : G(\phi) \geq g\}$$

toggle about the unit circle as follows:

- $g < 0 \iff [G < g] \cap \overline{\mathbf{D}} = \emptyset$.
- $[G = 0] = \mathbf{T}$.
- $g > 0 \iff [G \geq g] \subseteq \overline{\mathbf{D}}$.

Regarding the last item, $[G \geq g]$ is a disk contained in the closed unit disk with center and radius:

$$c := \overline{s}\frac{g}{1 + g|s|^2}, \qquad r := \frac{\sqrt{1 + g(|s|^2 - 1)}}{1 + g|s|^2}.$$

If $|s| > 1$ and $g \geq 0$, the radius is positive so that $[G \geq g]$ is nonempty. If $|s| < 1$, a sufficiently large $g \geq 0$ forces an imaginary radius or that $[G \geq g]$ is empty. The (g, s) pairs with $|s| < 1$ that admit a positive radius lie in the region

$$0 \leq g < \frac{1}{1 - |s|^2}. \tag{7.10}$$

For the output gain

$$G_L(S_L) = \frac{1 - |S_L|^2}{|1 - S_{22}S_L|^2} \geq G_{L,u},$$

a passive S_{22} gives the bounds

$$G_{L,u} \leq \frac{1}{1 - |S_{22}|^2}. \tag{7.11}$$

For the input gain

$$G_G(S_G, S_L) := \frac{1 - |S_G|^2}{|1 - S_1(S_L)S_G|^2} \geq G_{G,u},$$

a passive input reflectance $S_1(S_L)$ gives the bound

$$G_{G,u} < \frac{1}{1 - |S_1(S_L)|^2} \leq \frac{1}{1 - S_{1,u}^2}. \tag{7.12}$$

7.3.3 L^∞ Feasible Gains

We split the L^∞ feasible gain problem into determining L^∞ feasible input and output gains. That is, the amplifier designer wants the input and output gains to exceed their specified performance limits.

> **L^∞ Feasible Gains.** Given $G_{G,u}$, $G_{L,u} \in L^\infty(j\mathbf{R})$, determine if
>
> $$\emptyset \neq \overline{B}L^\infty(j\mathbf{R}) \cap [G_L \geq G_{L,u}],$$
>
> $$\emptyset \neq \overline{B}L^\infty(j\mathbf{R}, \mathbf{C}^2_\infty) \cap [G_G \geq G_{G,u}].$$

For the output gain, Inequality 7.11 gives the bound

$$0 \leq G_{L,u}(j\omega) \leq \frac{1}{1 - |S_{22}(j\omega)|^2}, \tag{7.13}$$

assuming $\|S_{22}\|_\infty < 1$. For both our amplifiers, the NE323484A and the NE321000, their scattering functions in Section 6.4 show S_{22} is passive: $\|S_{22}\|_\infty < 1$. For the nonpassive case, see [53]. The following question is reminiscent of the "toggling" of the stability functions.

Question 10. What happens to the gain when a wideband amplifier has $|S_{11}|$ or $|S_{22}|$ both passive and active?

Figure 7.20 shows the maximal user-requested gain that the NE323484A can deliver. This figure shows that if the amplifier designer wants a wideband constant $G_{L,u}$, then $G_{L,u} \leq 0.74$ dB. For comparison, Figure 7.21 shows the maximal user-requested gain that the NEC321000 can deliver. However, these bounds are useless without the stability constraints. Thus, we combine the feasible gains with stability.

> **L^∞ Stable and Feasible Gains.** Given $G_{G,u}$, $G_{L,u}$, $S_{1,u}$, and $S_{2,u}$ in $L^\infty(j\mathbf{R})$, determine if
>
> $$\emptyset \neq [G_L \geq G_{L,u}] \cap [|S_1| \leq S_{1,u}],$$
>
> $$\emptyset \neq [G_G \geq G_{G,u}] \cap [|S_1| \leq S_{1,u}] \cap [|S_2| \leq S_{2,u}].$$

Asking if the output gain is L^∞ stable and feasible is equivalent to asking if the gain disk and the stability sublevel sets intersect. Section 7.2 points out the stability sets may be a disk, half plane, or the exterior of a disk—all in various configurations with the unit disk. The output gain center and radius functions are

$$C_{g_L} := \overline{S_{22}} \frac{G_{L,u}}{1 + G_{L,u}|S_{22}|^2}, \quad R_{g_L} := \frac{\sqrt{1 + G_{L,u}(|S_{22}|^2 - 1)}}{1 + G_{L,u}|S_{22}|^2},$$

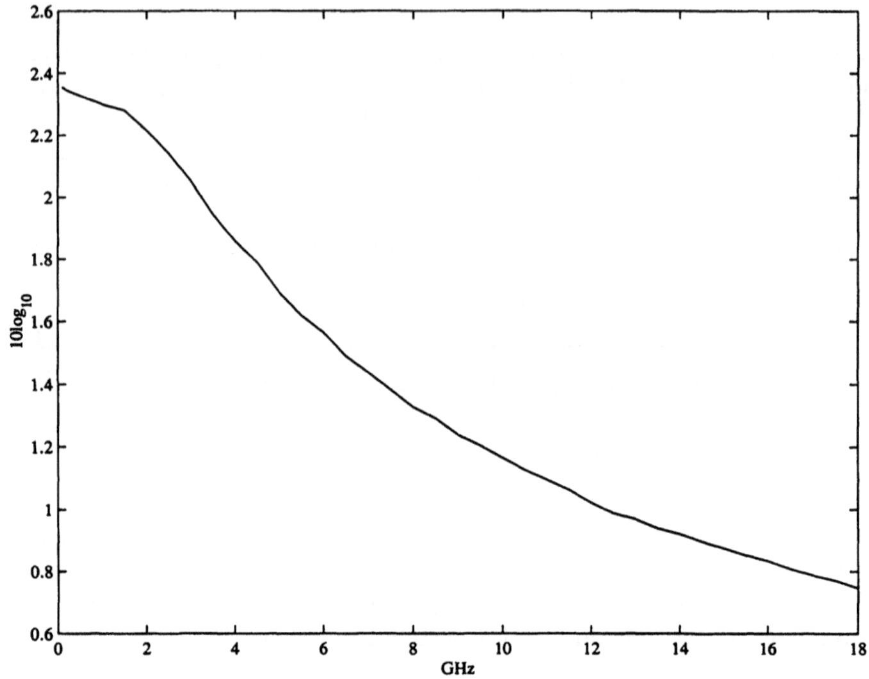

Fig. 7.20. $G_{L,u}$ bounds for the NE32484A.

assuming $\|S_{22}\|_\infty < 1$.

The question of whether the amplifier designer's requested $G_{L,u}$ is L^∞ stable and feasible is equivalent to asking if

$$\emptyset \neq \overline{D}(C_{g_L}, R_{g_L}) \cap [|S_1| \leq S_{1,u}]?$$

This question is answered by Lemmas 7.2.1 and 7.2.2 and tabulated in Table 7.3. The "practical" means that the half-plane stability sets are ignored. This completes the solution of L^∞ feasibility for the output gain.

Table 7.3. Practical tests for stable L^∞ feasibility of the output gain.

Stability Region	$G_{L,u}$ Feasibility	Nonempty Intersection
Interior	$\overline{D}(C_L, R_L) \cap \overline{D}(C_{g_L}, R_{g_L}) \neq \emptyset$	$\|C_{g_L} - C_L\| \leq R_L + R_{g_L}$
Exterior	$\overline{X}(C_L, R_L) \cap \overline{D}(C_{g_L}, R_{g_L}) \neq \emptyset$	$\|C_{g_L} - C_L\| \geq R_L - R_{g_L}$

Turning to the input gain

$$G_G(S_G, S_L) = \frac{1 - |S_G|^2}{|1 - S_1(S_L)S_G|^2},$$

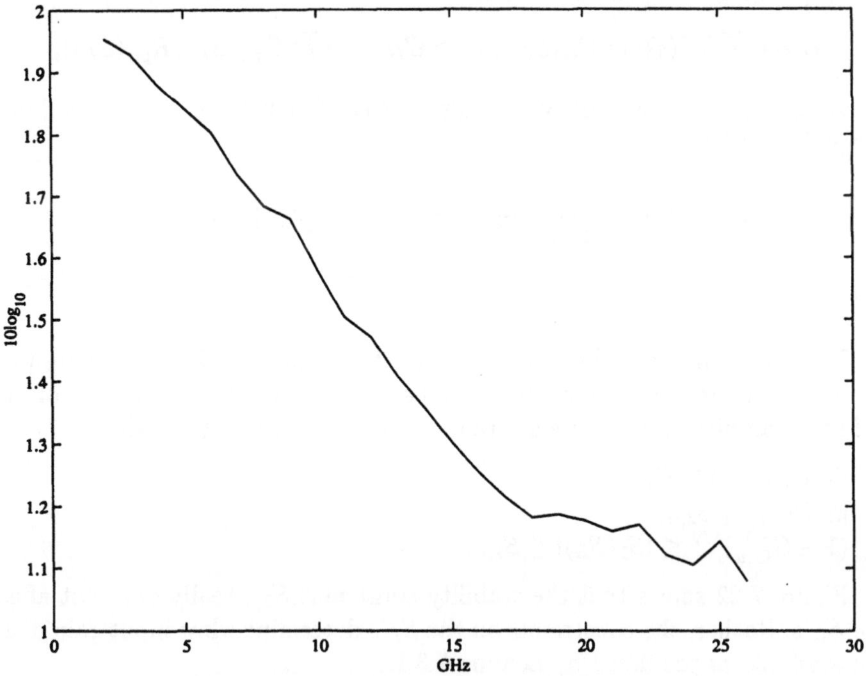

Fig. 7.21. $G_{L,u}$ bounds for the NEC321000.

input stability requires $\|S_1(S_L)\|_\infty \le 1$ for $S_L \in \overline{B}L^\infty(j\mathbf{R})$. Take the domain of the input gain as

$$\mathcal{D}_1 := \left\{ \begin{bmatrix} S_G \\ S_L \end{bmatrix} \in \overline{B}L^\infty(j\mathbf{R}, \mathbf{C}^2_\infty) : S_L \in [|S_1| \le S_{1,u}] \right\}.$$

Lemma 7.3.1 *Let $S_{1,u}$, $G_{G,u} \in L^\infty(j\mathbf{R})$ with $G_{G,u} \ge 1$ and $0 \le S_{1,u} \le 1$.*

$$\mathcal{D}_1 \bigcap [G_G \ge G_{G,u}] \subseteq \bigcup \overline{D}(C_{gG}(S_L), R_{gG}(S_L)) \times \{S_L\}$$

where the union is over all load reflectances

$$S_L \in [(1 - G_{G,u}^{-1})^{1/2} \le |S_1| \le S_{1,u}] \bigcap \overline{B}L^\infty(j\mathbf{R})$$

with associated center and radius functions

$$C_{gG}(S_L) := \overline{S_1(S_L)} \frac{G_{G,u}}{1 + G_{G,u}|S_1(S_L)|^2},$$

$$R_{gG}(S_L)) := \frac{\sqrt{1 + G_{G,u}(|S_1(S_L)|^2 - 1)}}{1 + G_{G,u}|S_1(S_L)|^2}.$$

Proof: Suppose $S_L \in [|S_1| \le S_{1,u}]$. Then

$$\{S_G \in \overline{B}L^\infty(j\mathbf{R}) : G_G(S_G, S_L) \ge G_{G,u}\} = \overline{D}(C_{g_G}(S_L), R_{g_G}(S_L)),$$

provided the radius is positive. A sufficient condition for a positive radius is Inequality 7.12:

$$G_{G,u} \le \frac{1}{1 - |S_1(S_L)|^2} \iff 1 - G_{G,u}^{-1} \le |S_1(S_L)|^2.$$

///

The following plots illustrate Lemma 7.3.1. Figures 7.22 and 7.23 are projections of a frequency slice of the superlevel set of the input gain of the NE32484 amplifier. The dots are random S_G, $S_L \in \overline{\mathbf{D}}$ that satisfy

- $G_G(S_G, S_L) \ge G_{G,u}$,
- $|S_2(S_G)| \le S_{2,u}$,
- $(1 - G_{G,u}^{-1})^{1/2} \le |S_1(S_L)| \le S_{1,u}$.

Figure 7.22 shows that the stability constraint $S_{2,u}$ really does not affect the S_G's. Rather, the constraints on the S_L's determine what input gain disks are available as predicted by Lemma 7.3.1.

Figure 7.23 shows how the available S_L's are lie between the gain and stability constraints. For comparison, Figures 7.24 and 7.25 show the effect of demanding more input gain. The available S_L's decrease as the set

$$[(1 - G_{G,u}^{-1})^{1/2} \le |S_1| \le S_{1,u}]$$

"pinches off." The decreased population of S_L's then decreases the available S_G's. By plotting the stability and gain constraints for the S_L's, the amplifier designer can graphically see the stability and gain constraints.

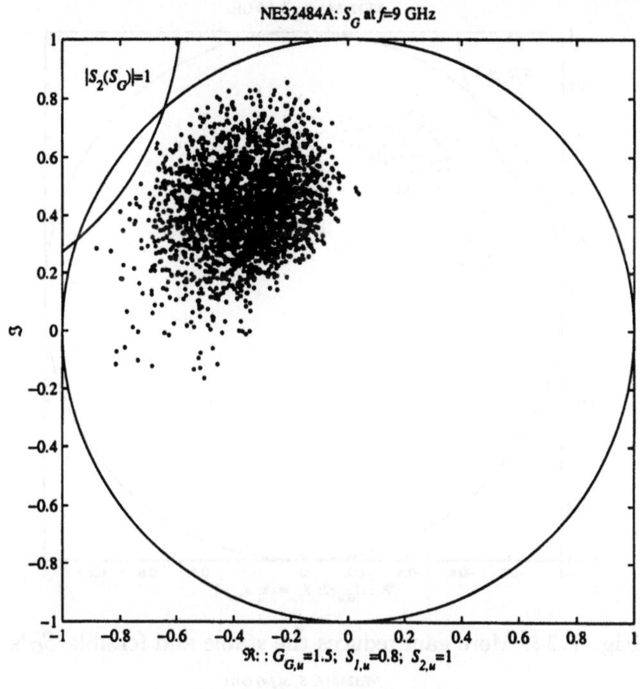

Fig. 7.22. Stable and feasible S_G's.

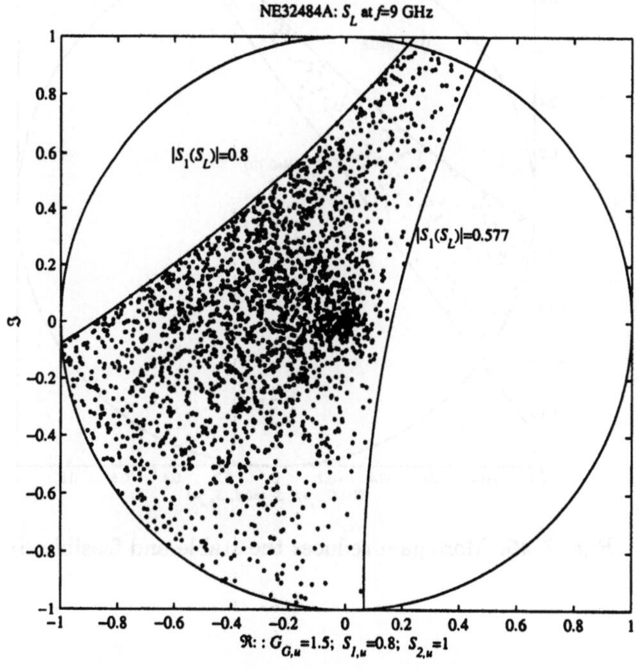

Fig. 7.23. Stable and feasible S_L's.

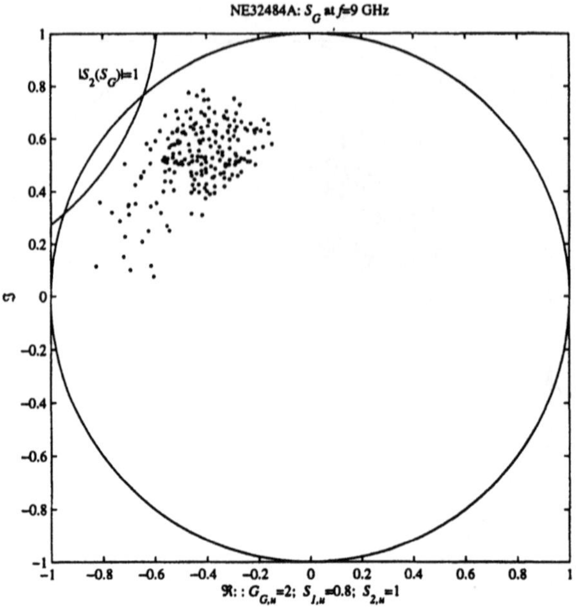

Fig. 7.24. More gain reduces the stable and feasible S_G's

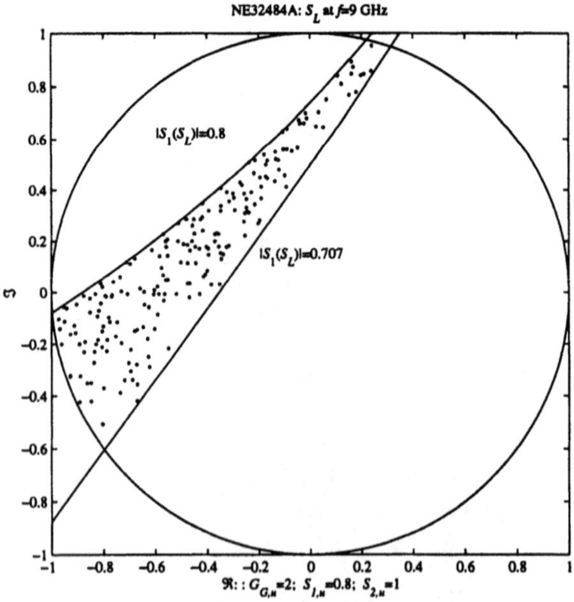

Fig. 7.25. More gain reduces the stable and feasible S_L's.

We close this section by raising the visualization issues for amplifier design. Recall Figure 7.2, which showed the gain and noise disks. For the amplifier designer, such a plot is an excellent design tool. It is featured in all the amplifier texts and shows at a glance whether the design constraints are L^∞ feasible—at a single frequency. Lemma 7.3.1 parameterizes the more complex set of L^∞ stable and feasible gains. The visualization of this set seems difficult. A simpler version asks about visualizing S_L's in Figures 7.23 and 7.25.

Question 11. What is an efficient way to present the available S_L's or

$$[(1 - G_{G,u}^{-1})^{1/2} \le |S_1| \le S_{1,u}]$$

as a function of frequency?

More generally, both input and output stability need to be constrained. Take the domain of the input gain as

$$\mathcal{D}_2 := \{[|S_2| \le S_{2,u}] \cap \overline{B}L^\infty(j\mathbf{R})\} \times \{[|S_1| \le S_{1,u}] \cap \overline{B}L^\infty(j\mathbf{R})\}.$$

This restricts the S_G's. The restricted S_G's reduce the available S_L's. If we want to see the frequency effects of $\mathcal{D}_2$, we must visualize a subset of $\mathbf{R} \times \mathbf{C}^2$.

Question 12. What is an efficient way to visualize $\mathcal{D}_2$?

7.3.4 H^∞ Feasible Gains

Once the amplifier designer has seen whether the gains are L^∞ feasible, the next step is to determine whether there are reflectances S_G and S_L in $H^\infty(\mathbf{C}_+)$ that meet the gain and stability constraints.

Stable H^∞ Feasible Gains. Given $G_{G,u}$, $G_{L,u}$, $S_{1,u}$, and $S_{2,u} \in L^\infty(j\mathbf{R})$, determine if

$$\emptyset \ne [G_L \ge G_{L,u}] \cap [|S_1| \le S_{1,u}] \cap \overline{B}H^\infty(\mathbf{C}_+),$$

$$\emptyset \ne [G_G \ge G_{G,u}] \cap [|S_1| \le S_{1,u}] \cap [|S_2| \le S_{2,u}] \cap \overline{B}H^\infty(\mathbf{C}_+).$$

Consider first the output gain. Assume L^∞ stability puts the output gain disk in the unit ball. Consequently, we can drop the unit ball constraint and ask

$$\emptyset \ne [G_L \ge G_{L,u}] \cap [|S_1| \le S_{1,u}] \cap H^\infty(\mathbf{C}_+)?$$

This is a nonconvex Nehari problem. As in Section 7.2, one approach is to find a disk that "threads" this intersection:

$$\overline{D}(C_{X,L}, R_{X,L}) \subseteq [G_L \ge G_{L,u}] \cap [|S_1| \le S_{1,u}]$$

and then ask for intersection with H^∞

Table 7.4. CXS tube for S_1 that threads the L^∞ stable and feasible region of the output gain.

Interior Region	$[\|S_1\| \le S_{1,u}] = \overline{D}(C_L, R_L)$	
$C_{X,L} = \{e_1 + e_2\}/2$	$e_1 = C_{g_L} + t_1'(C_L - C_{g_L})$	$t_1' := \min\{1 + t_2, t_1\}$
$R_{X,L} = \|e_1 - e_2\|/2$	$e_2 = C_{g_L} + t_2'(C_L - C_{g_L})$	$t_2' := \max\{-t_1, 1 - t_2\}$
Exterior Region	$[\|S_1\| \le S_{1,u}] = \overline{X}(C_L, R_L)$	
$C_{X,L} = \{e_1 + e_2\}/2$	$e_1 = C_{g_L} - t_1(C_L - C_{g_L})$	$t_1 := R_{g_L}\|C_L - C_{g_L}\|^{-1}$
$R_{X,L} = \|t_1 + t_2'\|/2$	$e_2 = C_{g_L} + t_2'(C_L - C_{g_L})$	$t_2' := \min\{t_1, 1 - t_2\}$
		$t_1 = R_{g_L}\|C_L - C_{g_L}\|^{-1}$ $t_2 = R_L\|C_L - C_{g_L}\|^{-1}$

$$\emptyset \neq \overline{D}(C_{X,L}, R_{X,L}) \cap H^\infty(\mathbf{C}_+)?$$

Table 7.4 computes the CXS tube from Lemmas 7.2.1 and 7.2.2.

For the input gain, the stability disks can be either disks or exteriors of disks. In the case of unilateral gain, the problem reduces to the same form as the output gain.

Question 13. Assume the input gain is modeled as unilateral gain:

$$G_G = G_{G,U} = \frac{1 - |S_G|^2}{|1 - S_{11}S_G|^2}.$$

Find an analytic form for any input disk $\overline{D}(C_{X,G}, R_{X,G})$ that threads $[G_{GU} \ge G_{G,u}] \cap [|S_2| \le S_{2,u}]$.

However, the analytic approach becomes unwieldy when faced with the input gain or the constraints of the full Amplifier Matching Problem. Therefore, we leave the development of the unilateral amplifier to the ambitious reader and turn to a numerical solution.

7.4 An H^∞ Multidisk Method

We close this chapter by gathering the preceding gain, noise, and stability results into a single H^∞ Multidisk Method. This is only one approach. Many others await development by the reader.

Assume the given amplifier is embedded in the input and output matching circuits of Figure 7.1. The amplifier designer has specified

- gain constraints $G_{G,u}$, $G_{L,u}$,
- a noise constraint F_u,
- stability constraints $S_{1,u}$, $S_{2,u}$.

Can we find lossless matching circuits with input and output reflectances S_G, $S_L \in \overline{B}H^\infty(\mathbf{C}_+)$ that satisfy the following?

AMP-1 Gain constraints: $S_L \in [G_L \geq G_{L,u}]$, $(S_G, S_L) \in [G_G \geq G_{G,u}]$.
AMP-2 Noise constraints: $S_G \in [F \leq F_u]$.
AMP-3 Stability constraints: $S_G \in [|S_2| \leq S_{2,u}]$, $S_L \in [|S_1| \leq S_{1,u}]$.

The first question is whether these superlevel and sublevel sets admit any intersection.

> **L^∞ Feasible Constraints.** Given $G_{G,u}$, $G_{L,u}$, $S_{1,u}$, $S_{2,u}$, and $F_u \in L^\infty(j\mathbf{R})$, determine if
>
> $$\begin{aligned} \emptyset \neq\ & \overline{B}L^\infty(j\mathbf{R}, \mathbf{C}^2_\infty) \cap [G_G \geq G_{G,u}] \\ & \cap \{[F \leq F_u] \times [|S_1| \leq S_{1,u}]\} \\ & \cap \{[|S_2| \leq S_{2,u}] \times [G_L \geq G_{L,u}]\}. \end{aligned}$$

If this intersection is nonempty, the next question is whether the intersection also intersects H^∞.

> **H^∞ Feasible Constraints.** Given $G_{G,u}$, $G_{L,u}$, $S_{1,u}$, $S_{2,u}$, and $F_u \in L^\infty(j\mathbf{R})$, determine if
>
> $$\begin{aligned} \emptyset \neq\ & \overline{B}H^\infty(\mathbf{C}_+, \mathbf{C}^2_\infty) \cap [G_G \geq G_{G,u}] \\ & \cap \{[F \leq F_u] \times [|S_1| \leq S_{1,u}]\} \\ & \cap \{[|S_2| \leq S_{2,u}] \times [G_L \geq G_{L,u}]\}. \end{aligned}$$

Both questions are open and are excellent points of departure for research. Specifically, the problems are to parameterize the intersection and extend Nehari's Theorem to this nonconvex set.

In this chapter, the H^∞ Multidisk Method finds input and output disks of circular cross section (CXS) that thread these constraints.

> **L^∞ CXS Feasible Constraints.** Given $G_{G,u}$, $G_{L,u}$, $S_{1,u}$, $S_{2,u}$, and $F_u \in L^\infty(j\mathbf{R})$, find the center and radius functions such that:
>
> $$\begin{aligned} & \overline{D}(C_{X,G}, R_{X,G}) \times \overline{D}(C_{X,L}, R_{X,L}) \\ & \subseteq \overline{B}L^\infty(j\mathbf{R}, \mathbf{C}^2_\infty) \cap [G_G \geq G_{G,u}] \\ & \cap \{[F \leq F_u] \times [|S_1| \leq S_{1,u}]\} \\ & \cap \{[|S_2| \leq S_{2,u}] \times [G_L \geq G_{L,u}]\}. \end{aligned}$$

Once the constraints are L^∞ CXS feasible, the amplifier designer then asks if these constraints are H^∞ feasible. Because the CXS tubes are L^∞ feasible, both tubes satisfy the amplifier designer's constraints and are contained in the unit ball of L^∞. Thus, testing for intersection with the unit ball of H^∞ reduces to

$$\emptyset \neq \overline{D}(C_{X,G}, R_{X,G}) \times \overline{D}(C_{X,L}, R_{X,L}) \subseteq H^\infty(j\mathbf{R}, \mathbf{C}^2_\infty)?$$

Because the disks are separable, only the scalar Nehari Theorem is needed to implement the H^∞ Multidisk Method.

> **H^∞ CXS Feasible Constraints.** Assume $G_{G,u}$, $G_{L,u}$, $S_{1,u}$, $S_{2,u}$, and $F_u \in L^\infty(j\mathbf{R})$ are L^∞ CXS feasible. These constraints are H^∞ feasible provided
>
> $$\emptyset \neq \overline{D}(C_{X,G}, R_{X,G}) \cap H^\infty(\mathbf{C}_+),$$
>
> $$\emptyset \neq \overline{D}(C_{X,L}, R_{X,L}) \cap H^\infty(\mathbf{C}_+).$$

The following examples illustrate this H^∞ CXS Multidisk Method. Consider the NE32484A amplifier of Section 6.4. Figure 6.20 reports for the series/stub matching that

- $G_T \geq 9.1$ dB.
- $F \leq 1.4$ dB.
- $|S_1| \leq 1$, $|S_1| \leq 0.7$.

Figure 7.26 shows the amplifier gains associated with this matching circuit. The amplifier gain G_0 rolls off from 14 dB to 8 dB as frequency sweeps from 0.1 to 18 GHz. The input gain G_G and the output gain G_L vary only by a fraction of a dB. However, the transducer power gain G_T always exceeds 9 dB. This is possible because the series/shunt matching stubs *accept a low frequency loss in G_G and G_L to realize high-frequency gains that offset the roll-off of G_0*. For the initial implementation of the H^∞ Multidisk Method, we relax the noise and stability requirements and focus on the gain.

We first baseline the H^∞ Multidisk Method using flat constraints. Focusing on the gain, we ask

Can the input and output gains exceed 0 dB?

Such a flat gain design realizes $G_T \geq 8$ dB. Equivalently, we ask if the following flat constraints are feasible:

- $G_{G,u} = 0$ dB, $G_{L,u} = 0$ dB.
- $F_u = 1.5$ dB.
- $S_{1,u} = 1$, $S_{2,u} = 0.8$.

The first step is to find large CXS disks that meet these flat constraints. By numerically optimizing the center and radius functions at each sample frequency, we obtain a large CXS disk that meets the flat constraints.

Figures 7.27 and 7.28 report on this numerical optimization. Each CXS disk is plotted in the unit circle as follows. For each frequency, the center is

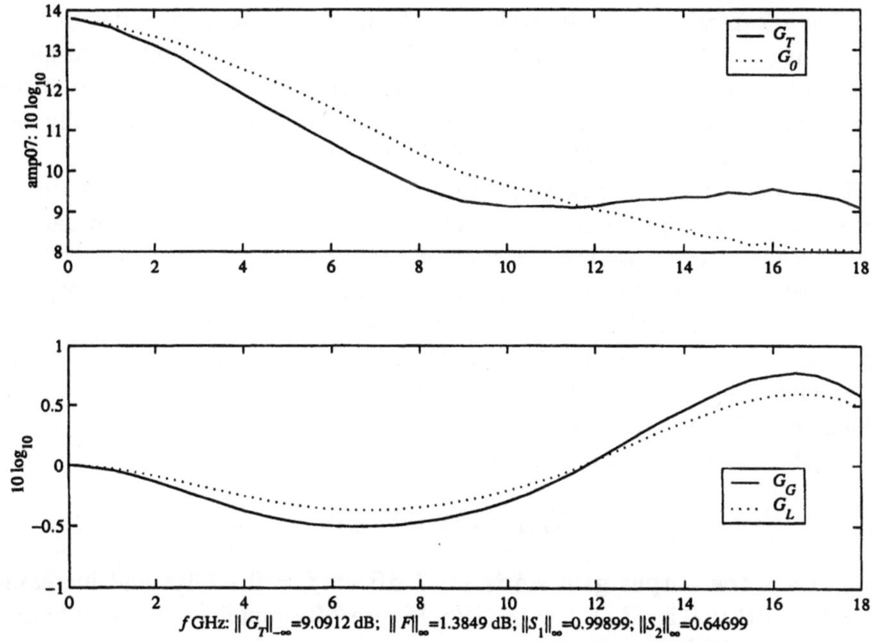

Fig. 7.26. Series/shunt amplifier gains of the NE32484A.

plotted as the heavy line. At each sample point, a circle of the corresponding radius is drawn. Both plots show that these CXS disks lie in the unit circle, so are L^∞ feasible. Thus, the second step is to see if these CXS disks are H^∞ feasible using the Nehari test.

To implement the Nehari test in Theorem 3.10.1, we first use the Cayley transform (Lemma 3.5.1) to pull the center and radius functions, defined only at sample frequencies on $\{j\omega_k\}$, onto the corresponding sample points $\{\exp(j\theta_k)\}$ on the unit circle $\mathbf{T}$. These samples of the center c_X and radius r_X functions are extended to the entire unit circle. The corresponding Fourier coefficients are estimated by applying an FFT of length N_{fft} to c_X and r_X. The Nehari test

$$\emptyset \neq \overline{D}(c_X, r_X) \cap H^\infty(\mathbf{T})? \iff \mathcal{T}_{\tilde{r}_X^2} - \mathcal{H}^*_{c_X}\mathcal{H}_{c_X}$$

is implemented by using the FFT estimates of the Fourier coefficients and truncating the matrices to a finite size. The truncation size is determined by the "Fourier cutoff" listed on the plots. In the discarded portions of the Hankel and Toeplitz matrices, all estimated Fourier coefficients do not exceed the Fourier cutoff.

Figures 7.29 and 7.30 report on the Nehari test. Both plots show that the smallest eigenvalues are barely different from zero. Accepting this level of numerical resolution, this H^∞ Multidisk Method informs the amplifier designer that these flat constraints are feasible. That is, there are matching circuits that can realize these flat constraints to obtain $G_T = 8$ dB.

This information puts the series/shunt matching circuit in a design context. Flat gains are H^∞ feasible at 8 dB gain whereas the sinusoidal gains of the series/shunt realize 9 dB gain (Figure 7.26). Consequently, 8 dB is the performance limit of the flat constraints. To get more gain, we naturally ask about extending the H^∞ Multidisk Method to handle nonflat gains.

The series/shunt stubs had 9 dB gain, even though Figure 7.26 shows input and output gain G_G and G_L are small, because the stubs trade high gain at low frequencies to boost the low gain at high frequencies. In terms of CXS tubes, we are asking for a large gain at the high frequencies. However, this gain demand causes the CXS tubes to "pinch off." To compensate for this tight high end, we relax the constraints at the low end. Thus, we need CXS tubes with a big radius at the low frequencies that let us interpolate the tight tube at the high frequencies.

On the forthcoming plots, the output gain is reported as

$$G_{L,u} = [-1 \;\; 0.5] \quad \text{dB}.$$

This means the output gain starts at -1 dB at $f = 0.1$ GHz and increases linearly in dB to 0.5 dB at $f = 18$ GHz. Figure 7.20 supplies the upper bound for G_L. This *dB-linear* gain constraint is

$$G_{L,u}(f) = -1 + 1.5\frac{f - 0.1}{18 - 0.1} \quad [\text{dB}].$$

A similar notation is used for the input gain G_G. These dB-linear gains reveal the design challenge of this amplifier. To get $G_T > 9.5$ dB at $f = 18$ GHz requires

$$G_{G,u}(18) + G_{L,u}(18) \geq 1.5 \quad [\text{dB}].$$

Figures 7.31 and 7.32 plot the associated CXS disks of one dB-linear design. At the low frequencies, the CXS disks are relatively large. At high frequencies, the CXS disks are "pinching off." This pinching off is reflected in the Nehari plot of Figures 7.33 and 7.34.

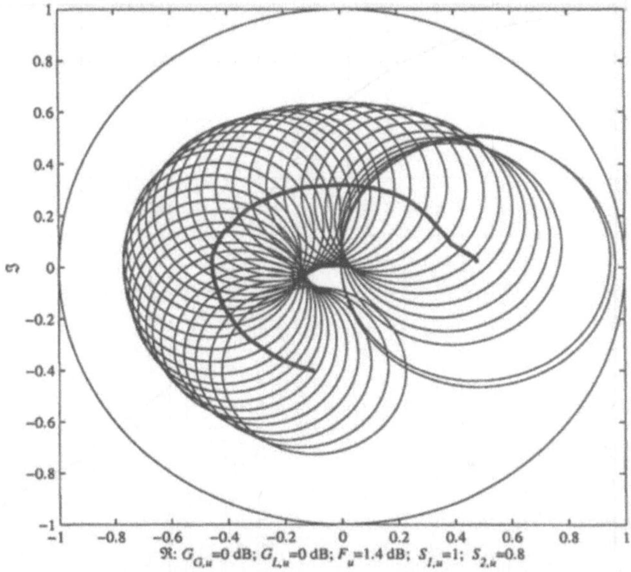

Fig. 7.27. NE32484A input CXS disks.

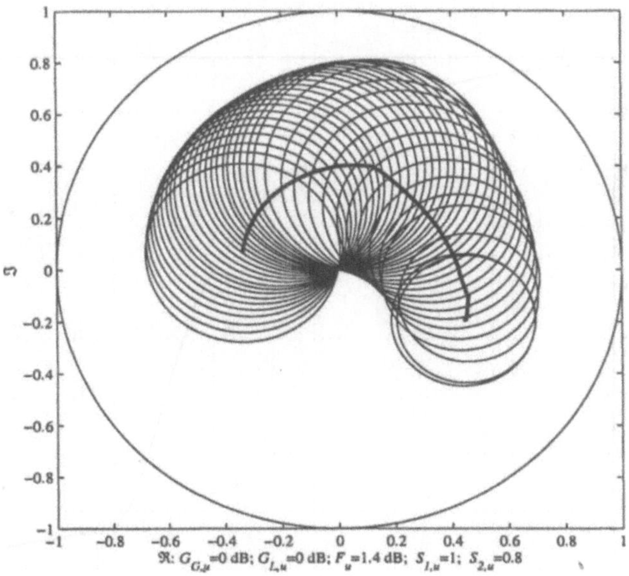

Fig. 7.28. NE32484A output CXS disks.

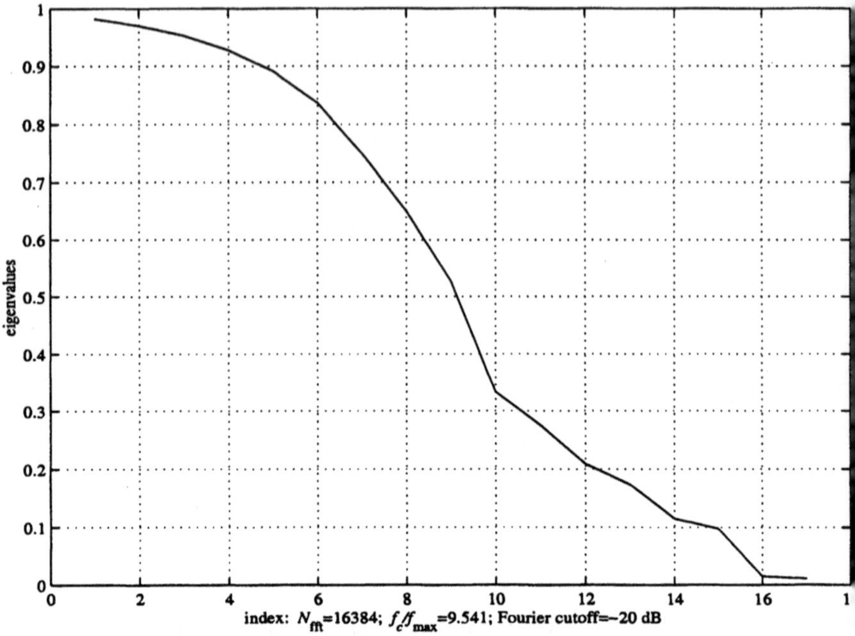

Fig. 7.29. Nehari test for the input CXS disk.

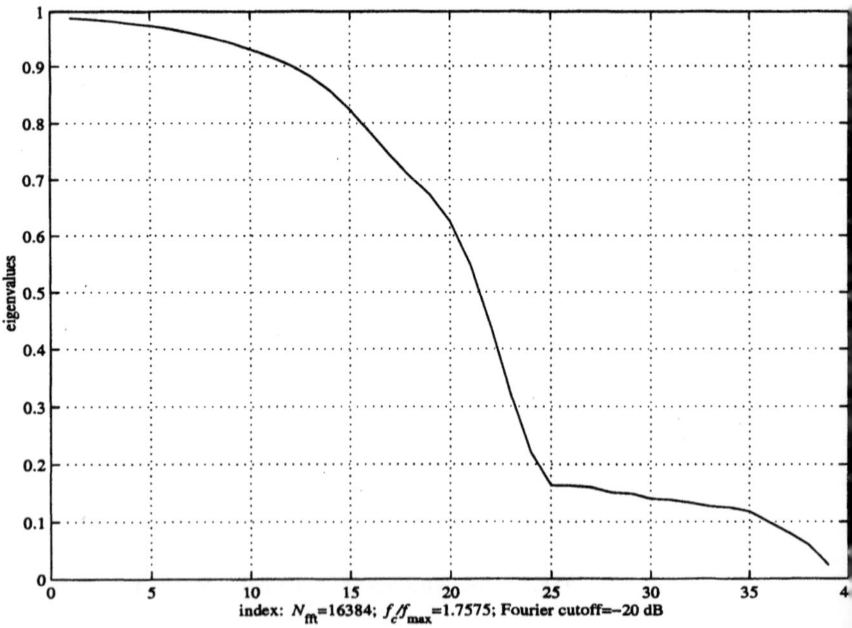

Fig. 7.30. Nehari test for the output CXS disk.

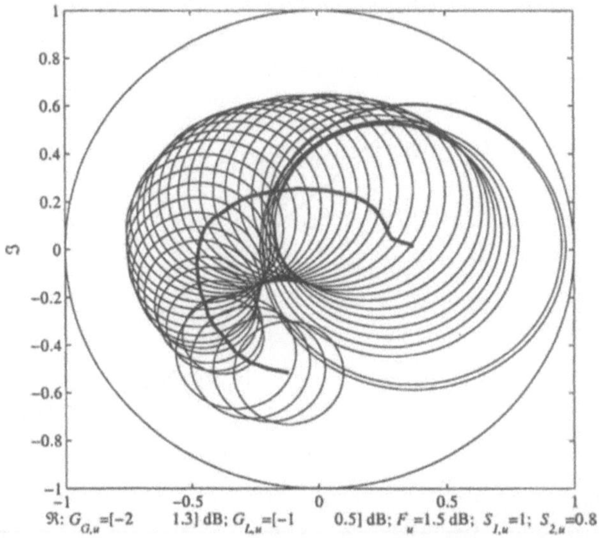

Fig. 7.31. NE32484A input CXS disks with dB-linear gains.

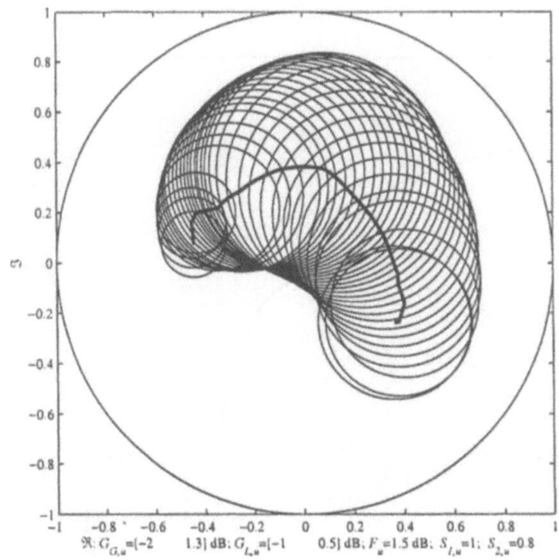

Fig. 7.32. NE32484A output CXS disks with dB-linear gains.

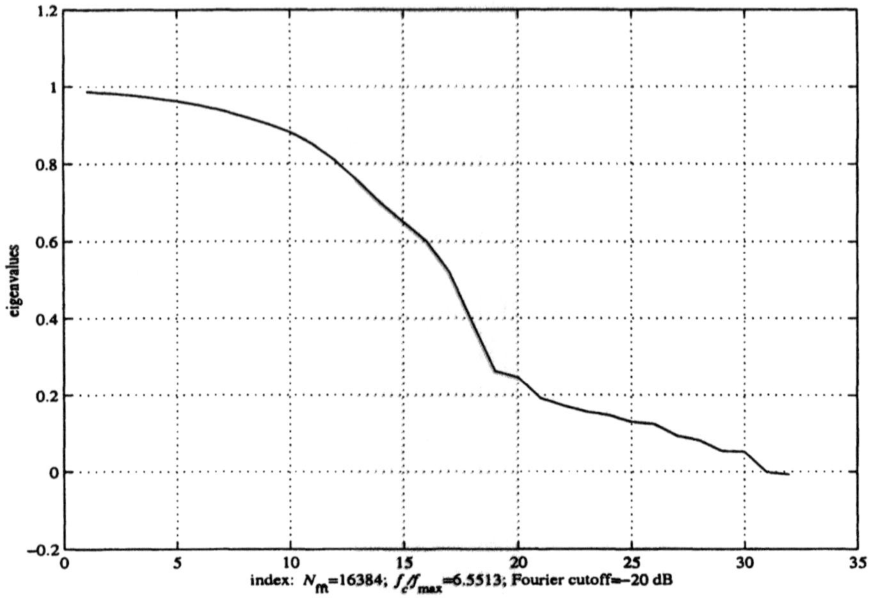

Fig. 7.33. Nehari test for the dB-linear input CXS disk.

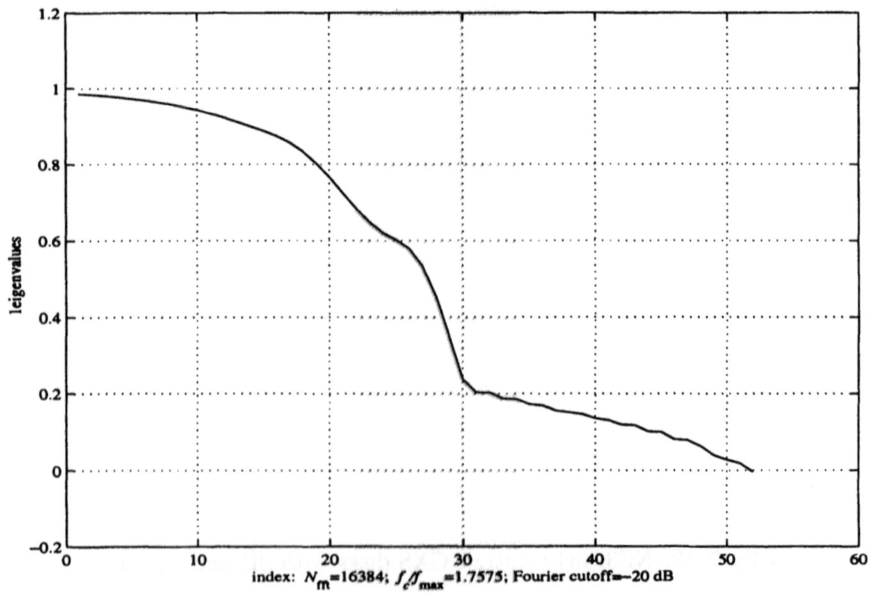

Fig. 7.34. Nehari test for the dB-linear output CXS disk.

The slight negativity in the Nehari test suggests we have reached a performance limit at this level of numerical resolution. Indeed, Figure 7.35 shows that the resulting G_T exhibits an "equal-ripple" phenomenon. Thus,

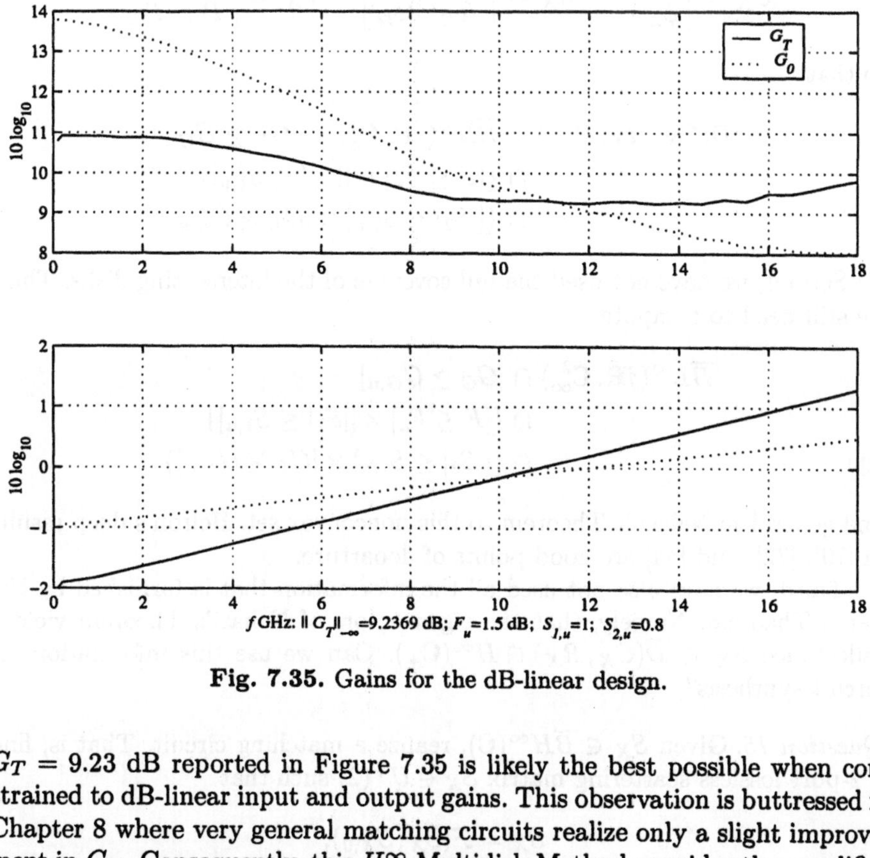

Fig. 7.35. Gains for the dB-linear design.

G_T = 9.23 dB reported in Figure 7.35 is likely the best possible when constrained to dB-linear input and output gains. This observation is buttressed in Chapter 8 where very general matching circuits realize only a slight improvement in G_T. Consequently, this H^∞ Multidisk Method provides the amplifier designer with the following performance limits:

- $G_T \geq 9.2$ dB.
- $F \leq 1.5$ dB.
- $|S_1| \leq 1$, $|S_2| \leq 0.8$.

In practice, the amplifier designer would start with the H^∞ Multidisk Method. to obtain these performance bounds. These bounds show that series/shunt matching is nearly optimal. Moreover, that the input stability problem is not unique to the series/shunt circuit, but afflicts all such matching circuits (Chapter 9 shows how to overcome such stability problems).

We conclude by making explicit some research opportunities. First, we have not used the full power of Helton's Theorem 3.10.1. That is, the CXS disks need not be separable but can be coupled as follows.

Question 14. Find center, radius, and scaling functions

$$C_X = \begin{bmatrix} \phi_G \\ \phi_L \end{bmatrix}, \quad R_X = \begin{bmatrix} R_{11} & R_{12} \\ R_{21} & R_{22} \end{bmatrix}, \quad P_X = \begin{bmatrix} P_{11} & P_{12} \\ P_{21} & P_{22} \end{bmatrix}$$

so that

$$\begin{aligned} \overline{D}(C_X, R_X, P_X,) \subseteq \overline{B}L^\infty(j\mathbf{R}, \mathbf{C}^2_\infty) &\cap [G_G \geq G_{G,u}] \\ &\cap \{[F \leq F_u] \times [|S_1| \leq S_{1,u}]\} \\ &\cap \{[|S_2| \leq S_{2,u}] \times [G_L \geq G_{L,u}]\}? \end{aligned}$$

Second, we have not used the full coverage of the intersecting disks. Thus, we still need to compute

$$\begin{aligned} \overline{B}L^\infty(j\mathbf{R}, \mathbf{C}^2_\infty) &\cap [G_G \geq G_{G,u}] \\ &\cap \{[F \leq F_u] \times [|S_1| \leq S_{1,u}]\} \\ &\cap \{[|S_2| \leq S_{2,u}] \times [G_L \geq G_{L,u}]\} \end{aligned}$$

and generalize Nehari's Theorem to this nonconvex set. Helton's deep results in [70], [72], and [44] are good points of departure.

Third, we have also not used all the information that is furnished by Nehari's Theorem. Namely, that the eigenvectors of Nehari's Theorem yield a reflectance $S_X \in \overline{D}(C_X, R_X) \cap H^\infty(\mathbf{C}_+)$. Can we use this information for circuit synthesis?

Question 15. Given $S_X \in \overline{B}H^\infty(\mathbf{C})$, realize a matching circuit. That is, find a 2-port lossless scattering matrix $\mathcal{S}_X \in U^+(2)$ such that

$$S_X = \mathcal{F}(\mathcal{S}_X, S_{X,0}),$$

where the load is typically $S_{X,0} = 0$.

8

State-Space Methods for Single Amplifiers

The H^∞ Methods bound the amplifier's performance available from *any* matching circuit. In fact, a good portion of Chapter 7 is devoted to making these H^∞ bounds tight. In contrast, most engineering approaches specify a matching circuit and optimize over the component values. This is the approach taken in Chapter 6. The State-Space Method stands between these two extremes by optimizing over all possible matching circuits of a specified degree. The reactive components are specified but the optimization takes place over all component values and circuit topologies. The state-space representation of Chapter 4 completely parameterizes the lumped, lossless N-ports from the orthogonal group. The state-space representation also parameterizes a subset of the lumped-distributed N-ports whose distributed elements are transmission lines of the same electrical length. Both representations admit generalizations to single-stage amplifiers, multistage amplifiers, and amplifier representation theory.

8.1 Parameterizing $U^+(N,d)$

Chapter 3 introduced the class of *real inner* functions

$$U^+(N) := \{S \in \Re\overline{B}H^\infty(\mathbf{C}_+, \mathbf{C}^{N\times N}) : S(j\omega)^H S(j\omega) = I_N \quad \text{a.e.}\}.$$

This class contains all scattering matrices representing any lossless N-port. Chapter 4 analyzed the subclass of lumped, lossless N-ports denoted by

$$U^+(N,d) := \{S \in U^+(N) : \deg_{\text{SM}}[S(p)] \le d\}.$$

Theorem 4.2.1 provides the *state-space representation* for $S \in U^+(N,d)$:

$$S(p) = \mathcal{F}(S_a, S_L; p) := S_{a,11} + S_{a,12}S_L(p)(I_d - S_{a,22}S_L(p))^{-1}S_{a,21},$$

where the *augmented load* $S_L(p)$ consists of N_L inductors and N_C capacitors

$$S_L(p) = \begin{bmatrix} qI_{N_L} & 0 \\ 0 & -qI_{N_C} \end{bmatrix}, \quad \left(q = \frac{p-1}{p+1}\right),$$

and $N_L + N_C = d$. The *augmented scattering matrix*

$$S_a = \begin{bmatrix} S_{a,11} & S_{a,12} \\ S_{a,21} & S_{a,22} \end{bmatrix} \begin{matrix} N \\ d \end{matrix}$$
$$\quad N \quad d$$

is a constant, real, orthogonal matrix.

Let $\mathcal{O}[M]$ denote the *orthogonal group* of real $M \times M$ orthogonal matrices:

$$\mathcal{O}[M] := \{S_a \in \mathbf{R}^{M \times M} : S_a^T S_a = I_M\}.$$

The orthogonal group decomposes into two *connected components*: the *special orthogonal group*

$$S\mathcal{O}[M] := \{S_a \in \mathcal{O}[M] : \det[S_a] = 1\}$$

and its complement consisting of the orthogonal matrices with determinant -1. Fix $\Sigma \in \mathcal{O}[N]$ with $\det[\Sigma] = -1$. The complement of the special orthogonal group is

$$S\mathcal{O}[M]\Sigma := \{S_a \Sigma : S_a \in \mathcal{O}[M]\}.$$

Consequently, the orthogonal group decomposes as

$$\mathcal{O}[M] = S\mathcal{O}[M] \bigcup S\mathcal{O}[M]\Sigma.$$

Because $U^+(N, d)$ allows gyrators, and the cascade of a gyrator with an inductor is a capacitor (Chapter 1 or [8, pages 22–25]), the distinction between the inductors and capacitor may be ignored. Thus, the augmented load may be *fixed*:

$$S_L(p) = qI_d.$$

Because the augmented load is fixed, any $S \in U^+(N, d)$ is parameterized as

$$S(p) = \mathcal{F}(S_a, S_L; p) := S_{a,11} + S_{a,12}(q^{-1}I_d - S_{a,22})^{-1} S_{a,21}.$$

Thus, the lumped, lossless N-ports are parameterized[1] by the orthogonal group

$$U^+(N, d) = \mathcal{F}(\mathcal{O}[N + d], S_L) \quad (S_L(p) = qI_d).$$

Any time such a parameterization occurs, it is good practice to ask about its structure.

---

[1] Theorem 4.2.1 parameterizes the scattering matrices of degree d from $\mathcal{O}[N + d]$ whereas $U^+(N, d)$ is closed because it includes all real, lossless scattering matrices $S(p)$ of degree $\det_{SM}[S(p)] \le d$. If $\det_{SM}[S(p)] = d' < d$ and $S(p) = \mathcal{F}(S_a, qI_{d'})$ with $S_a \in \mathcal{O}[N + d']$, this minimal representation may be extended to a nonminimal representation

$$S(p) = \mathcal{F}(S_a \oplus I_{d-d'}, qI_d),$$

where $S_a \oplus I_{d-d'}$ belongs to $\mathcal{O}[N + d]$.

Question 16. Is $U^+(N, d)$ a stratification of dimension $\dim[\mathcal{O}[N+d]] = (N+d)(N+d-1)/2$?

We will need to track the connected components of the orthogonal group. Introduce

$$U_+^+(N,d) := \mathcal{F}(SO[N+d], S_L),$$
$$U_-^+(N,d) := \mathcal{F}(SO[N+d]\Sigma, S_L).$$

Thus, the special orthogonal group parameterizes the scattering matrices of $U^+(N, d)$ as

$$\begin{aligned} U^+(N,d) &= U_+^+(N,d) \bigcup U_-^+(N,d) \\ &:= \mathcal{F}(SO[N+d], S_L) \bigcup \mathcal{F}(SO[N+d]\Sigma, S_L). \end{aligned}$$

8.2 Parameterization by $U^+(2, d_G) \times U^+(2, d_L)$

Figure 8.1 shows that the Amplifier Matching Problem is an optimization problem over $U^+(2, d_G) \times U^+(2, d_L)$. By Theorem 4.2.1, we are optimizing

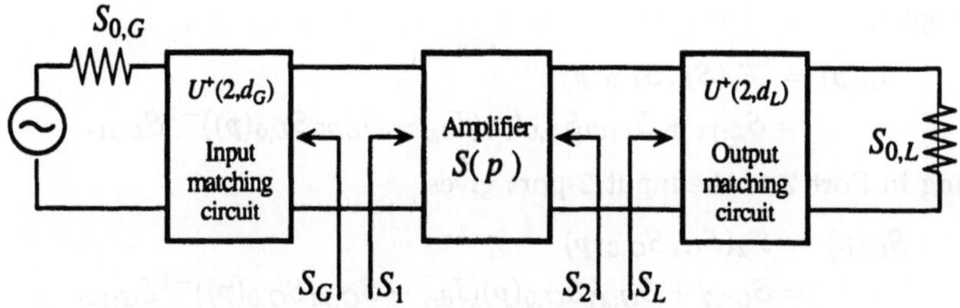

Fig. 8.1. Matching over all lumped, lossless 2-ports of degree (d_G, d_L).

over all lumped, lossless 2-ports of degree not exceeding d_G on the input and degree not exceeding d_L on the output.

Let $S_{G,a} \in \mathcal{O}[d_G+2]$ denote the augmented scattering matrix for the input matching circuit:

$$S_{G,a} = \begin{bmatrix} S_{G,a;11} & S_{G,a;12} \\ S_{G,a;21} & S_{G,a;22} \end{bmatrix} \begin{matrix} 2 \\ d \end{matrix}.$$
$$\qquad\qquad\quad 2 \qquad d$$

Likewise, let $S_{L,a} \in \mathcal{O}[d_L+2]$ denote the augmented scattering matrix for the output matching circuit:

$$S_{L,a} = \begin{bmatrix} S_{L,a;11} & S_{L,a;12} \\ S_{L,a;21} & S_{L,a;22} \end{bmatrix} \begin{matrix} 2 \\ d \end{matrix}.$$
$$\qquad\qquad\quad 2 \qquad d$$

Because we are not restricted to reciprocal 2-ports, the augmented loads may be written as follows:

$$S_{G,a}(p) := \frac{p-1}{p+1} I_{d_G},$$
$$S_{L,a}(p) := \frac{p-1}{p+1} I_{d_L}.$$

Define the mapping $\mathcal{F} : \mathcal{O}[2+d_G] \to U^+(2,d_G)$ as

$$\begin{aligned} S_G(p) &= \mathcal{F}(\mathcal{S}_{G,a}, S_{G,a}; p) \\ &:= \mathcal{S}_{G,a;11} + \mathcal{S}_{G,a;12} S_{G,a}(p)(I_{d_G} - \mathcal{S}_{G,a;22} S_{G,a}(p))^{-1} \mathcal{S}_{G,a;21}. \end{aligned}$$

With a slight abuse of notation, define the mapping $\mathcal{F} : \mathcal{O}[2+d_L] \to U^+(2,d_L)$ as

$$\begin{aligned} S_L(p) &= \mathcal{F}(\mathcal{S}_{L,a}, S_{L,a}; p) \\ &:= \mathcal{S}_{L,a;11} + \mathcal{S}_{L,a;12} S_{L,a}(p)(I_{d_L} - \mathcal{S}_{L,a;22} S_{L,a}(p))^{-1} \mathcal{S}_{L,a;21}. \end{aligned}$$

Just as in Section 8.1, these mappings parameterize the input and output 2-ports:

$$U^+(2,d_G) = \mathcal{F}(\mathcal{O}[2+d_G], S_{G,a}),$$
$$U^+(2,d_L) = \mathcal{F}(\mathcal{O}[2+d_L], S_{L,a})$$

independent of any circuit topology. These 2-ports parameterize the generator and load reflectances shown in Figure 8.1. Looking in Port 1 of the output 2-port gives

$$\begin{aligned} S_L(p) &= \mathcal{F}_1(\mathcal{S}_L, S_{L,0}; p) \\ &:= \mathcal{S}_{L;11} + \mathcal{S}_{L;12} S_{L,0}(p)(I_{d_L} - \mathcal{S}_{L;22} S_{L,0}(p))^{-1} \mathcal{S}_{L;21}. \end{aligned}$$

Looking in Port 2 of the input 2-port gives

$$\begin{aligned} S_G(p) &= \mathcal{F}_2(\mathcal{S}_G, S_{G,0}; p) \\ &:= \mathcal{S}_{G;22} + \mathcal{S}_{G;21} S_{G,0}(p)(I_{d_G} - \mathcal{S}_{G;11} S_{G,0}(p))^{-1} \mathcal{S}_{G;12}. \end{aligned}$$

Finally, we need to keep track of the connected components of the orthogonal groups. Fix $\Sigma_G \in \mathcal{O}[2+d_G]$ with $\det[\Sigma_G] = -1$ and set

$$U_+^+(2,d_G) := \mathcal{F}(S\mathcal{O}[2+d_G], S_G),$$
$$U_-^+(N,d_G) := \mathcal{F}(S\mathcal{O}[2+d_G]\Sigma_G, S_G).$$

Fix $\Sigma_L \in \mathcal{O}[2+d_L]$ with $\det[\Sigma_L] = -1$ and set

$$U_+^+(2,d_L) := \mathcal{F}(S\mathcal{O}[2+d_L], S_L),$$
$$U_-^+(N,d_L) := \mathcal{F}(S\mathcal{O}[2+d_L]\Sigma_L, S_L).$$

The input and output 2-ports decompose as

$$U^+(2,d_G) = U_+^+(2,d_G) \bigcup U_-^+(2,d_G),$$

$$U^+(2,d_L) = U_+^+(2,d_L) \bigcup U_-^+(2,d_L).$$

8.3 Lumped Matching

As in Chapter 6, the multiple objectives in amplifier matching are to maximize the transducer power gain, to minimize the noise figure, and to maintain stability over the lumped 2-ports of fixed degree.

AMPLIFIER MATCHING OVER THE LUMPED 2-PORTS. Find $S_G \in \mathcal{F}_2(U^+(2,d_G), S_{0,G})$ and $S_L \in \mathcal{F}_1(U^+(2,d_L), S_{0,L})$ to simultaneously
AMP-1 Maximize the transducer power gain: $\|G_T(S_G, S_L)\|_{-\infty}$.
AMP-2 Minimize the noise figure: $\|F(S_G)\|_\infty$.
AMP-3 Guarantee stability: $\|S_1(S_L)\|_\infty$, $\|S_2(S_G)\|_\infty \le 1$.

The Goal Attainment Method is the multiobjective optimizer applied to this matching problem. Set

$$X := \mathcal{F}_2(U^+(2,d_G), S_{0,G}) \times \mathcal{F}_1(U^+(2,d_L), S_{0,L})$$

and define the multiobjective function $\gamma : X \to \mathbf{R}^4$ as

$$\gamma(S_G, S_L) := \begin{bmatrix} -\|G_T(S_G,S_L)\|_{-\infty} \\ \|F(S_G)\|_\infty \\ \|S_1(S_L)\|_\infty \\ \|S_2(S_G)\|_\infty \end{bmatrix}.$$

As discussed in Chapter 6, the user selects a design goal $\gamma_u < \gamma(X)$ and a weight vector $\mathbf{w} \in \mathbf{R}^4_+$. The Goal Attainment Method is

$$\min\{t \in \mathbf{R}\}$$

subject to $(S_G, S_L) \in X$ and

$$\gamma(S_G,S_L) - t\mathbf{w} = \begin{bmatrix} -\|G_T(S_G,S_L)\|_{-\infty} \\ \|F(S_G)\|_\infty \\ \|S_1(S_L)\|_\infty \\ \|S_2(S_G)\|_\infty \end{bmatrix} - t \begin{bmatrix} w_G \\ w_F \\ 0 \\ 0 \end{bmatrix} \le \begin{bmatrix} -G_{T,u} \\ F_u \\ S_{1,u} \\ S_{2,u} \end{bmatrix} = \gamma_u.$$

The zeros in the weight vector enforce stability. Section 8.2 determined that the 2-ports decompose as

$$U^+(2,d_G) = U^+_+(2,d_G) \bigcup U^+_-(2,d_G),$$

$$U^+(2,d_L) = U^+_+(2,d_L) \bigcup U^+_-(2,d_L).$$

The implication for matching is that the optimizer must be applied to all four $(\pm,\pm)$ combinations. If it could be determined that $U^+_+(2,d) = U^+_-(2,d)$ (not likely), or that the $(\pm,\pm)$ combinations did not matter for matching (likely) the cost of a search could be reduced by a factor of 4. Thus, an excellent research question is the following:

Question 17. How does $U^+_\pm(2,d)$ reside in $U^+(2,d)$?

Although the orthogonal group has two components, we suspect that $U^+_+(2,d_G)$ and $U^+_-(2,d_G)$ are connected on the scattering matrices of degree strictly less that d_G.

Question 18. Is $U^+(2,d)$ connected?

Regardless of connectivity, the numerical optimizer needs to know if it matters which piece of $U^+(2,d)$ is selected.

Question 19. Is there a significant difference between optimization over $U^+_+(2,d)$ or $U^+_-(2,d)$?

Closely allied with the selection of $U^+_\pm(2,d)$ is the engineering interpretation of the "±" subsets. Although $U^+_\pm(2,d)$ represents the lumped, lossless 2-ports, it would be interesting to know the following:

Question 20. What is the physical meaning of the "±" in $U^+_\pm(2,d)$?

The following matching examples make these questions concrete. The amplifier is the NE32484A discussed in Section 6.4. The NE32484A is a wideband (0.1–18 GHz), low-noise, high-gain communication amplifier. Figures 8.2 and 8.3 present the NE32484A's matching performance over all lumped, lossless 2-ports of degree 1 and 2, respectively. The upper panel shows the gain-noise plane in dB. The lower panel shows the stability in absolute value. The domain of $\gamma(S_G, S_L)$ is restricted to

$$X := U^+_+(2,d) \times U^+_+(2,d).$$

To sketch the four-dimensional image of $\gamma(X)$, a random collection of reflectances has been plotted on both panels. The upper panel shows the triangular shape discussed in Section 6.3. Thus, the "almost minimal region" still exists. The lower panel shows that input stability is still a problem for this wideband amplifier. Therefore, the Goal Attainment Method is started from the point with the smallest $\|S_1(S_L)\|_\infty$. This point is marked with the "+." The "$\diamond$" marks the terminal point of the Goal Attainment Method.

We conjecture that the "(±, ±)" subsets do not affect the multiobjective optimization. To explore this assertion, Tables 8.1 and 8.2 presents the matching results for each (±, ±) combination. The (+, +) results are from Figures 8.2 and 8.3. The selection of the (±, ±) subsets makes little difference in the matching and answers Question 19 negatively.

Once matching over the lumped 2-ports is in place for a given degree, it is natural to ask for the matching performance over a range of degrees.

Question 21. What is the amplifier's performance as a function of (d_G, d_L)?

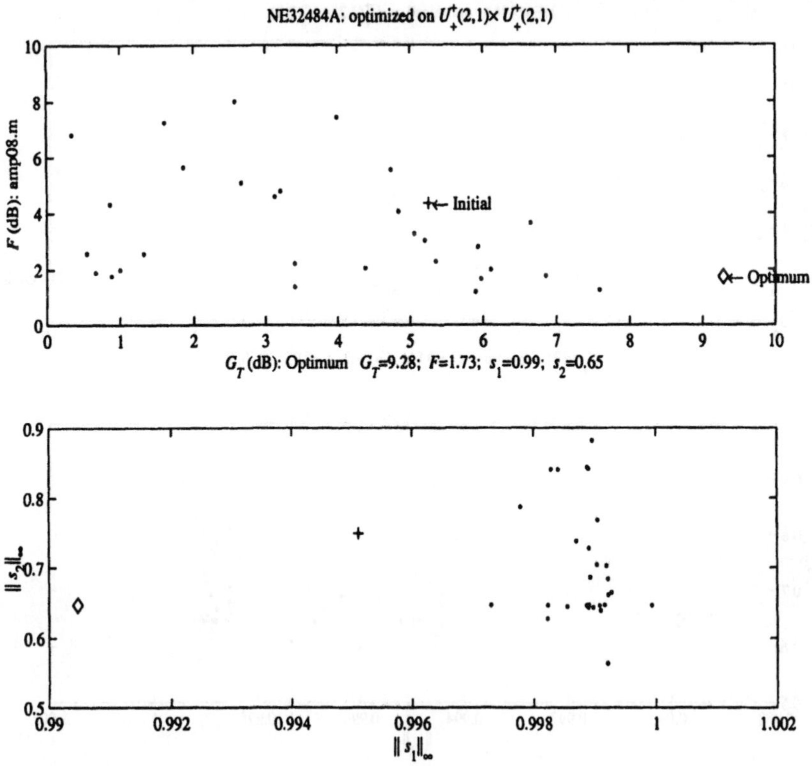

Fig. 8.2. Matching over the (+,+) component of the lumped, lossless 2-ports of degree $(d_G, d_L) = (1,1)$.

Table 8.1. Amplifier performance for $U_\pm^+(2,1) \times U_\pm^+(2,1)$.

Component	Gain (dB)	Noise (dB)	$\|S_1\|_\infty$ (abs)	$\|S_2\|_\infty$ (abs)
$(+,+)$	9.28	1.73	0.99	0.65
$(+,-)$	9.28	1.73	0.99	0.65
$(-,+)$	9.34	1.81	0.99	0.65
$(-,-)$	9.34	1.81	0.99	0.65

Figure 8.4 illustrates this question by matching over $U_+^+(2,d) \times U_+^+(2,d)$ for $d = 0, \ldots, 6$. For each degree, 10 randomly selected reflectances (S_G, S_L) started the Goal Attainment Method. For most of these starting reflectances, the Goal Attainment Method successfully terminated. The resulting gains and noise figures are plotted as a function of degree. Typically, increasing the degree gets more gain but increases the noise figure. However, the NE32484A amplifier functions are known only at 37 frequencies over 0.–18 GHz.

Previous work on impedance matching revealed an *overfitting phenomena* [2]. A matching 2-port with a sufficiently large number of parameters can get

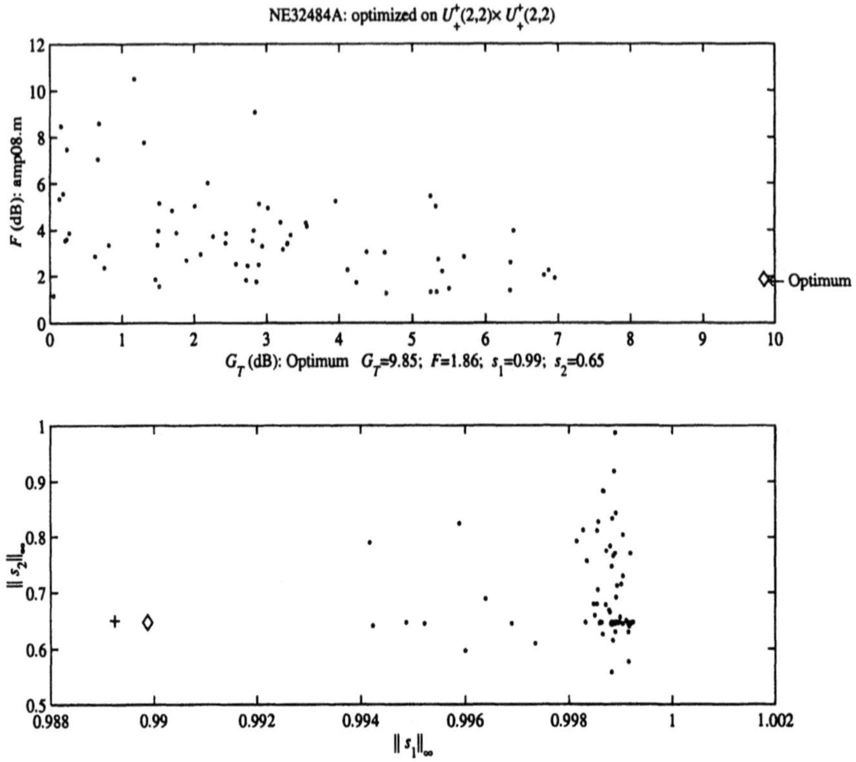

Fig. 8.3. Matching over the (+,+) component of the lumped, lossless 2-ports of degree $(d_G, d_L) = (2, 2)$.

Table 8.2. Amplifier performance for $U_{\pm}^{+}(2,2) \times U_{\pm}^{+}(2,2)$.

Component	Gain (dB)	Noise (dB)	$\|S_1\|_\infty$ (abs)	$\|S_2\|_\infty$ (abs)
(+,+)	9.85	1.86	0.99	0.65
(+,−)	9.77	1.83	0.99	0.65
(−,+)	9.69	2.01	0.99	0.65
(−,−)	9.69	2.01	0.99	0.65

a very good match at all sample frequencies. In fact, there exist bogus engineering papers claiming perfect matching without realizing that the match is good only on the sample frequencies. Off the sample frequencies, the matching performance is worse than the H^∞ bound. In Figure 8.4, only 37 frequency samples are available for the NE32484A amplifier. The number of parameters in $U^+(2,d) \times U^+(2,d)$ is $(d+2)(d+1)$. When $d = 6$, there are 56 parameters. These parameters may allow an optimizer to exceed the H^∞ bounds on the sample points.

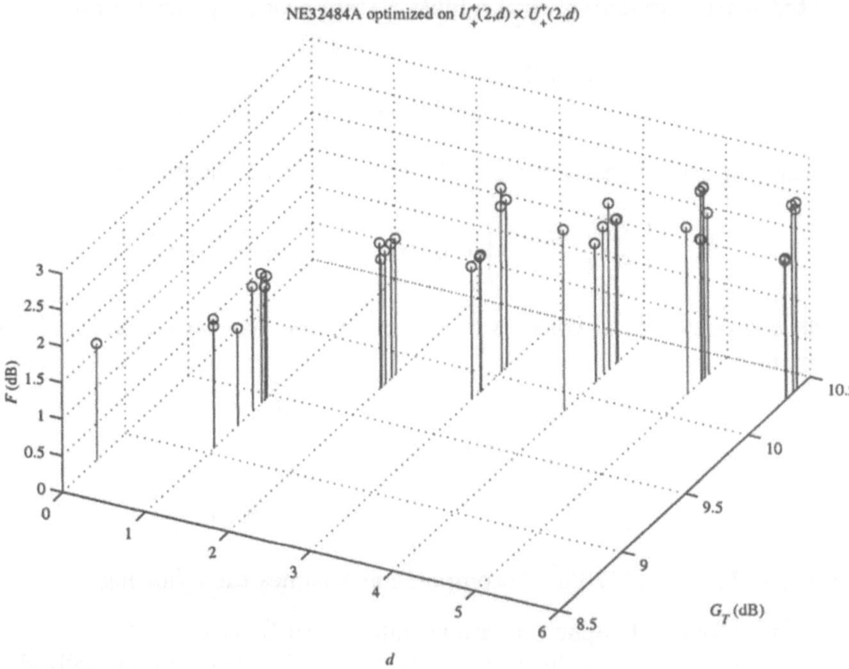

Fig. 8.4. Gain-noise performance as a function of degree $d = 0, \ldots, 6$.

Question 22. Is their an overfitting phenomenon for amplifier matching?

Chapter 10 makes a systematic presentation of the research topics that arise when optimizing over finite frequency samples. The H^∞ Methods may warn the amplifier designer when overfitting occurs and too optimistic matching performance is computed. For the circuits in this chapter, no overfitting analysis undertaken. Rather, the H^∞ Methods will simply benchmark the performance. Before we make this comparison, the state-space Method is generalized to lumped-distributed matching circuits.

8.4 Lumped-Distributed Matching

Section 6.3.5 shows the matching limitations of lumped 2-ports with distrubuted 2-ports. Section 4.4 offered a state-space representation that includes distributed components. The basic idea is to attach distributed elements to the augmented multiport along with the inductors and capacitors. Figure 4.6 illustrates this idea and is the basis for Wohler's [128, pages 168–172] claim of characterizing lumped-distributed N-ports. Section 4.4 makes explicit the problems with this characterization claim. Therefore, a class of lumped-distributed, lossless N-ports is defined directly from Figure 4.6. The associated scattering matrix $S(p)$ admits a state-space representation

$$\begin{aligned} S(p) &= \mathcal{F}(S_a, S_L; p) \\ &= S_{a,11} + S_{a,12} S_L(p)(I_d - S_{a,22} S_L(p))^{-1} S_{a,21}, \end{aligned}$$

where the augmented scattering matrix is a constant orthogonal matrix:

$$S_a = \begin{bmatrix} S_{a,11} & S_{a,12} \\ S_{a,21} & S_{a,22} \end{bmatrix} \in \mathcal{O}[N + N_L + N_C + 2N_U],$$

and the augmented load consists of N_L inductors, N_C capacitors, and N_U unit elements

$$S_L(p) = \begin{bmatrix} q \times I_{N_L} & & 0 \\ 0 & -q \times I_{N_C} & 0 \\ 0 & 0 & I_{N_U} \otimes \begin{bmatrix} 0 & e^{-\tau p} \\ e^{-\tau p} & 0 \end{bmatrix} \end{bmatrix},$$

for $q = (p-1)(p+1)^{-1}$. This decomposition assumes the following.

- The inductors and capacitors are normalized to $Z_0 = 1$.
- All unit elements have the same electrical length τ and are normalized to the characteristic impedance Z_{c,n_u} of each transmission line.

The image of the orthogonal group is tracked with the "±" subscript:

$$U^+_{\tau,+}(N, N_L, N_C, N_U) := \mathcal{F}(SO[N + N_L + N_C + 2N_U], S_L),$$

$$U^+_{\tau,-}(N, N_L, N_C, N_U) := \mathcal{F}(SO[N + N_L + N_C + 2N_U]\Sigma, S_L),$$

where $\Sigma \in \mathcal{O}[N + N_L + N_C + 2N_U]$ has $\det[\Sigma] = -1$. Define the class of lumped-distributed, lossless N-ports as follows:

$$U^+_\tau(N, N_L, N_C, N_U) := U^+_{\tau,+}(N, N_L, N_C, N_U) \bigcup U^+_{\tau,-}(N, N_L, N_C, N_U).$$

The remarks that close Section 4.4 are made specific to amplifier matching using the unit elements. In the augmented multiport, it is the transformers that sweep over all inductor and capacitor values. Thus, the augmented load can fix the values of the inductors and capacitors at 1 and augmented multiport parameterizes the lumped elements. However, only the characteristic impedance Z_c of a unit element is modified by the transformer—not its delay τ. The implication for the state-space representation is that the augmented multiport can only parameterize the characteristic impedance Z_c of each unit element. For this reason, the τ must index these scattering matrices.

Finally, as in Section 6.3.4, we choose τ be small enough so that

$$\tau\omega < \frac{\pi}{2}$$

for all ω in the frequency range $[\omega_{\min}, \omega_{\max}]$. To guarantee this inequality, set

$$\tau = \frac{1}{2} \times \frac{\pi}{2\omega_{\max}}.$$

Figure 8.5 shows the amplifier matching using a single unit element in the input 2-port and the output matching 2-port. Comparing Figure 8.5 with Figure 8.4 shows that the performance of the single unit element exceeds the performance of the lumped 2-ports of degree 6. Figure 8.6 adds a second unit element to each 2-port to obtain still better performance.

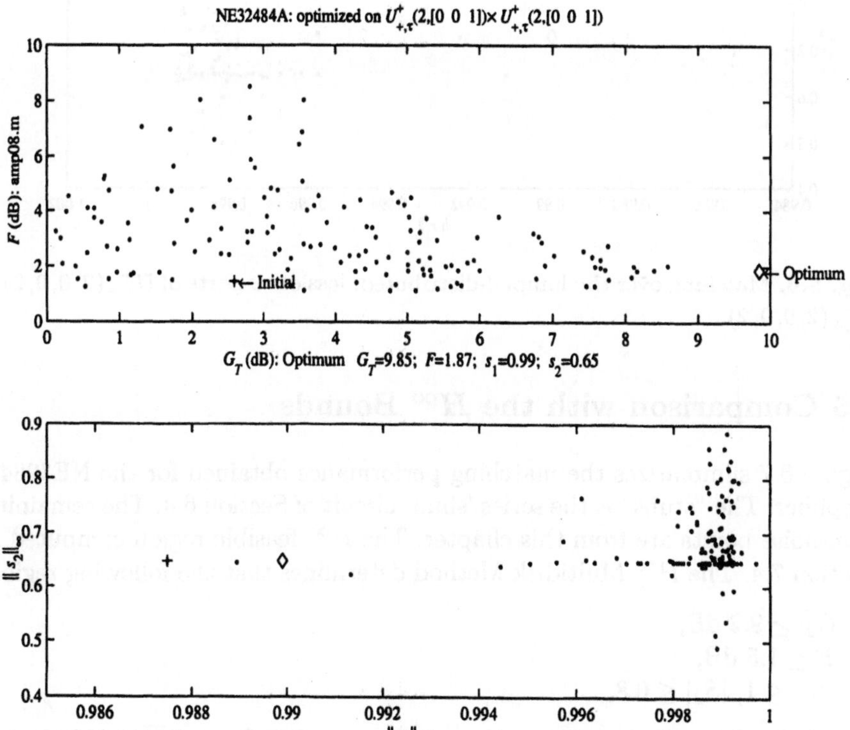

Fig. 8.5. Matching over the lumped-distributed, lossless 2-ports of $U^{+}_{\tau,+}(2,0,0,1) \times U^{+}_{\tau,+}(2,0,0,1)$.

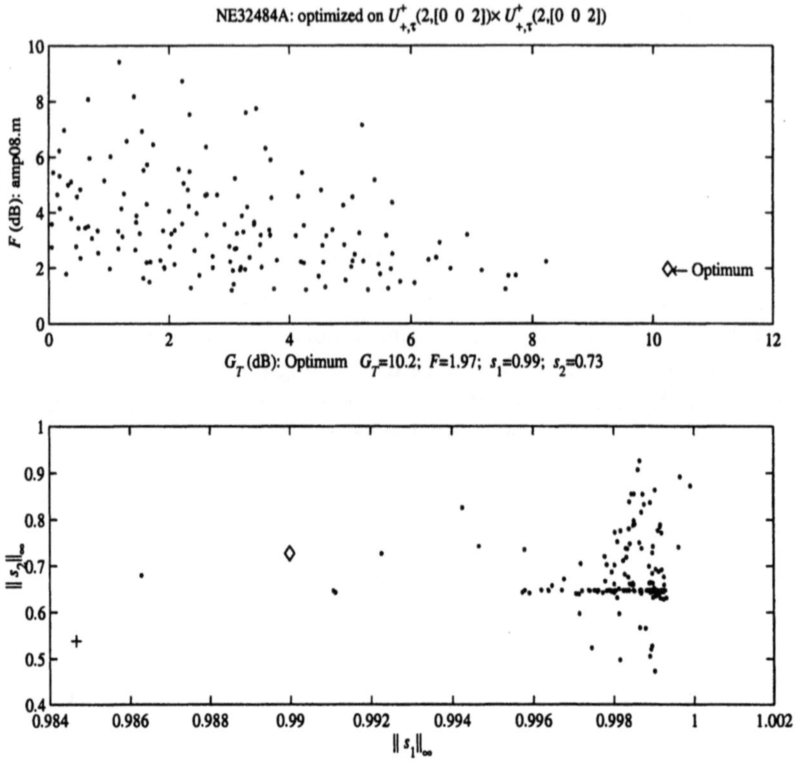

Fig. 8.6. Matching over the lumped-distributed, lossless 2-ports of $U^{+}_{\tau,+}(2,0,0,2) \times U^{+}_{\tau,+}(2,0,0,2)$.

8.5 Comparison with the H^∞ Bounds

Figure 8.7 summarizes the matching performance obtained for the NE3284A amplifier. The "stubs" is the series/shunt circuit of Section 6.4. The remaining gain-noise points are from this chapter. The H^∞ feasible region computed in Section 7.4. The H^∞ Multidisk Method determines that the following region:

- $G_T \geq 9.2$ dB,
- $F \leq 1.5$ dB,
- $|S_1| \leq 1$, $|S_2| \leq 0.8$,

is H^∞ feasible. That is, there are reflectances S_G, $S_L \in \Re\overline{B}H^\infty(\mathbf{C}_+)$ that meet the following performance bounds.

- $G_T(S_G, S_L; j\omega) \geq 9.2$ dB.
- $F(S_G; j\omega) \leq 1.5$ dB.
- $|S_1(S_L; j\omega)| \leq 1$, $|S_2(S_G; j\omega)| \leq 0.8$.

For this particular choice of bounds, the stubs circuit is nearly optimal. However, the matching circuits show there exists a trade-off between gain and

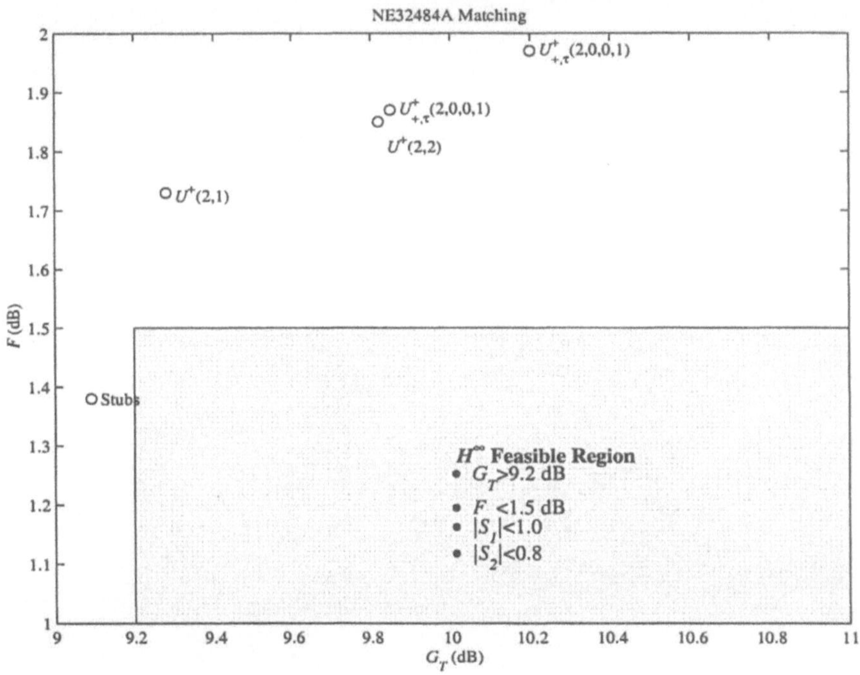

Fig. 8.7. NE32484 gain-noise performance.

noise. For example, if the amplifier designer could accept more noise, say $F = 1.7$ dB, the H^∞ Multidisk Method could be run again to compute another H^∞ feasible region with more gain. This new region could benchmark the more complex circuits with greater gain. More generally, by incrementing the noise and recomputing the gain, the amplifier designer can map out a gain-noise performance curve and associated H^∞ feasible regions.

However, we have come to the performance limits available by the input/output matching circuits. In particular, the stability margins obtained by both the circuits and H^∞ Methods are problematic. Chapter 9 introduces a general state-space approach that overcomes this stability problem, that extends to multiple amplifiers, and opens up a new collection of H^∞ topics.

9

State-Space Methods for Multiple Amplifiers

The amplifier literature contains many schemes to connect multiple amplifiers. For example, the classical cascode circuit [25] of Figure 9.1 directly connects two transistors and is still a topic in amplifier matching [90].

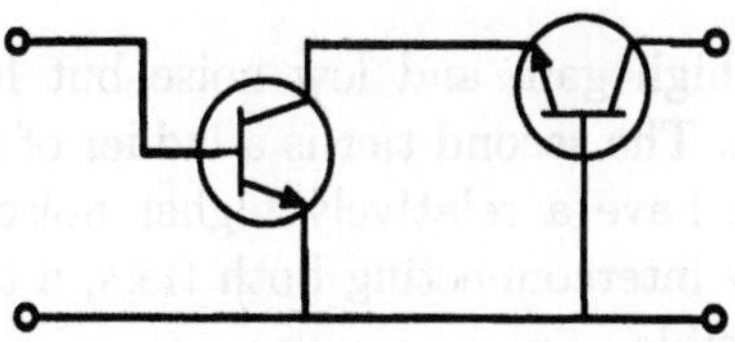

Fig. 9.1. The CE-CB or cascode circuit.

More generally, the cascade of multiple amplifiers constitutes a large portion of the amplifier literature and continues as an active research topic [120]. For example, Figure 9.2 shows two amplifiers in cascade connected by three lossless matching circuits. With three matching circuits, the Amplifier Matching Problem is an optimization problem over $U^+(2) \times U^+(2) \times U^+(2)$. Recall

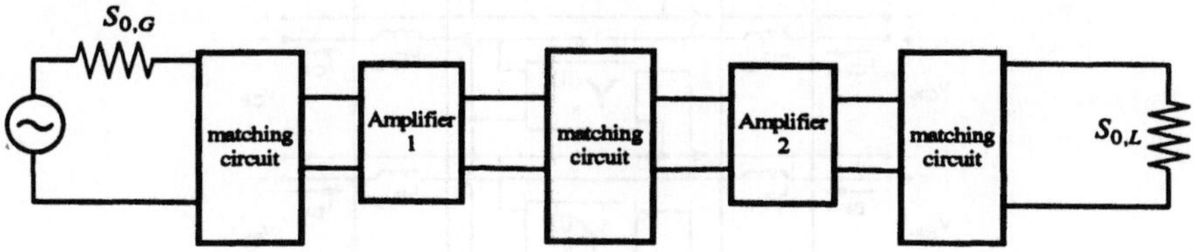

Fig. 9.2. Cascade of two amplifiers.

from Chapter 4 that $U^+(2)$ denotes the collection of all lossless 2×2 scattering matrices.

Still more generally, Figure 9.3 shows the schematic for a two-tier matrix amplifier [102]. The first tier is a ladder of high electron-mobility transistors

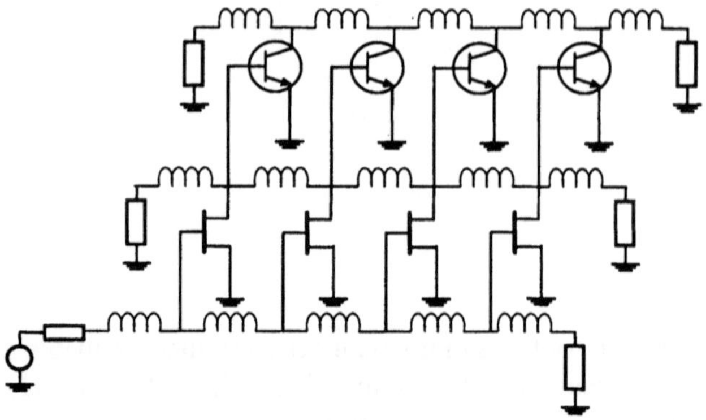

Fig. 9.3. A two-tier matrix amplifier (courtesy of the IEEE).

(HEMT) that exhibit high gain and low noise but have a relatively large DC power consumption. The second tier is a ladder of heterojunction bipolar transistors (HBT) that have a relatively higher noise figure but lower DC power consumption. By interconnecting both tiers, a trade-off between gain, noise, and power is possible.

Figure 9.4 illustrates a generalization of the two-tier matrix amplifier to the three-tier amplifier cell [100]. An amplifier is constructed by cascading these cells and terminating all but two ports.

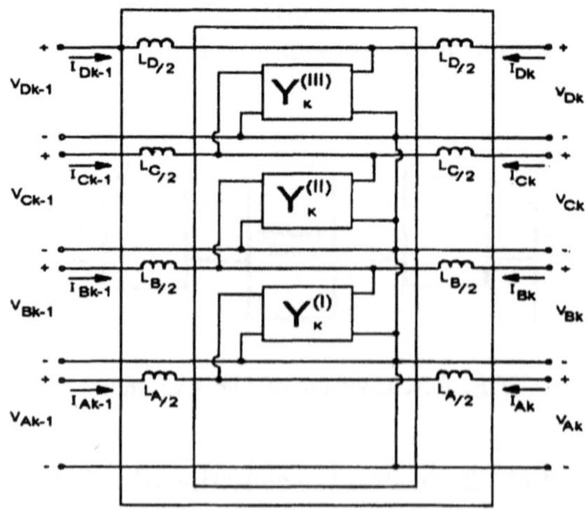

Fig. 9.4. A three-tier matrix amplifier cell [100] (courtesy of the IEEE).

A generalization to the multitier amplifier is the "circuit technique of coupling one or more input or output lines of a distributed amplifier into several output or input lines, respectively" [31]. Figure 9.5 illustrates this technique where nine cascode pairs implement an active wideband distributed impedance divider.

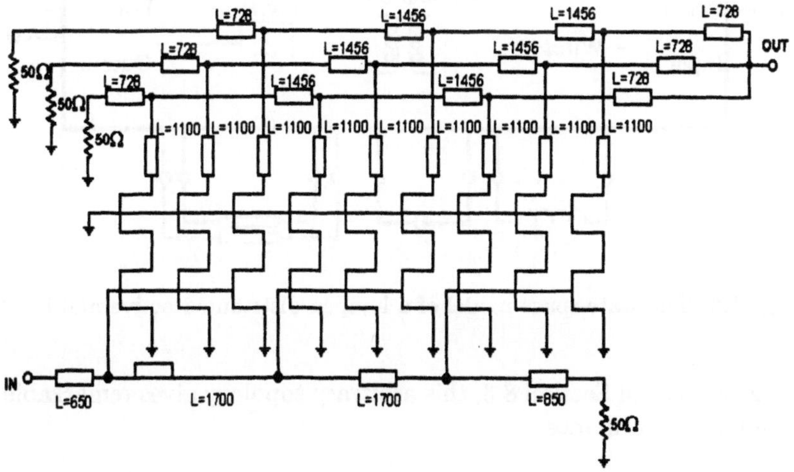

Fig. 9.5. An active wideband impedance divider [31] (courtesy of the IEEE).

The mathematical question naturally follows and is the most general amplifier question of this book.

Question 23. What is the optimal performance of N_A amplifiers embedded in an arbitrary matching topology?

Only a few papers address this question [107], [23]. Figure 9.6 illustrates the state-space solution. The amplifiers are attached to a lossless, lumped-distributed multiport. The reactive elements and the amplifiers are fixed. The matching is computed by sweeping over the augmented scattering matrices of the augmented multiport. Such a model of an active multiport includes all the *negative-feedback* topologies that can stabilize the attached amplifiers. A nice history of negative feedback is in [89].

To verify these assertions, we start by matching a single amplifier. Figure 9.7 shows the matching performance of one amplifier over the lumped, lossless multiports of degree 4—no transmission lines. The stability axis constitutes the horizontal plane. The transducer power is the vertical axis. The initial stable guesses are marked as dots. A "most" stable point is marked with a circle and starts the multiobjective optimizer. The diamond marks the resulting local optima. Compared to the standard matching using input and

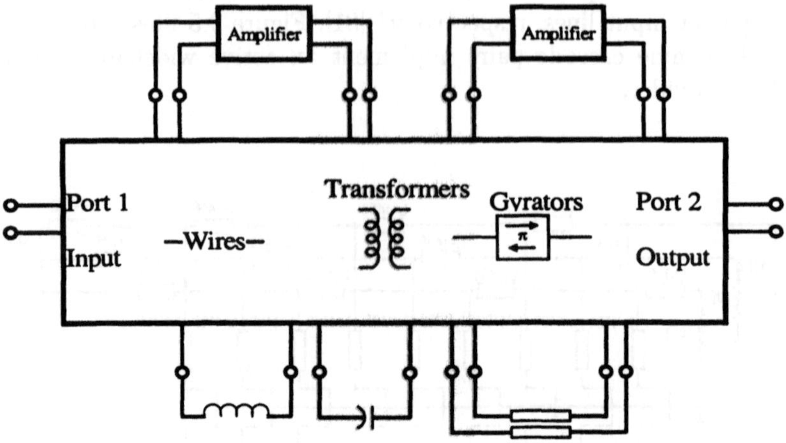

Fig. 9.6. The state-space model of a lumped-distributed active multiport.

output 2-ports as in Section 8.3, this arbitrary topology gives remarkable gain and stability performance.

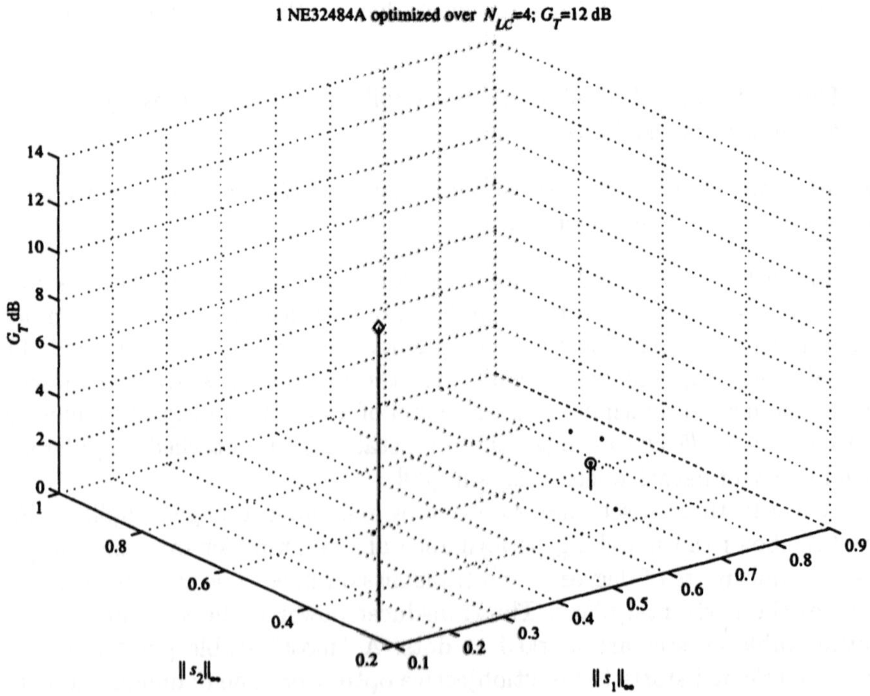

Fig. 9.7. Matching one amplifier over the lumped, lossless multiports of degree 4.

Figure 9.8 adds a second amplifier. We are now matching two amplifiers over the lumped, lossless multiports of degree 2. The gain has almost doubled but the stability is back to the usual marginal level associated with this amplifier (Figures 8.2, 8.3). Both Figures 9.7 and 9.8 only optimize the gain

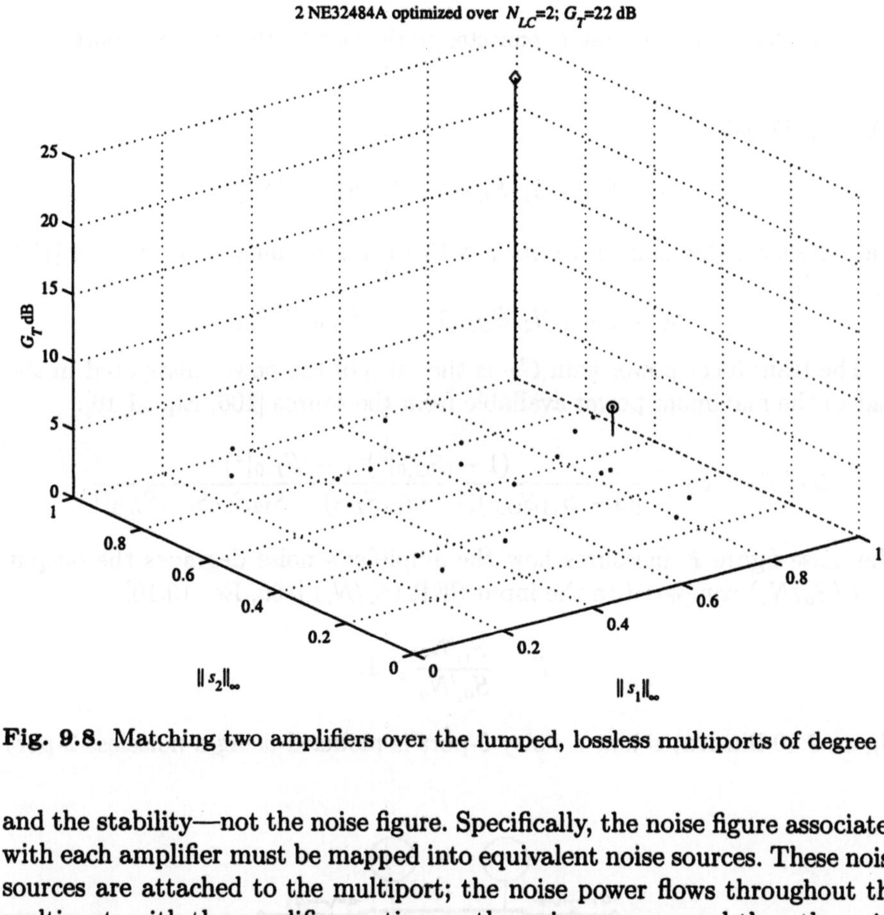

Fig. 9.8. Matching two amplifiers over the lumped, lossless multiports of degree 2.

and the stability—not the noise figure. Specifically, the noise figure associated with each amplifier must be mapped into equivalent noise sources. These noise sources are attached to the multiport; the noise power flows throughout the multiport, with the amplifiers acting on the noise powers; and then the noise levels at the input and output ports determine the overall noise figure.

9.1 Functions of an Active 2-Port

Although Figure 9.6 is the representation that we will ultimately use, the noise figure computations necessitate a return to the basic 2-port. Figure 9.9 shows the general setup for the amplifier functions. A 2-port, either passive or active, delivers the signal entering Port 1 to the load that terminates Port 2. The amplifier functions are defined in terms of the 2-port's scattering matrix. S_1 denotes the reflectance looking into Port 1 with Port 2 terminated in $S_{L,0}$

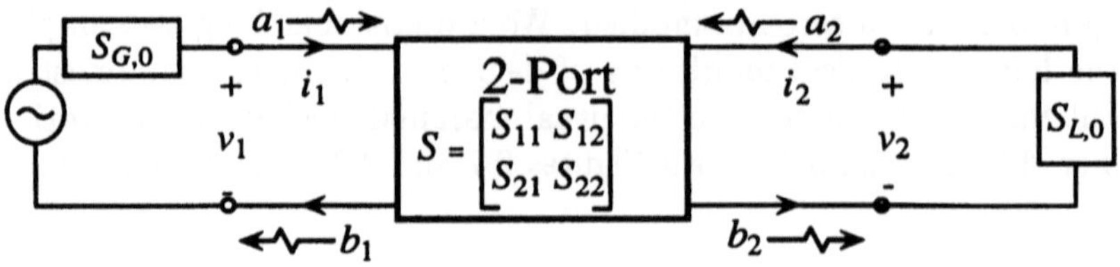

Fig. 9.9. Signal generator connected to the load by the (active) 2-port.

[106, Eq. 11.6a]:

$$S_1 = S_{11} + S_{12}S_{L,0}(1 - S_{22}S_{L,0})^{-1}S_{21}.$$

The reflectance looking into Port 2 with Port 1 terminated in $S_{G,0}$ is [106, Eq. 11.6b]:

$$S_2 = S_{22} + S_{21}S_{G,0}(1 - S_{11}S_{G,0})^{-1}S_{12}.$$

The transducer power gain G_T is the ratio of the power dissipated in the load to the maximum power available from the source [106, Eq. 11.16]:

$$G_T(S) = |S_{21}|^2 \frac{(1 - |S_{G,0}|^2)(1 - |S_{L,0}|^2)}{|(1 - S_{11}S_{G,0})(1 - S_{22}S_{L,0}) - S_{12}S_{21}S_{G,0}S_{L,0}|^2}.$$

The noise figure F measures how the amplifier's noise degrades the output SNR (S_o/N_o) compared to the input SNR (S_i/N_i) [106, Eq. 10.10]:

$$F = \frac{S_i/N_i}{S_o/N_o} \geq 1.$$

Figure 9.10 illustrates how a noisy 2-port is modeled using a noiseless 2-port

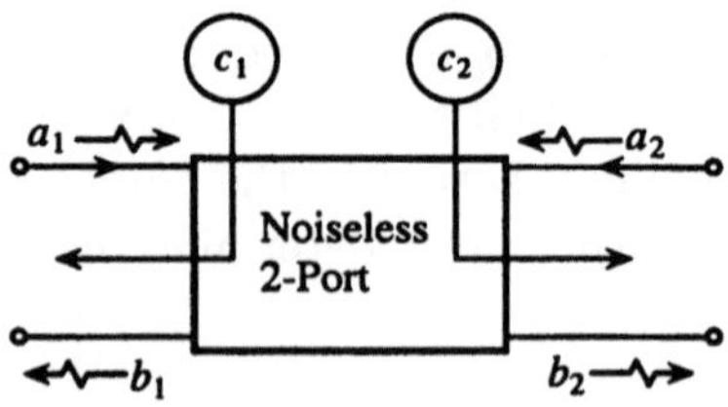

Fig. 9.10. A noisy 2-port model.

and two noise waves c_1 and c_2 emanating from Ports 1 and 2, respectively. If the noiseless 2-port has scattering matrix S, then [54, Eq. 3]

$$\mathbf{b} = \begin{bmatrix} b_1 \\ b_2 \end{bmatrix} = \begin{bmatrix} S_{11} & S_{12} \\ S_{21} & S_{22} \end{bmatrix} \begin{bmatrix} a_1 \\ a_2 \end{bmatrix} + \begin{bmatrix} c_1 \\ c_2 \end{bmatrix} = S\mathbf{a} + \mathbf{c}.$$

The noise waves are zero-mean, wide-sense stationary random processes with noise correlation matrix [54, Eq. 4]

$$C_S = E[\mathbf{c}\mathbf{c}^H] = \begin{bmatrix} E[|c_1|^2] & E[c_1\overline{c_2}] \\ E[c_2\overline{c_1}] & E[|c_2|^2] \end{bmatrix},$$

where E denotes the expectation operator. The noise figure for the 2-port is [126, Eqs. 12–13]

$$F(S) = 1 + \frac{\alpha^H C_S \alpha}{k_B T_0 (1 - |S_{G,0}|^2)}, \tag{9.1}$$

where $k_B = 1.380 \times 10^{-23}$ J/K is Boltzmann's constant [106, page 550], $T_0 =$ 290 Kelvin is the standard temperature [106, page 556], and α denotes the vector [126, Eq. 13]

$$\alpha = \begin{bmatrix} S_{G,0} \\ (1 - S_{G,0}S_{11})S_{21}^{-1} \end{bmatrix}. \tag{9.2}$$

Thus, the scattering matrix and noise covariance matrix of the 2-port determine the amplifier functions.

9.2 Active Multiport Scattering

To compute the scattering matrix of the 2-port, assume the 2-port admits a state-space representation shown in Figure 9.6. The transmission lines are omitted for now. Group the inductors, capacitors, resistors, and amplifiers into the *augmented load* with scattering matrix

$$S_L(p) = \begin{bmatrix} qI_{N_L} & 0 & 0 & 0 & \cdots & 0 \\ 0 & -qI_{N_C} & 0 & 0 & \cdots & 0 \\ 0 & 0 & 0_{N_R} & 0 & \cdots & 0 \\ 0 & 0 & 0 & S_{\mathrm{amp}}(p) & \cdots & 0 \\ \vdots & \vdots & \vdots & \vdots & \ddots & \vdots \\ 0 & 0 & 0 & 0 & \cdots & S_{\mathrm{amp}}(p) \end{bmatrix}, \tag{9.3}$$

where $q = (p-1)/(p+1)$. The remaining nonreactive multiport contains only wires, transformers, and gyrators. Thus, the nonreactive multiport has an *augmented scattering* matrix S_a that is real, lossless, and constant. Then S_a belongs to the orthogonal group:

$$S_a \in \mathcal{O}[2+d'] := \{S_a \in \mathbf{R}^{(2+d')\times(2+d')} : S_a^T S_a = I_{2+d'}\},$$

where $d' = N_L + N_C + N_R + 2N_A$. Number the ports to correspond to the matrix indices of S_a as follows: Ports 1 and 2 are the input and output as in Figure 9.6; the next N_L ports attach the inductors; the next N_C ports attach the capacitors; the next N_R ports attach the resistors; the remaining $2N_A$

are where the amplifiers attach. Using this numbering scheme, partition the augmented scattering matrix as

$$S_a = \begin{bmatrix} S_{a;11} & S_{a;12} \\ S_{a;21} & S_{a;22} \end{bmatrix} \begin{matrix} 2 \\ d' \end{matrix} .$$
$$\qquad\quad 2 \quad\; d'$$

The scattering matrix $S(p)$ for the lumped, active 2-port can be written

$$\begin{aligned} S(p) &= \begin{bmatrix} S_{11}(p) & S_{12}(p) \\ S_{21}(p) & S_{22}(p) \end{bmatrix} \\ &= S_{a,11} + S_{a,12} S_L(p) \{ I_{d'} - S_{a,22} S_L(p) \}^{-1} S_{a,21}. \end{aligned} \tag{9.4}$$

Let $\mathcal{A}(N_L, N_C, N_R, N_A)$ denote the collection of the scattering matrices of Equation 9.4. That is, $\mathcal{A}(N_L, N_C, N_R, N_A)$ models all the active 2-ports that may be realized by attaching the lumped elements and amplifiers to a network of wires, gyrators, and transformers. Because the load is fixed, this class of active 2-ports admits the following state-space parameterization.

STATE-SPACE PARAMETERIZATION OF $\mathcal{A}(N_L, N_C, N_R, N_A)$. Let

$$d' = N_L + N_C + N_R + 2N_A.$$

Define the mapping $\mathcal{F} : \mathcal{O}[2 + d'] \longrightarrow \mathcal{A}(N_L, N_C, N_R, N_A)$ as

$$\begin{aligned} S(p) &= \mathcal{F}(S_a; p) \\ &:= S_{a,11} + S_{a,12} S_L(p) \{ I_{d'} - S_{a,22} S_L(p) \}^{-1} S_{a,21}, \end{aligned}$$

where $S_L(p)$ is given by Equation 9.3.

Because the augmented load contains the active $S_{\mathrm{amp}}(p)$, there is no guarantee that $\{I_{d'} - S_{a,22} S_L(p)\}^{-1}$ exists for $p \in \mathbf{C}_+$, even though $\|S_{a,22}\| \leq 1$. For example, suppose the amplifier is a negative resistor $-R$ [92, Chapter 9], [53, Section 388–397], [106, Section 11.5]. Suppose this amplifier terminated a shunt capacitor. The input impedance is

$$z(p) = \mathcal{G}\left(\begin{bmatrix} 1 & 0 \\ pC & 1 \end{bmatrix}, -R; p \right) = \frac{-R}{1 - pRC},$$

which has a pole at $(RC)^{-1}$ in $\mathbf{C}_+$. From this single example, we conjecture that

$$\begin{aligned} \mathcal{A}(N_L, N_C, N_R, N_A) &\subset H_k^\infty(\mathbf{C}_+, \mathbf{C}^{2\times 2}) \\ &=: \bigcup_{d=0}^{k} \{ \phi^{-1} H : H \in H^\infty(\mathbf{C}_+, \mathbf{C}^{2\times 2}), \phi \in U^+(1, d) \}. \end{aligned}$$

Once a space is settled upon, the topological structure of these active 2-ports relevant to optimization can be explored.

Question 24. Is $\mathcal{A}(N_L, N_C, N_R, N_A)$ closed, compact, or connected?

Allied with these mathematical questions is the question of how these scattering matrices correspond to a physical circuit. Recall from Chapter 4 that Theorem 4.2.1 maps out this correspondence for the lumped, passive multiports. The same question arises for these active multiports.

Question 25. What active 2-ports are modeled by $\mathcal{A}(N_L, N_C, N_R, N_A)$?

The final question concerns implementation. Because the orthogonal group splits across the special orthogonal group:

$$\mathcal{O}[2+d'] = \mathcal{SO}[2+d'] \bigcup \mathcal{SO}[2+d']\Sigma,$$

where $\Sigma \in \mathcal{O}[2+d']$ with $\det[\Sigma] = -1$, both mappings

$$\mathcal{F} : \mathcal{SO}[2+d'] \longrightarrow \mathcal{A}(N_L, N_C, N_R, N_A),$$

$$\mathcal{F} : \mathcal{SO}[2+d']\Sigma \longrightarrow \mathcal{A}(N_L, N_C, N_R, N_A)$$

may cover the same set. Do we need both components of the orthogonal group?

Question 26. Does $\mathcal{F}(\mathcal{SO}[2+d']) = \mathcal{A}(N_L, N_C, N_R, N_A)$?

The scattering matrix and noise covariance matrix of the 2-port determine the amplifier functions. This section completes the discussion of the scattering matrix of the active 2-port. To finish the computations for the amplifier functions, we need the noise covariance matrix of the 2-port.

9.3 Multiport Noise

To compute the noise figure from the state-space representation, observe that both the amplifiers and resistors are noise sources. These noise sources contribute to the noise correlation matrix C_S of Equation 9.1 as follows [126]:

$$C_S(S_a; p) = \Lambda(S_a; p)\, C_L(p)\, \Lambda(S_a; p)^H, \tag{9.5}$$

where Λ is the mapping on $\mathcal{O}[2+d']$

$$\Lambda(S_a; p) := S_{a,12} + S_{a,12} S_L(p)(I_{d'} - S_{a,22} S_L(p))^{-1} S_{a,22},$$

and $C_L(p)$ is the noise covariance matrix of the augmented load:

$$C_L(p) = \begin{bmatrix} 0_{N_L} & & & & & \\ & 0_{N_C} & & & & \\ & & k_B T_0 I_{N_R} & & & \\ & & & C_{\mathrm{amp}}(p) & & \\ & & & & \ddots & \\ & & & & & C_{\mathrm{amp}}(p) \end{bmatrix},$$

with N_A copies of the noise covariance matrix $C_{\mathrm{amp}}(p)$ of the amplifier strung along the diagonal. This completes the discussion of noise covariance matrix. With the scattering matrix determined in the preceding section, we are ready to compute the amplifier functions of these active 2-ports

9.4 Optimization on $\mathcal{A}(N_L, N_C, N_R, N_A)$

With respect to the state-space parameterization of $\mathcal{A}(N_L, N_C, N_R, N_A)$, the multiobjective amplifier function of Chapter 6 has the form:

$$\gamma(S_a) = \begin{bmatrix} -\|G_T(\mathcal{F}(S_a))\|_{-\infty} \\ \|F(\mathcal{F}'(S_a))\|_{\infty} \\ \|S_1(\mathcal{F}(S_a))\|_{\infty} \\ \|S_2(\mathcal{F}(S_a))\|_{\infty} \end{bmatrix}.$$

The prime in $\mathcal{F}'(S_a)$ denotes that the noise figure is not only a function of the scattering matrix $S(p) = \mathcal{F}(S_a)$ but also a function of $\Lambda(S_a)$. Thus, the Multiobjective Amplifier Problem is

$$\min\{\gamma(S_a) : S_a \in \mathcal{O}[d' + 2]\},$$

which can be solved by the Goal Attainment Method discussed in Chapter 6. Figure 9.11 illustrates the geometry. The amplifier designer wants to find a

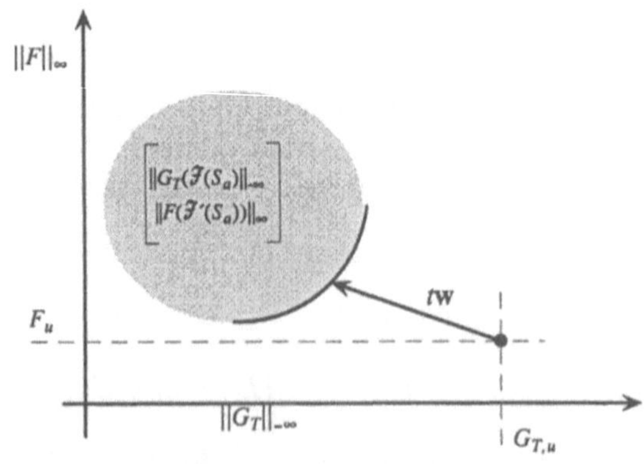

Fig. 9.11. Goal Attainment Method in the gain-noise plane.

multiport that gets "as close as possible" to a desired gain $G_{T,u}$, noise figure F_u, and stability constraints $S_{1,u}$, $S_{2,u}$. Starting from the desired gain $G_{T,u}$ and noise F_u, the amplifier designer chooses a weight vector $\mathbf{w}$ that is extended by **tw** until it just touches the closest Pareto point. (For clarity, the plot suppresses the stability functions.) What follows is high-level implementation for estimating matching performance.

$\mathcal{A}(N_L, N_C, N_R, N_A)$ PERFORMANCE ESTIMATE: *Bounds the performance of N_A user-specified amplifiers connected to any passive lumped multiport of Figure 9.6 containing N_L inductors, N_C capacitors, and N_R resistors.*

Data

- *The sample frequencies $\Omega = \{\omega_n : n = 1, \ldots N_\Omega\}$.*
- *The reflectance of the generator at the sample frequencies $S_{G,0}(j\omega_n)$.*
- *The reflectance of the load at the sample frequencies $S_{L,0}(j\omega_n)$.*
- *The scattering function of the user-selected amplifier $S_{\mathrm{amp}}(j\omega_n)$.*
- *The noise covariance matrix $C_{\mathrm{amp}}(j\omega_n)$ of the amplifier.*

Input

- *The number of inductors N_L, capacitors N_C, resistors N_R, and amplifiers N_A.*
- *The stability bounds $S_{1,u}$, $S_{2,u} \leq 1$.*
- *The gain and noise design goals $G_{T,u}$ and F_u.*
- *The gain and noise weights w_G and w_F.*

Computation

- *Determine if the stability bounds are feasible. Set $d' = N_L + N_C + N_R + 2N_A$. Randomly sample the orthogonal group $\mathcal{O}[2+d']$ to find any augmented scattering matrix $S_a \in \mathcal{O}[2+d']$ such that*

$$\gamma(S_a) = \begin{bmatrix} -\|G_T(\mathcal{F}(S_a))\|_{-\infty} \\ \|F(\mathcal{F}'(S_a))\|_\infty \\ \|S_1(\mathcal{F}(S_a))\|_\infty \\ \|S_2(\mathcal{F}(S_a))\|_\infty \end{bmatrix} \leq \begin{bmatrix} -1 \\ 10 \\ S_{1,u} \\ S_{2,u} \end{bmatrix}.$$

  These feasible S_a's meet the stability requirements with negligible constraints on the gain and noise: gain exceeds 0 dB; noise is bounded by 10 dB.
- *If the stability bounds are feasible, set the user's design goal and weight vector as*

$$\gamma_u = \begin{bmatrix} -G_{T,u} \\ F_u \\ S_{1,u} \\ S_{2,u} \end{bmatrix} \text{ and } \mathbf{w} = \begin{bmatrix} w_G \\ w_F \\ 0 \\ 0 \end{bmatrix}.$$

  The zeros in the weight vector force the stability constraints.
- *Starting from a feasible S_a with $\gamma(S_a)$ closest to γ_u,*

$$\min\{t \in \mathbf{R}\}$$

subject to the constraints $S_a \in \mathcal{O}[2+d']$ *and*

$$\gamma(S_a) - t\mathbf{w} \le \gamma_u.$$

Output *A minimizing* $S_{a,\mathrm{opt}} \in \mathcal{O}[2+d']$ *and the associated amplifier performance:*

$$\gamma(N_L, N_C, N_R, N_A) = \gamma(S_{a,\mathrm{opt}}).$$

By running this algorithm with various weights, Figure 9.11 shows that the Pareto set is mapped out. This technique of varying the weights is used in the next section to map the Pareto set of wideband amplifiers.

9.5 Wideband Matching

Matching from $\mathcal{A}(N_L, N_C, N_R, N_A)$ is demonstrated using the NE321000 amplifier. As discussed in Section 6.4, the NE321000 is a low-noise heterojunction FET transistor used for communications that operates over 2–26 GHz.

We start by connecting only one NE321000 to a lumped, lossless multiport as shown in Figure 9.6. The resistors and the distributed elements are omitted. That is, the amplifier's performance is determined by optimizing over $\mathcal{A}(N_L, N_C, N_R, N_A)$ with $N_R = 0$ and $N_A = 1$. Figure 9.12 shows the resulting performance with the stability bounds

$$S_{1,u}, S_{2,u} \le 0.8.$$

The degree of the multiports is $d = N_L + N_C$. The curves show the performance of the general matching multiports of degree d. The curves are generated by fixing the degree and letting the weights w_G and w_F vary to map out the Pareto points. As the amplifier designer demands more and more gain, the noise figure starts to increase but far more slowly than the gain. However, no matching circuit can continue to deliver arbitrary gain. For each degree d, the figure shows there is a gain "wall" past which any slight increase in gain come a terrific increase in noise.

Figure 9.12 lets the amplifier designer explore these arbitrary amplifier topologies—including all feedback circuits to enforce the stability margin. For comparison, the performance of the series/shunt cascade of Figure 6.16 is plotted as the bullet. The series/shunt cascade has no stability margin: $\|S_1\|_\infty = 1$. Indeed, stability has been a problem dogging our designs through Chapters 6, 7, and 8. The power of the general multiport topology is that we can enforce nontrivial stability constraints. In particular, Figure 9.12 informs the amplifier designer that there is a $d = 1$ circuit that attains better performance than the series/shunt circuit with a much better stability margin. Consequently, the amplifier designer is directed to using fewer reactive elements with a feedback circuit topology to enforce stability.

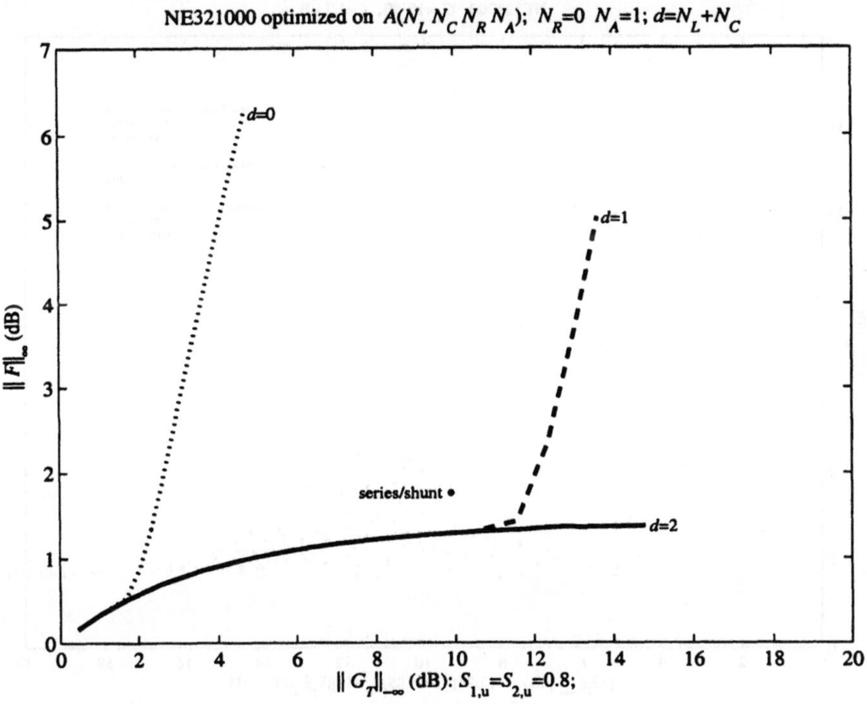

Fig. 9.12. Performance of the NE321000 connected to lumped, lossless matching multiports of arbitrary topology.

The next matching example uses two NE321000 amplifiers. We start by extending the series/shunt circuit. Figure 9.13 has the series/shunt cascade for input and output matching circuits. The intermediate matching circuit is

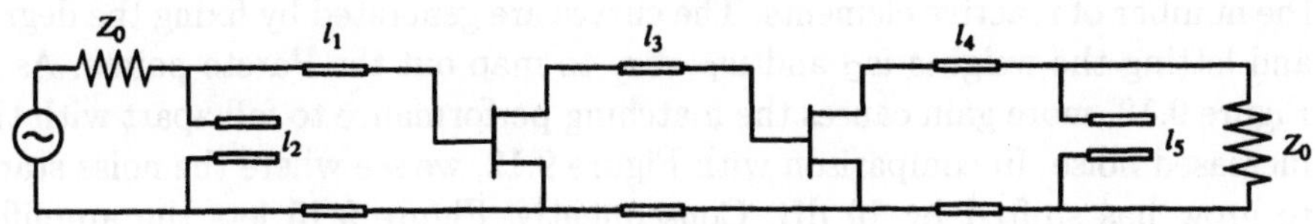

Fig. 9.13. Two FETs and matching circuits.

the series line (no shunt). Figure 9.14 presents the performance of this cascade circuit in the gain-noise plane. The dots mark the gain-noise performance for random selections of line lengths. The dot labeled "optimum" marks the result of applying the Goal Attainment Method. The optimal lengths are reported in millimeters and wavelength λ at 10 GHz. As expected, the second amplifier

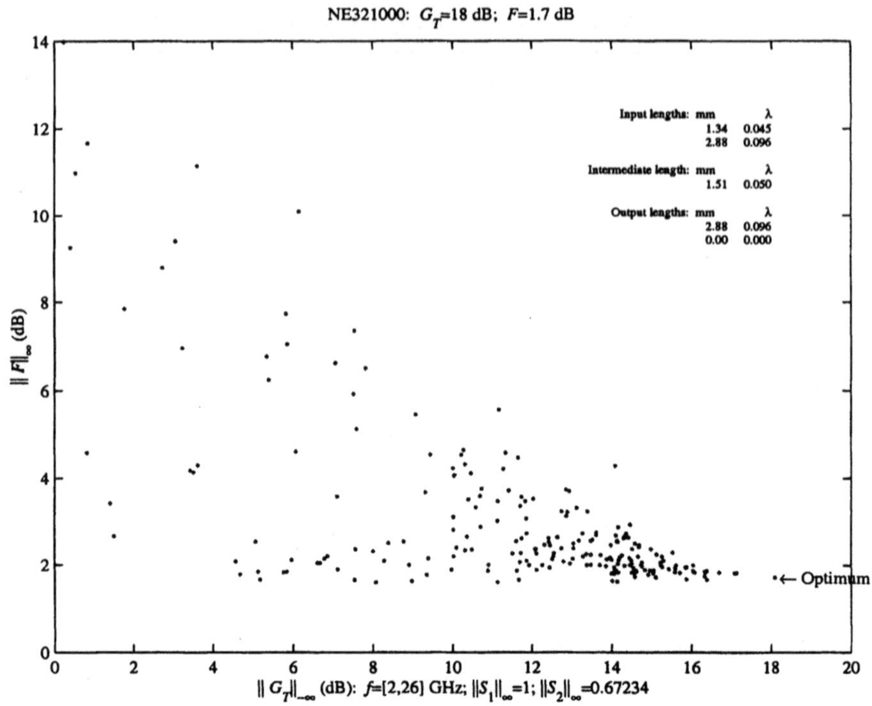

Fig. 9.14. Gain and noise trade-offs for two NE321000's using the series/shunt cascade of Figure 9.13.

boosts the gain over 18 dB. The second amplifier also adds more noise. Also as expected, the stability margin is still very small.

Figure 9.15 compares the performance of the 2-FET series/shunt cascade, marked with the bullet, against the performance available from the lumped, lossless matching multiports using the same two FETs. The degree d counts the number of reactive elements. The curves are generated by fixing the degree and letting the weights w_G and w_F vary to map out the Pareto points. As in Figure 9.12, more gain causes the matching performance to fall apart with the increased noise. In comparison with Figure 9.12, we see where the noise starts to grow has shifted by 10 dB. Consequently, Figure 9.15 lets the amplifier designer know that a $d = 1$ circuit exists with larger stability margin, larger gain, and less noise that the cascade circuit. As before, the amplifier designer is directed to use fewer reactive elements and to explore alternative topologies.

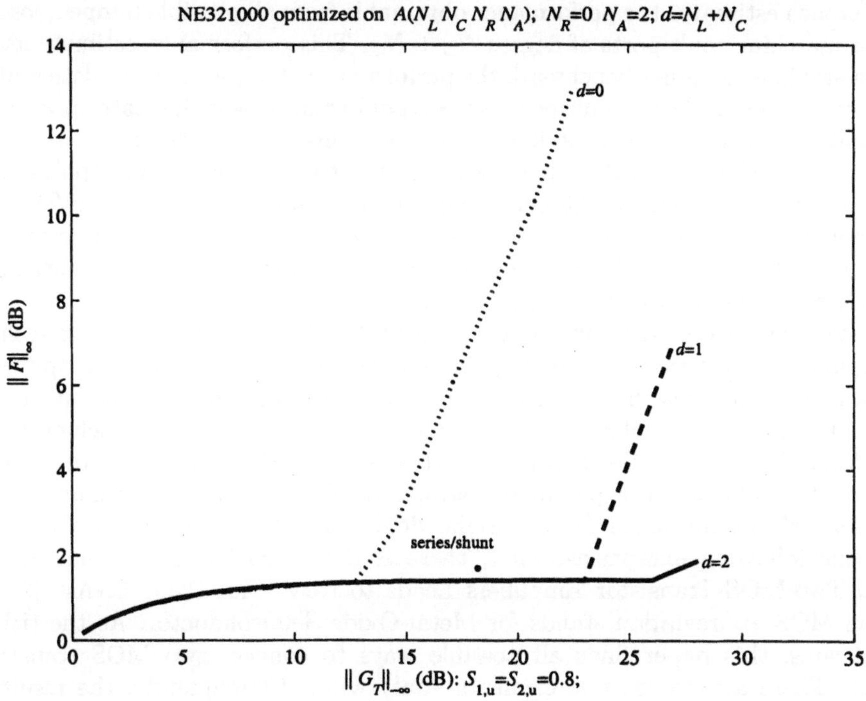

Fig. 9.15. Performance of two NE321000 FETs connected to lumped, lossless matching multiports of arbitrary topology.

9.6 Comparison with the Real-Frequency Technique

As amplifier performance demands continue to increase, the amplifier designer must migrate from single-stage amplifiers, to multi-stage amplifiers [16], to distributed amplifiers [31], and to matrix amplifiers [102]. As a consequence, the matching multiport becomes more complex and difficult to analyze. A means of computing the most general performance is valuable to the amplifier designer for the ranking of candidate multiports [94]:

> Multistage amplifiers are urgently needed with the advance in technologies, due to the fact [that the] signal-stage cascode amplifier is no longer suitable in low-voltage designs ... In fact, there are not many discussions on the comparison of the existing compensation technologies.

Most of the amplifier Computer-Aided Design (CAD) programs are device and topology specific. The high-end CAD programs incorporate detailed models of the biasing circuitry and many parasitic effects. In contrast, the state-space approach estimates the performance obtainable from all possible lumped, lossless matching multiports of degree $N_L + N_C$. This performance estimate lets the amplifier designer benchmark the performance of a specific candidate multiport. If the candidate multiport shows a performance near the state-space estimate, then the amplifier designer can accept this near-optimal performance. Otherwise, the electrical engineer is motivated to search for other topologies.

For example, Section 9.5 benchmarked the standard cascade series/shunt topology. Low-order, lumped, lossless multiports do exist that deliver more gain and less noise while enforcing the stability constraint of 0.8. In contrast, both cascade circuits were almost unstable.

We close by setting up comparisons with other general matching techniques. The electrical engineering community has developed many amplifier matching programs. Some are quite detailed and are kept current with online updates of the latest data from the manufacturers. Others include the nontrivial effects of the biasing circuitry on the specified transistor operational points and the packaging parasitics. These programs compute the amplifier's performance for a specific device and topology. We are aware of the following exceptions. First, there is the journal paper "Generating *All* Two-MOS-Transistor Amplifiers Leads to New Wide-Band LNAs" [23]. The MOS abbreviation stands for Metal-Oxide Semiconductor. As the title indicates, this paper finds all possible ways to connect two MOS transistors. There are no reactive elements—only wires. Consequently, the resulting configurations are special cases subsumed by the state-space estimate of $\mathcal{A}(N_L = 0, N_C = 0, N_R = 0, N_A = 2)$. That is, simply attach two transistors to a lossless multiport of degree 0 and estimate optimal performance.

The second exception is the application of the Real-Frequency Technique[1] to amplifier design. The development of the Real-Frequency Technique closest to the state-space estimate is by Jung and Wu [86]. This paper starts from the cascade of Figure 9.16. This Real-Frequency Technique consists of the

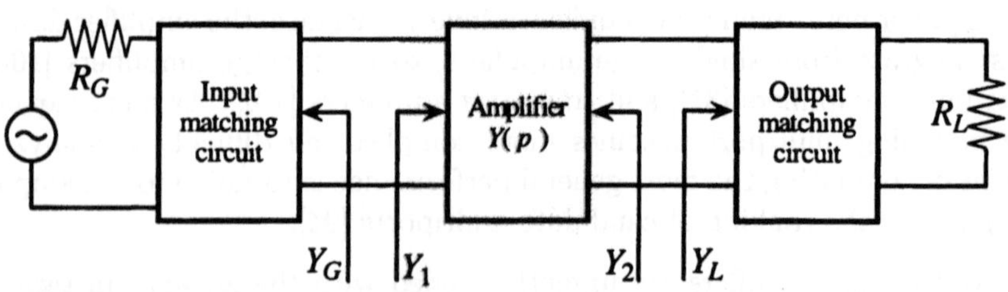

Fig. 9.16. Generic amplifier, matching circuits, and admittances.

[1] The reader is warned that there exist two amplifier-design approaches that identify themselves as the "Real-Frequency Technique." The approach we do not discuss uses the Belevitch representation of ladders [143], [88].

following steps:

- Model the generator's admittance Y_G and load admittance Y_L by analytic splines.
- Perform a multiobjective optimization over these splined admittances.
- Fit an input matching circuit to Y_G and an output matching circuit to Y_L.

The key idea is using the analytic splines to model the admittances. Write the load admittance as

$$Y_L(j\omega) = G_L(j\omega) + B_L(j\omega),$$

where $G_L(j\omega)$ denotes the conductance and $B_L(j\omega)$ denotes the susceptance. Let $\Omega = \{\omega_n\}$ denote the collection of sample frequencies. Introduce the space of piecewise linear splines $\mathcal{G}_L(j\Omega)$ on $j\Omega$. An element $G_L \in \mathcal{G}_L(j\Omega)$ is linear on each interval $[\omega_n, \omega_{n+1}]$ and continuous across each interval. Denote the Hilbert transform of G_L as [103, Eq. 10-28]

$$\mathcal{H}[G_L; j\omega] := -\frac{1}{\pi}\int_{-\infty}^{\infty} \frac{G_L(j\omega)}{\omega - \omega'} d\omega'.$$

Define the space of analytic splines that model the load admittances as follows:

$$\mathcal{Y}_L(j\Omega) := \{G_L + jB_L : G_L \in \mathcal{G}_L(j\Omega),\ G_L(j\omega) \geq 0,\ B_L = \mathcal{H}[G_L]\}.$$

Likewise, let $\mathcal{Y}_G(j\Omega)$ denote the corresponding model of the generator's admittances. If the amplifier has an admittance matrix

$$Y = \begin{bmatrix} Y_{11} & Y_{12} \\ Y_{21} & Y_{22} \end{bmatrix},$$

the transducer power gain can be written as [86]:

$$G_T(Y_G, Y_L) = \frac{4|Y_{21}|^2 G_G G_L}{|(Y_{11} + Y_G)(Y_{22} + Y_L) - Y_{12}Y_{21}|^2}.$$

The noise figure can be written as [106, Eq. 11.54]

$$F(Y_G) = F_{\min} + \frac{R_N}{G_G}|Y_G - Y_{\text{opt}}|^2.$$

That is, the amplifier functions can be computed using the admittances Y_G and Y_L. A multiobjective optimization may be applied to

$$\gamma(Y_G, Y_L) = \begin{bmatrix} \|G_T(Y_G, Y_L)\|_{-\infty} \\ \|F(Y_G)\|_\infty \\ \|S_1(Y_L)\|_\infty \\ \|S_2(Y_G)\|_\infty \end{bmatrix}$$

where $Y_G \in \mathcal{Y}_G(j\Omega)$ and $Y_L \in \mathcal{Y}_L(j\Omega)$ to compute Pareto optimal admittances $Y_{G,\mathrm{opt}}$ and $Y_{L,\mathrm{opt}}$. Once optimal admittances $Y_{G,\mathrm{opt}}$ and $Y_{L,\mathrm{opt}}$ have been computed, the last step of the Real-Frequency Technique is to fit a circuit model, typically a ladder, to $Y_{G,\mathrm{opt}}$ and $Y_{L,\mathrm{opt}}$. This can be quite difficult and requires the specification of a topology.

The closest point of comparison between the Real-Frequency Technique and the state-space estimate is that both compute a Pareto minimum. However, the multiobjective optimization is computed over completely different spaces. The Real-Frequency Technique minimizes $\gamma(Y_G, Y_L)$ over $Y_G \in \mathcal{Y}_G(j\Omega)$ and $Y_L \in \mathcal{Y}_L(j\Omega)$. The state-space approach minimizes $\gamma(S)$ over $S \in \mathcal{A}(N_L, N_C, N_R = 0, N_A)$. Consequently, the matching multiports are different. The Real-Frequency Technique is currently restricted to the cascade connections of Figure 9.16, although the generalization to multiports is straightforward. In contrast, the state-space approach uses the general multiport of Figure 9.6 omitting transmission lines or resistors. In summary, there exist no approaches to optimize multiple amplifier performance over all the lumped, lossless multiports independent of circuit topology other than the state-space estimate presented here.

10

Research Topics

This book discussed several approaches to amplifier matching:

- Standard Multiobjective Optimization,
- H^∞ Multidisk Methods,
- H^∞ Multiobjective Approaches,
- State-Space Methods.

The development of these methods generated many questions. The Circuit-Scattering Correspondence of Chapter 4 led to questions regarding the existence and representation of lossless multiports. In particular, foundational questions regarding distributed multiports are challenging mathematical problems.

Darlington's Theorem of Chapter 5 asked how various orbits fill up the unit ball of H^∞. How the orbits reside in the unit ball is pivotal for the H^∞ methods because Darlington's Theorem turns optimizing over $U^+(2,\infty)$ to a convex, scalar-valued problem that admits a computational solution by Nehari's Theorem.

Chapter 6 started with the standard multiobjective optimization solution to the Amplifier Matching Problem. Recent developments regarding the utopic point and Pareto optima set await application to amplifier design. The H^∞ context, the utopic point, computing and characterizing all the Pareto optima, and generalizing the Goal Attainment Methods, reflect the current limits of H^∞ research.

The H^∞ Multidisk Method developed in Chapter 7 computes solid and tight bounds for the Amplifier Matching Problem. That development generated many questions regarding the associated "disks in function space," the distinction between the disk algebra and H^∞, and the specialization to the amplifier functions.

Chapters 8 and 9 developed and generalized the State-Space Methods. In Chapter 9 we finally see how to overcome the stability problem that has been dogging our amplifiers since Chapter 6. It is the full multiport approach that stabilizes these wideband amplifiers and readily generalizes to multiple amplifiers. The multiple-amplifier matching problem is an optimization problem over $U^+(N,d)$. Its H^∞ extention is the computation of the Pareto optima in $U^+(N)$. The generalization of a Darlington Theorem awaits as does the replication of the entire H^∞ program for the multiple-amplifier matching problem.

Our final chapter organizes these questions into several research topics. What is fascinating is how this simple application—optimizing an amplifier's performance—can direct attention to unexplored analysis.

10.1 Lumped-Distributed Matching

Chapter 9 left transmission lines out of the matching problem. This section remedies that omission. We are given N_A noisy amplifiers. These amplifiers are to be connected to a lumped-distributed, lossless multiport. Figure 10.1 illustrates the matching multiport. A lossless multiport is to be found that simultaneously makes the following trade-offs:

MultiAmp-1 Maximize the transducer power gain G_T.
MultiAmp-2 Minimize the noise figure F.
MultiAmp-3 Guarantee stability: $|S_1| \leq 1$, $|S_2| \leq 1$.

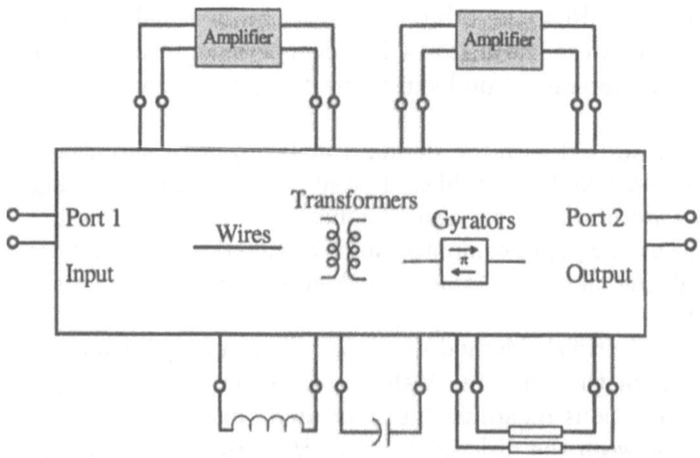

Fig. 10.1. Lossless multiport of lumped elements and uniform transmission lines for matching noisy amplifiers.

The State-Space Method parameterizes the scattering matrix $S(p)$ of the active 2-port. Group the N_L inductors, N_R resistors, N_C capacitors, N_U uniform transmission lines with delay τ, and N_A amplifiers into the *augmented load*

$$S_L(p) = \begin{bmatrix} qI_{N_L} & & & & & \\ & -qI_{N_C} & & & & \\ & & S_{\mathrm{UE}}(p) & & & \\ & & & \ddots & & \\ & & & & S_{\mathrm{amp}}(p) & \\ & & & & & \ddots \end{bmatrix}, \quad \left(q = \frac{p-1}{p+1}\right).$$

The scattering matrix of the transmission lines have the form of Example 4.3.1:

$$S_{\mathrm{UE}}(p) = \begin{bmatrix} 0 & e^{-\tau p} \\ e^{-\tau p} & 0 \end{bmatrix},$$

with delay τ fixed. The amplifier's scattering matrix

$$S_{\mathrm{amp}}(p) = \begin{bmatrix} S_{\mathrm{amp},11}(p) & S_{\mathrm{amp},12}(p) \\ S_{\mathrm{amp},21}(p) & S_{\mathrm{amp},22}(p) \end{bmatrix}$$

is known from measurements. Although no resistors are used ($N_R = 0$), the resistor indexing is kept for further generalizations. The remaining nonreactive multiport—wires, transformers, gyrators—is represented by its *augmented scattering* matrix S_a that belongs to the orthogonal group:

$$S_a \in \mathcal{O}[2 + d'], \qquad d' = N_L + N_C + N_R + 2N_U + 2N_A.$$

Index the input and output ports as Ports 1 and 2, respectively. Index the remaining ports to correspond to the decomposition in the augmented load. Partition the augmented scattering matrix as

$$S_a = \begin{bmatrix} S_{a;11} & S_{a;12} \\ S_{a;21} & S_{a;22} \end{bmatrix} \begin{matrix} 2 \\ d' \end{matrix}. \\ \qquad\quad\; 2 \qquad d'$$

The scattering matrix $S(p)$ for the lumped-distributed, active 2-port can be written

$$\begin{aligned} S(p) &= \begin{bmatrix} S_{11}(p) & S_{12}(p) \\ S_{21}(p) & S_{22}(p) \end{bmatrix} \\ &= S_{a,11} + S_{a,12} S_L(p) \{I_{d'} - S_{a,22} S_L(p)\}^{-1} S_{a,21} \\ &=: \mathcal{F}(S_a, S_L; p). \end{aligned}$$

Let $A_\tau(N_L, N_C, N_R, N_U, N_A)$ denote this collection of the scattering matrices:

$$A_\tau(N_L, N_C, N_R, N_U, N_A) := \{\mathcal{F}(S_a, S_L) : S_a \in \mathcal{O}[2 + d']\}.$$

Then $A_\tau(N_L, N_C, N_R, N_U, N_A)$ *ostensibly* models all the active 2-ports that may be realized by attaching lumped elements, transmission lines, and amplifiers to a network of wires, gyrators, and transformers. Section 4.4 makes explicit some of the mathematical traps that lurk in distributed system models. Nevertheless, if we accept this collection of scattering matrices as a set that our optimizers operate on, its topological properties are relevant to any optimization program. Is $A_\tau(N_L, N_C, N_R, N_U, N_A)$

- closed?
- compact?
- connected?

There are also deep questions regarding correspondence between the physical multiports and these scattering matrices.

- Can every scattering matrix $S(p)$ of $A_\tau(N_L, N_C, N_R, N_U, N_A)$ be realized by a multiport of Figure 10.1?
- Does every multiport of Figure 10.1 have a scattering matrix $S(p)$ that belongs to $A_\tau(N_L, N_C, N_R, N_U, N_A)$?

That is, we ask how Theorem 4.2.1 on passive multiports generalizes to active multiports. These scattering matrices readily generalize to passive multiports. Passive multiports allow the amplifier designer to trade noise for stability.

10.2 Passive Matching

Throughout this book, we have noticed that none of our wideband amplifiers is unconditionally stable. Unconditionally stable amplifiers allow the stability constraint to be dropped. There are two techniques for stabilizing an amplifier [121]:

- add resistors, or
- use feedback.

Because resistors are noise sources, adding resistors trades gain and noise for stability. Both the resistors and the feedback can be subsumed by matching with a *passive* multiport as shown in Figure 10.2. The generalization of passive matching with an arbitrary topology follows from $A_\tau(N_L, N_C, N_R, N_U, N_A)$ coupled with the noise model of Equations 9.1, 9.2, and 9.5. This matching problem has a relatively small literature. The model paper is by Bruccoleri, Klumperink, and Nauta that systematically determines all possible topologies for 2-MOS-transistor wideband amplifiers with noise [23].

As in Section 10.1, $A_\tau(N_L, N_C, N_R, N_U, N_A)$ raises a host of mathematical and engineering questions. Rather than repeat that discussion, Darlington's Theorem offers another point of view. Recall that Darlington's Theorem identifies the orbit of the zero reflectance under the action of $U^+(2,\infty)$:

$$\overline{\mathcal{F}(U^+(2,\infty),0)} = \Re \overline{B} A_1(\mathbf{C}_+).$$

Theorem 4.2.1 can be interpreted as a Darlington's Theorem. Given the reactive augmented load

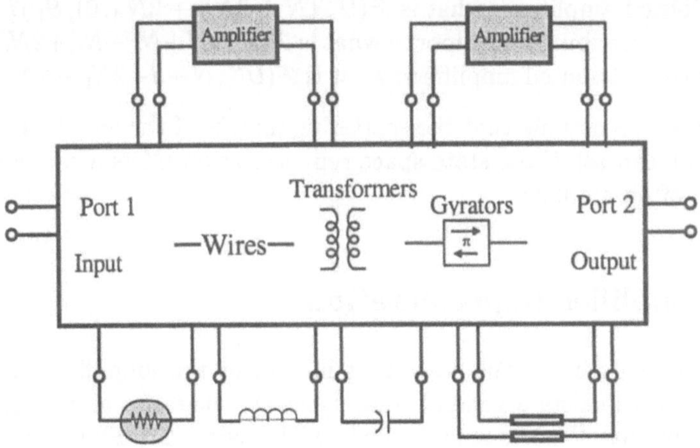

Fig. 10.2. Passive multiport of lumped elements, uniform transmission lines, and resistors (noise sources) for matching noisy amplifiers.

$$S_L(p) = \begin{bmatrix} qI_{N_L} & 0 \\ 0 & -qI_{N_C} \end{bmatrix}, \quad \left(q = \frac{p-1}{p+1} \right).$$

Theorem 4.2.1 computes the orbit of $S_L(p)$ under the action of the lossless multiports of degree 0 as the lossless multiports of degree not exceeding $d = N_L + N_C$:

$$\mathcal{F}(U^+(N+d,0), S_L) = U^+(N,d).$$

Adding resistors to the augmented load

$$S_L(p) = \begin{bmatrix} qI_{N_L} & 0 & 0 \\ 0 & -qI_{N_C} & 0 \\ 0 & 0 & 0_{N_R} \end{bmatrix}$$

picks up all the rational passive multiports of degree not exceeding $d = N_L + N_C$ and normal rank of at least N_R. *What is the corresponding orbit when the augmented load contains amplifiers?* For concreteness, let the augmented load be

$$S_L(p) = \begin{bmatrix} qI_{N_L} & & & & & & \\ & -qI_{N_C} & & & & & \\ & & 0_{N_R} & & & & \\ & & & S_{\mathrm{UE}}(p) & & & \\ & & & & \ddots & & \\ & & & & & S_{\mathrm{amp}}(p) & \\ & & & & & & \ddots \end{bmatrix}$$

By swapping components out of the augmented load, we can raise questions regarding various orbits:

- Distributed amplifiers: what is $\mathcal{F}(U^+(N + 2N_U + 2N_A, 0), S_L)$?
- Lumped-distributed multiports: what is $\mathcal{F}(U^+(N+N_L+N_C+2N_U, 0), S_L)$?
- Lumped-distributed amplifiers: what is $\mathcal{F}(U^+(N+d+2N_U+2N_A, 0), S_L)$?

These questions assume that the scattering matrix of the amplifier is a given. Instead, we can ask if the state-space representation offers a canonical representation of an amplifier.

10.3 Amplifier Representations

Throughout this book, the scattering matrix of the amplifier was a given. We ask if there exists a general state-space structure for amplifiers as there exists for lumped N-ports as embodied by Theorem 4.2.1. Figure 10.3 presents an equivalent circuit for a tunnel diode that motivates our questions. The

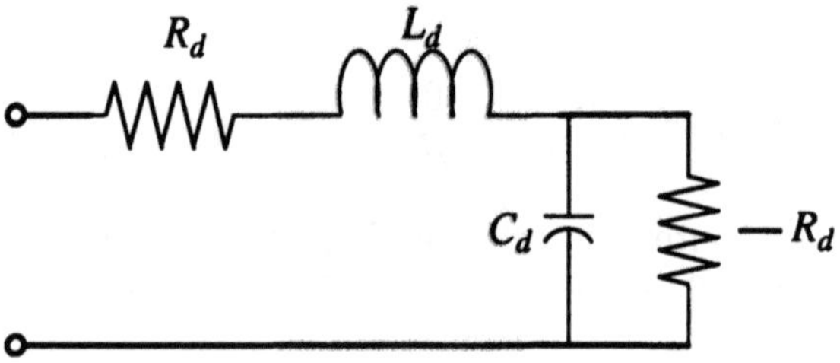

Fig. 10.3. Equivalent circuit of a tunnel diode [92, pages 445–447].

tunnel diode is a special case of the *negative-resistance* amplifiers [128, Section 3.4], [106], [53]. The theory of negative-resistance amplifiers was developed by Kuh and Rohrer [92] in the late 1960s but not fully explored according to Helton [66, page 61]. Figure 10.3 models a tunnel diode as a negative resistor and a positive resistor attached to a lossless 3-port. Does this circuit show us how to generalize Theorem 4.2.1 from passive to active multiports? Theorem 4.2.1 states that having no resistors gives a lossless multiport. Adding positive resistors gives a passive or lossy multiport. Can multiport activity be modeled by negative resistors?

Figure 10.4 illustrates the general case. The nonreactive and lossless multiport is terminated in the lumped, lossless elements (inductors and capacitors). The lossy resistors force passivity whereas the negative resistors force activity. By adding the negative resistors, how much of the class of active lumped multiports are we modeling? How much of Theorem 4.2.1 also generalizes? That is, can we count the number of negative resistors also to give a lower bound on the number of these elements?

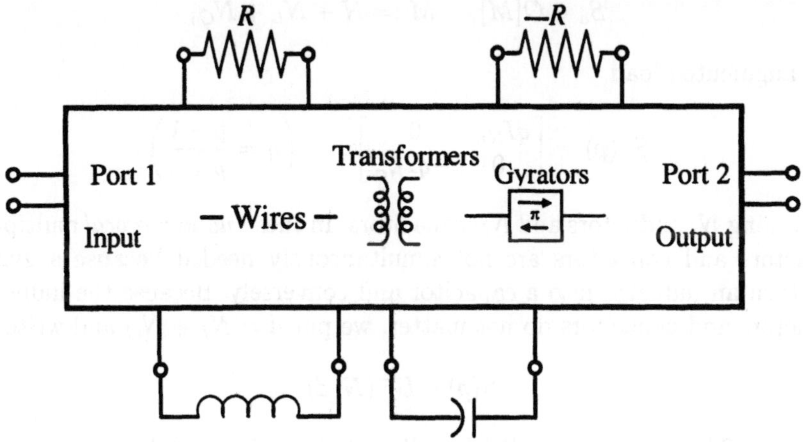

Fig. 10.4. A lumped, active 2-port.

Question 27. Does every lumped N-port with scattering matrix $S(p)$ admit a state-space representation

$$S(p) = \mathcal{F}(S_a, S_L; p) = S_{a,11} + S_{a,12}S_L(p)(I - S_{a,22}S_L(p))^{-1}S_{a,21}$$

where the augmented scattering matrix

$$S_a = \begin{bmatrix} S_{a,11} & S_{a,12} \\ S_{a,21} & S_{a,22} \end{bmatrix}$$

is a real, orthogonal, and constant matrix, and the augmented load has the form

$$S_L(p) = \begin{bmatrix} 0_{N_{R+}} & & & \\ & -2I_{R-} & & \\ & & qI_{N_L} & \\ & & & -qI_{N_C} \end{bmatrix}, \quad \left(q = \frac{p-1}{p+1}\right)?$$

The "-2" is one normalization for the negative resistors.

10.4 Reciprocal Matching

Throughout this book, state-space matching has always used lumped, lossless multiports containing gyrators. For these nonreciprocal multiports with scattering matrix $S(p)$, Theorem 4.2.1 supplies the state-space representation

$$S(p) = \mathcal{F}(S_a, S_L; p) = S_{a,11} + S_{a,12}S_L(p)(I_d - S_{a,22}S_L(p))^{-1}S_{a,21}$$

with augmented scattering matrix S_a in the the orthogonal group:

$$S_a \in \mathcal{O}[M], \quad M := N + N_L + N_C,$$

and augmented load

$$S_L(p) = \begin{bmatrix} qI_{N_L} & 0 \\ 0 & -qI_{N_C} \end{bmatrix}, \quad \left(q = \frac{p-1}{p+1}\right)$$

containing N_L inductors and N_C capacitors. In these *nonreciprocal* multiports, inductors and capacitors are not simultaneously needed because a gyrator can turn an inductor into a capacitor and conversely. Because the individual inductors and capacitors do not matter, we put $d = N_L + N_C$ and write

$$S(p) \in U^+(N, d).$$

However, Theorem 4.2.1 explicitly calls out that a lumped, lossless, *reciprocal* multiport—no gyrators—admits a state-space representation characterized by a *symmetric* augmented scattering matrix: $S_a = S_a^T$. In such a reciprocal multiport, the number of inductors and capacitors do matter. In this case, we write

$$S(p) \in U^+(N, [N_L, N_C], T),$$

where the "T" denotes the transpose in the symmetric matrix definition. The symmetric matrices that lie in the orthogonal group are a fascinating subset. Denote the real symmetric matrices as

$$\text{Symm}[M] := \{S \in \mathbf{R}^{M \times M} : S^T = S\}.$$

The symmetric orthogonal matrices are the intersection

$$\text{Symm}[N] \cap \mathcal{O}[M].$$

All lossless reciprocal N-ports containing N_L inductors and N_C capacitors may be parameterized as

$$U^+(N, [N_L, N_C], T) := \mathcal{F}(\text{Symm}[N] \cap \mathcal{O}[M], S_L).$$

The orthogonal group splits into two disconnected manifolds. Intersection with the symmetric matrices chops these manifolds into many disjoint submanifolds. With some work, the number and dimension of these submanifolds may be computed from the following result.

Corollary 10.4.1 *Let $S_a \in \mathcal{O}[M]$. The following are equivalent:*

- *S_a is symmetric.*
- *The eigenvalues of S_a are ± 1.*

Because $\mathcal{O}[M] \cap \text{Symm}[M]$ has many components, and $\mathcal{F}$ maps these components into $U^+(N, [N_L, N_C], T)$, the following question is relevant for reducing the computational load.

Question 28. How do these components cover $U^+(N, [N_L, N_C], T)$?

Because $\mathcal{F}$ maps several components of $\mathcal{O}[M] \cap \mathrm{Symm}[M]$ onto a single component of these 2-ports, what is an efficient way to undertake optimization over these lumped, lossless, reciprocal 2-ports?

In this section and the preceding sections, we computed a "matching circuit" as an element of the orthogonal group. Although wonderfully satisfying for a mathematician, such an "answer" is less than thrilling for the electrical engineer. The next section shows that getting a real matching circuit is rather difficult.

10.5 Getting the Matching Circuit

Any time we have presented the H^∞ results to any electrical engineers, their immediate question is always

Where is the matching circuit?

The most conservative answer is that the H^∞ theory does not supply a matching circuit—the H^∞ theory computes the *best possible performance* over all the lossless matching circuits. It is very rare in numerical optimization to know the global minimum. The H^∞ theory equips the amplifier designer with the best possible performance to benchmark to assess candidate matching circuits.

Finer detail regarding the degree of the matching circuits can supplement the H^∞ bounds using state-space matching of Chapters 8 and 9. For concreteness, suppose we are matching a single amplifier with input and output matching circuits selected from $U^+(2, d)$. By plotting the matching performance as a function of the degree d, we can see *how fast* the amplifier's performance approaches the H^∞ bound. Thus, the H^∞ theory equips the amplifier designer with an additional trade-off between performance and circuit complexity.

Finally, suppose we have found a near-optimal Pareto point. That is, we know the input and output scattering matrices $\mathcal{S}_G$, $\mathcal{S}_L \in U^+(2, d)$, and their state-space representations. Do we have a matching circuit? Although Theorem 4.2.1 guarantees that $\mathcal{S}_G$ and $\mathcal{S}_L$ admit realizations as lumped, lossless 2-ports, the resulting circuits are of academic interest rather than practical use.

To illustrate these *practical circuit synthesis problems* for the mathematical reader, let $\mathcal{Z}_L(p)$ denote the 2×2 impedance matrix of $\mathcal{S}_L(p)$. Suppose further that the 2-port is reciprocal. The partial-fraction or *Foster expansion* of this lossless 2-port is [6, Theorem 5.1], [99, Section 7-2], [119, Section 5-4]:

$$\mathcal{Z}_L(p) = \frac{K_0}{p} + 2p \sum_{m=1}^{M} \frac{K_m}{p^2 + \omega_m^2} + pK_\infty,$$

where $0 < \omega_1 < \ldots < \omega_M < \infty$ and each residue matrix K_m is real and positive semidefinite. A typical term in the expansion has the form [6, Eq. 5.9]:

$$Z_m(p) := K_m f(p)$$

and is called a Foster matrix. Figure 10.5 shows a 2-port realization of a Foster matrix. The impedance $\mathcal{Z}_L(p)$ is realized by connecting these elementary 2-

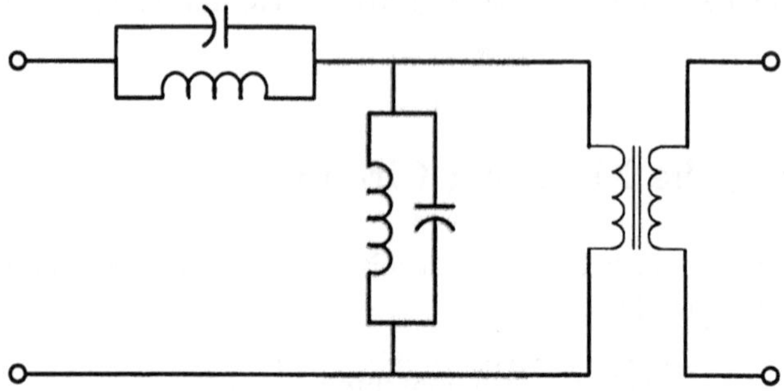

Fig. 10.5. A realization of a Foster matrix [119].

ports in series. Unfortunately, the Foster expansion is impractical. From Temes and LaParta [119]:

> The circuit ... needs many more reactances than the degree of $\mathcal{Z}_L(p)$ justifies. It also needs, in general, $M+2$ ideal transformers. Hence, it is not suitable for practical realization. It importance lies in its *feasibility*.

Transformers are currently inexpedient for wideband circuit design. To a lesser extent, the same "wire" technology limits the use of the inductor—putting lots of these magnetic devices on the same board can introduce cross-coupling. Nevertheless, the partial-fraction expansion illustrates one approach to circuit synthesis: find a mapping between an expansion of $S(p)$ and a useful class of 2-ports. The electrical engineering community has a huge literature exploring such mappings [6], [8], [99], [130], [55].

A more basic approach is to specify a circuit topology, specify the reactive elements, and then optimize over admissible element values as in Chapter 6. Equivalently, the element values parameterize a class $\mathcal{U} \subset U^+(2)$ of 2-ports. Basic mathematical questions relevant to optimization—existence, uniqueness, characterization, manifold structure of $\mathcal{U}$—are rarely considered.

For example, the most common matching circuit is a ladder. The scattering matrix $S(p)$ of a low-pass ladder is (Section 1.12)

$$S(p) = \frac{1}{g(p)} \begin{bmatrix} h(p) & 1 \\ 1 & -h_*(p) \end{bmatrix}.$$

Let $U^+(2,d,\mathrm{LP})$ denote the collection of these scattering matrices of degree d. Consider the class of all possible ladders:

$$U^+(2,\mathrm{LP}) := \overline{\bigcup_{d\geq 0} U^+(2,d,\mathrm{LP})}.$$

Suppose the amplifier designer has specified a 2-port $\mathcal{S}_L \in U^+(2)$. Can the ladders approximate the performance of the given 2-port? Specific questions are

- Compute the distance between $\mathcal{S}_L$ and $U^+(2,\mathrm{LP})$.
- Characterize best approximants of $\mathcal{S}_L$ of $U^+(2,\mathrm{LP})$.
- Determine the rate of convergence from $U^+(2,d,\mathrm{LP})$.
- Determine the state-space representation of the low-pass ladders.

The same questions apply for the high-pass ladders and more general ladders, such as the mid-series and mid-shunt ladders.

Terminated reflectances and Darlington's Theorem provide another approach to matching. Matching over each 2-port $\mathcal{S}_L$ is equivalent to matching over the terminated reflectances:

$$S_L = \mathcal{F}_1(\mathcal{S}_L, S_{L,0}).$$

If the loads have zero reflectances, Belevitch's Theorem (Section 1.12) gives that

$$S_L = \frac{h_L}{g_L}.$$

In the case of the low-pass ladders $U^+(2,d,\mathrm{LP})$, the discussion of Section 1.12 determines that such an S_L is drawn from the following class of rational functions:

$$\left\{\frac{h_L}{g_L} : h_L h_{L,*} = g_L g_{L,*} - 1; \deg[g] \leq d; \text{ and } g(p) \text{ strict Hurwitz}\right\}.$$

A similar class exists for the terminated impedance of the high-pass ladders. More generally, the mid-series and mid-shunt ladders have a terminated impedance characterized by the Fujisawa Theorems dating from the 1950s [6, pages 124-129], [134]. The last substantial advance in ladder theory is Fialkow's 1979 paper [47]. The optimization properties of these rational function classes are largely unexplored.

The *cascade approach* to matching is really Schur's Algorithm. On the unit disk, Schur's Algorithm has the form [7]:

- Input: $h_0 := h \in \overline{B}H^\infty(\mathbf{D})$.
- $\gamma_1 := h_0(0)$.
- for $m = 1$ to ∞ while $|\gamma_m| < 1$
  $h_m(z) := \frac{1}{z}\frac{h_{m-1}(z)-\gamma_m}{1-\overline{\gamma_m}h_{m-1}(z)}$

$\gamma_{m+1} := h_m(0)$

- end

The Schur parameters $\{\gamma_k\}$ characterize the elements of the unit ball of H^∞ and lead to continued fraction expansions [7] (i.e., realizations by ladders). To turn the Schur algorithm into a circuit synthesis scheme, start with the observation that

$$h_{m-1}(z) = \frac{\gamma_m + z h_m(z)}{1 + z\overline{\gamma_m} h_m(z)} = \mathcal{G}\left(\begin{bmatrix} z & \gamma_m \\ z\overline{\gamma_m} & 1 \end{bmatrix}, h_m; z\right).$$

The *Schur section*

$$\Theta_m(z) = \frac{1}{\sqrt{1-|\gamma_n|^2}} \begin{bmatrix} z & \gamma_m \\ z\overline{\gamma_m} & 1 \end{bmatrix} \quad (z = e^{+j\theta})$$

is J-unitary on the unit circle

$$\Theta_m(z)^H J \Theta(z) = J = \begin{bmatrix} 1 & 0 \\ 0 & -1 \end{bmatrix},$$

so can be identified as a chain scattering matrix of a lossless 2-port (Section 1.9). Transplanted to the right half plane, a Schur section takes the form [129], [35]:

$$\Theta_m(p) := \frac{1}{\sqrt{1-|\gamma_n|^2}} \begin{bmatrix} u_n(p) & \gamma_m \\ u_n(p)\overline{\gamma_m} & 1 \end{bmatrix}, \qquad u_n(p) = \frac{p - p_n}{p + \overline{p}_n},$$

where $p_n \in \mathbf{C}_+$ is a complex transmission zero. The Schur section represents a *nonreal*, lumped, lossless 2-port of degree 1. A cascade of Schur sections generates the input reflectance

$$S_1(p) = \mathcal{G}(\Theta_1 \ldots \Theta_d, S_L; p).$$

Sweeping over all Schur sections generates the *Schur Orbit of S_L*. With some work, each Θ_n can be turned into a Darlington section so that an actual circuit is theoretically possible [35].

However, this cascade is still infested with transformers [17]. Transformer-free circuit synthesis was substantially advanced by Fialkow's characterization of those impedances $Z(p)$ obtained by terminating lumped, lossless 2-ports *without transformers* [47]:

> *Any such an impedance $Z(p)$ can be realized as either a series connection or a parallel connection of two LC-R ladders for a total of four parallel LC-R ladders.*

Figure 10.6 illustrates a parallel ladder circuit. The 2-port is realized by three RC ladders connected in parallel. Although the 2-port is lossy, there are

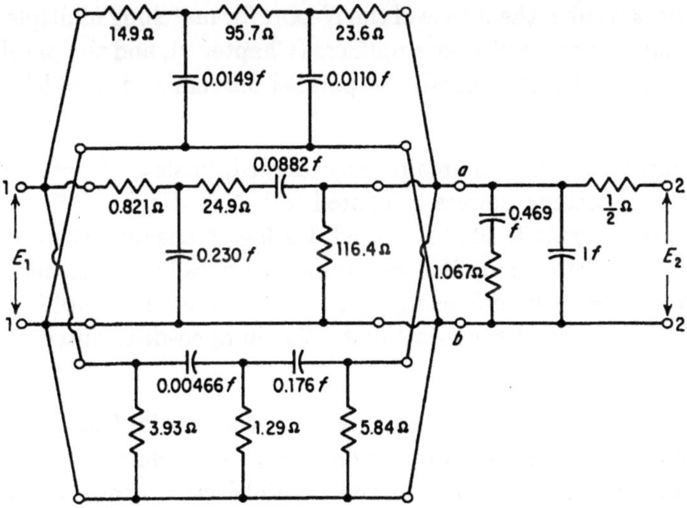

Fig. 10.6. 3 parallel RC ladders [130, Fig. 8-3] (courtesy of Macmillan).

no transformers and inductors. It remains to be seen how parallel RC ladders trade off amplifier gain and stability against noise.

Recent results in microwave filters use a parallel cascade of transmission lines to efficiently synthesize filters. Figure 10.7 shows one such parallel cascade structure. The circuit is the graphic labeled (a) and consists of a transmission line, a shunt resonance, followed by another transmission line. Two physical realizations are presented in graphics (b) and (c).

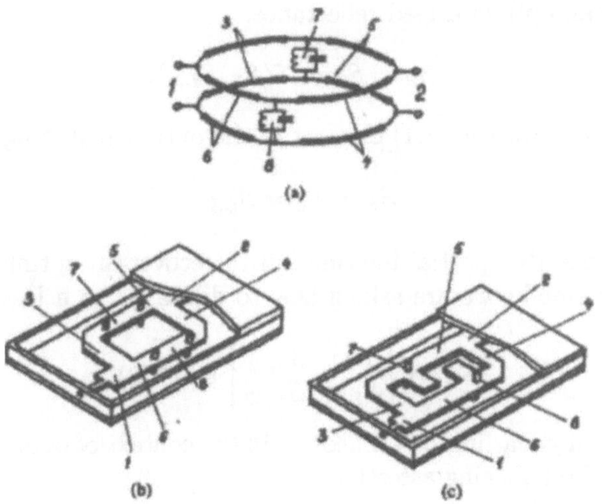

Fig. 10.7. (a) Parallel cascade; (b), (c) two physical realizations [101] (courtesy of the IEEE).

Fialkow's results, the utility of the N-port for matching multiple amplifiers, ladders visible in the multitier amplifiers (Chapter 9), and the parallel cascade raise the possibility of generalizing parallel ladders and parallel cascades to N-ports:

- Can parallel LC ladders realize any lumped, lossless N-port? If not, what are their matching powers compared to $U^+(N,d)$?
- Do Fialkow's results lift to lumped, lossless, transformerless N-ports?
- Does the multitier amplifier generalize the parallel ladder network?
- What are the trade-offs using parallel RC ladders for amplifier matching?
- Do Fialkow's results generalize to the lumped-distributed parallel cascades?

This section demonstrates that circuit synthesis is far from solved: the academic circuits are infested with transformers; many classes of rational circuits remain unexplored, especially in the multiobjective context of the Amplifier Matching Problem; most synthesis techniques require rational functions; the mathematical techniques need to be specialized to more practical circuit classes; and the "2-port" limitation of the synthesis literature needs to be lifted to N-ports.

What can the H^∞ theory offer to this problem of circuit synthesis? A piece of the Nehari Theorem that has not been exploited is that the optimum reflectance is also available. Specifically,

$$S_\times \in \overline{D}(C_\times, R_\times) \cap H^\infty(\mathbf{C}_+)$$

is computable from the eigenvector of the Toeplitz and Hankel operators [140], [63], [65], [66]. For example, in the Amplifier Matching Problem of Figure 2.1, $S_\times$ might be an optimum load reflectance:

$$S_\times = S_L = \mathcal{F}(S_L, S_{L,0}).$$

If $S_{L,0} = 0$, we know the (1,1) element of the output matching 2-port:

$$S_\times = S_L = S_{L,11}.$$

How can we use this partial information to recover an optimal matching 2-port? More formally, we are asking how to dilate $S_\times$ to a lossless scattering matrix:

$$S_\times \mapsto \begin{bmatrix} S_\times & S_{L,12} \\ S_{L,21} & S_{L,22} \end{bmatrix} \in U^+(2).$$

This is essentially Darlington's Theorem in the context of operator theory [73] and leads to the following assertion:

Circuit synthesis is really a problem in matrix dilations.

However, not every reflectance can dilate to a lossless 2-port. Wohlers [128, pages 100–101] shows that the 1-port with impedance

$$z(p) = \arctan(p)$$

cannot dilate to an $S \in U^{+}(2)$. The following Douglas-Helton result characterizes those elements in the unit ball of H^{∞} that came from a lossless N-port.

Theorem 10.5.1 [39], [40] *Let $S(p) \in \overline{B}H^{\infty}(\mathbf{C}_{+}, \mathbf{C}^{N\times N})$ be a real matrix function. The following are equivalent:*

(a) $S(p)$ admits a real inner dilation $\mathbf{S}(p) = \begin{bmatrix} S(p) & S_{12}(p) \\ S_{21}(p) & S_{22}(p) \end{bmatrix}$.

(b) $S(p)$ has a meromorphic pseudo-continuation of bounded type to the open left half plane $\mathbf{C}_{-}$. That is, there exists a $\phi \in H^{\infty}(\mathbf{C}_{-})$ and an $H \in H^{\infty}(\mathbf{C}_{-}, \mathbf{C}^{N\times N})$ such that

$$\lim_{\substack{\sigma>0 \\ \sigma\to 0}} S(\sigma + j\omega) = \lim_{\substack{\sigma>0 \\ \sigma\to 0}} \frac{H}{\phi}(-\sigma + j\omega) \quad \text{a.e.}$$

(c) There is an inner function $\phi \in H^{\infty}(\mathbf{C}_{+})$ such that $\phi S^{H} \in H^{\infty}(\mathbf{C}_{+}, \mathbf{C}^{N\times N})$.

An excellent research topic would be a constructive implementation of this dilation result using sampled data or the output from Nehari's Theorem. However, such a dilation still does not specify a matching circuit. All that the dilation produces is an optimal scattering matrix $\mathcal{S}_L \in U^{+}(2)$. Can this scattering matrix, though not unique, point us to a circuit class? For example, suppose we compute the distance to low-pass ladders:

$$\inf\{\|\mathcal{S}_L - \mathcal{S}\|_{\infty} : \mathcal{S} \in U^{+}(2, \mathrm{LP})\}.$$

Suppose further we find that $\mathcal{S}_L(p)$ is closer to the low-pass ladders $U^{+}(2, \mathrm{LP})$ than any other class of circuits. This minimal distance suggests that the low-pass ladders are good candidates for matching circuits. Thus, Nehari's Theorem may also guide us to a good matching circuit topology.

We conclude this section by highlighting selected electrical engineering results that apply to the dilation and synthesis problems. The seminal application of analytic function theory—Nevanlinna-Pick interpolation—to matching is Youla and Saito's 1967 paper "Interpolation by Positive-Real Functions" [133]. Youla has repeatedly returned to the matching problem. Youla's 1971 exposition on matching was a masterful summary of the problem and techniques [134]. In 1984, Youla, Carlin, and Yarman characterized when one 2N-port could be extracted from another [137]. In 1997, Youla, Winter, and Pillai examined the orbits of a load using Nevanlinna-Pick interpolation (and pointed out a common error in the literature). Finally, the wonderfully concise 1966 paper by Zeheb and Lempel [141] is immediately accessible to the mathematician by explictly displaying a *transformerless* parallel-ladder network that interpolates the scattering function. This "network interpolator" should allow exploration of Nevanlinna-Pick interpolation and scattering function approximation with a potentially practical synthesis method.

10.6 Working with Sampled Data

Question 22 pointed out the "overfitting phenomena." With only a few dozen frequency samples of the scattering matrix, fiddling with a sufficiently large number of circuit parameters should produce arbitrarily good performance—at the sample points. For concreteness, consider computing the stability bounds. In practice, the input and output reflectances are computed at sample frequencies:

$$\|S_1\|_{\infty,j\Omega} := \max\{|S_1(j\omega_k)| : \omega_k \in \Omega\},$$

where $\Omega = \{0 \leq \omega_1 < \ldots < \omega_K\}$. The multiobjective minimizers enforce stability on the sample points:

$$\|S_1\|_{\infty,j\Omega}, \|S_2\|_{\infty,j\Omega} \leq 1.$$

However, the stability constraint (MultiAmp-3 of Section 10.1) holds over the entire frequency axis.

Question 29. How can we guarantee stability from sampled data?

The same problem shows up in the gain and noise functions. We only know these functions at the sample points of the scattering function of the amplifier. The deep question is really

Question 30. How do we recover the scattering matrix from its samples?

Several schemes come to mind but the amplifier designer should be aware that additional information regarding the amplifier is invoked. For example, we can spline the sampled scattering matrix $\{S(j\omega_k)\}$ to "recover" the original scattering matrix on all of $j\mathbf{R}$. This assumes that we have a physical justification for the type of spline and the end point conditions. Another way to recover the amplifier is to fit the sampled data with a state-space model. Again, this requires a physical justification for selecting the model. A more robust approach is a derivative bound. Find matching circuits meeting these bounds:

$$\|S_1\|_{\infty,j\Omega} \leq S_{1,u},$$
$$\|S_1'\|_{\infty,j\Omega} \leq S_{1,u}',$$

where prime denotes the finite difference. Then

$$S_{1,u} + S_{1,u}'\|\Omega\| < 1$$

enforces stability (up to first-order) on the sample interval $[\omega_1, \omega_K]$. The interface between the discrete samples and the continuous frequency affords the applied mathematician fertile ground for explorations. The next section also takes up this theme.

10.7 Estimating the Nehari Error

The problem is to determine if

$$\mathcal{T}_{r^2} \geq \mathcal{H}_c^* \mathcal{H}_c,$$

when all we know are noisy samples of the center and radius functions measured at a finite number of frequencies. Of the several approaches to this problem [66], we use the simple Spline-FFT Method.

THE SPLINE-FFT NEHARI ALGORITHM *Given samples* $\{(jw_k, C(j\omega_k))\}$ *and* $\{(jw_k, R(j\omega_k))\}$, *where* $0 \leq \omega_1 < \omega_2 < \ldots < \omega_K < \infty$.

SF-1 Cayley transform the samples from $j\mathbf{R}$ *to the unit circle* $\mathbf{T}$:

$$c(e^{j\theta_k}) := C \circ \mathbf{c}^{-1}(e^{j\theta_k})$$
$$r(e^{j\theta_k}) := R \circ \mathbf{c}^{-1}(e^{j\theta_k}).$$

SF-2 Use a spline to extend $\{e^{j\theta_k}, c(e^{j\theta_k})\}$ *and* $\{e^{j\theta_k}, r(e^{j\theta_k})\}$ *to functions on the unit circle* $\mathbf{T}$.

SF-3 Approximate the Fourier coefficients using the FFT:

$$\widehat{c}(N;n) := \frac{1}{N}\sum_{n'=0}^{N-1} e^{-j2\pi n n'/N} c(e^{+j2\pi n'/N}),$$

$$\widehat{r}(N;n) := \frac{1}{N}\sum_{n'=0}^{N-1} e^{-j2\pi n n'/N} r(e^{+j2\pi n'/N}).$$

SF-4 Make the truncated Toeplitz and Hankel matrices:

$$\mathcal{T}_{r^2,M,N} = \left[\widehat{r^2}(N; m_1 - m_2)\right]_{m_1,m_2=0}^{M-1},$$
$$\mathcal{H}_{c,M,N} = \left[\widehat{c}(N; -(m_1+m_2))\right]_{m_1,m_2=0}^{M-1}$$

SF-5 Find the smallest eigenvalue of

$$A_{M,N} := \mathcal{T}_{r^2,M,N} - \mathcal{H}_{c,M,N}^H \mathcal{H}_{c,M,N}.$$

There are two main "knobs" for this estimate: the FFT size N and the truncation size M. We are aware of the following sources of error:

- The samples are corrupted by measurement errors.
- The spline extends the sampled data to functions defined on the unit circle.
- The Fourier coefficients are computed from an FFT of size N.
- The operator A is computed from $M \times M$ truncations.

Question 31. Are these all the sources of error (neglecting roundoff)?

Only the last two items are under control of the algorithm through the FFT size N and the truncation M.

Question 32. How can the Spline-FFT Nehari Algorithm automatically adapt M and N to obtain a user-specified error for the eigenvalue estimates?

10.8 Additional Constraints

The Amplifier Matching Problem in Chapter 2 listed the competing objectives:

AMP-1 maximize the gain,
AMP-2 minimize the noise,
AMP-3 guarantee stability.

Additional constraints may also placed on the matching circuits. Control of the input and output Voltage Standing-Wave Ratio (VSWR) is one example [117], [90]. Referring to either the basic amplifier circuit of Figure 2.1 or the more general N-ports of Chapter 9, the goal is to have the input reflectance S_{in} and the output reflectance S_{out} match Z_0. Here S_{in} is the reflectance looking into the input port of the input matching circuit [117]:

$$|S_{\text{in}}|^2 = 1 - \frac{(1-|S_1|^2)(1-|S_G|^2)}{|1-S_1 S_G|^2}.$$

Likewise, S_{out} is the reflectance looking into the output port of the output matching circuit [117]:

$$|S_{\text{out}}|^2 = 1 - \frac{(1-|S_2|^2)(1-|S_L|^2)}{|1-S_2 S_L|^2}.$$

Because the VSWR is [106, page 67]

$$\text{VSWR} = \frac{1+|S|}{1-|S|},$$

a upper bound on the VSWR sets an upper bound on the reflectance. Equivalently, the VSWR constraint is formalized as follows:

AMP-4 $|S_{\text{in}}| < S_{\text{in},u}$ and $|S_{\text{out}}| < S_{\text{out},u}$.

Adding the AMP-4 constraint simply adds another objective function in multiobjective amplifier optimization and adds another disk to the H^∞ Multidisk Method.

10.9 Nonlinear Matching

Throughout this book, we have always used the small-signal model of the amplifier. The small-signal model permits the use of the *linear* scattering operator

$$\mathbf{b}(p) = S_{\text{amp}}(p)\mathbf{a}(p)$$

that maps the incoming waves to the reflected waves. From Gonzalez [53, page 352]:

the small-signal S parameters are not useful for power amplifier design because power amplifiers usually operate in nonlinear regions. The small-signal S parameters can be used in large-signal amplifiers operating in class A (i.e., linear output power). However, for classes AB, B, or C, the small-signal S parameters are not suitable for design purposes. A set of large-signal S parameters is needed to characterize the transistor for power applications. Unfortunately, the measurement of large-signal S parameters is difficult and not properly defined. Therefore, an alternative set of large-signal parameters is needed to characterize the transistor. This can be done by providing information of source and load reflection coefficients as a function of output power and gain.

Figure 10.8 displays this information for a Dexcel 3501A chip biased for an output power of 18.5 dBm at 12 GHz [121, pages 105–106]. The contours or *sublevel sets* show the output power as a function of S_G and S_L. Maximum

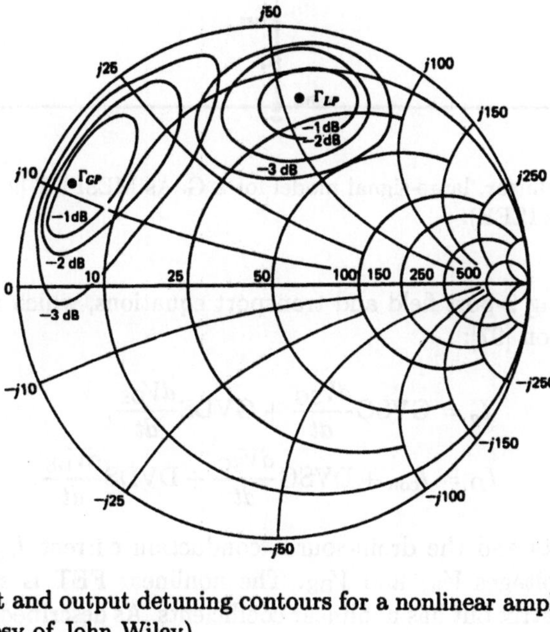

Fig. 10.8. Input and output detuning contours for a nonlinear amplifier [121, pages 105–106] (courtesy of John Wiley).

power occurs at the reflectances labeled Γ_{GP} and Γ_{LP}. The contours mark the 1 dB, 2 dB, ..., reductions in output power corresponding to deviations of S_G and S_L from Γ_{GP} and Γ_{LP}, respectively. What is interesting is that these sublevel sets are not disks but approximately convex. These shapes make it reasonable to conjecture that the H^∞ techniques for such sublevel sets apply to these nonlinear amplifiers [74].

Models of nonlinear amplifiers fall into two categories. The bulk of the literature contains empirical measurements. A smaller portion actually derives a model from physical principles, typically leading to a nonlinear partial differential equation model [97], [118]. Figure 10.9 illustrates the equivalent circuit obtained from such a physical analysis for a Gallium-Arsenide MESFET (MEtal-Semiconductor Field-Effect Transistor). The basic FET is governed

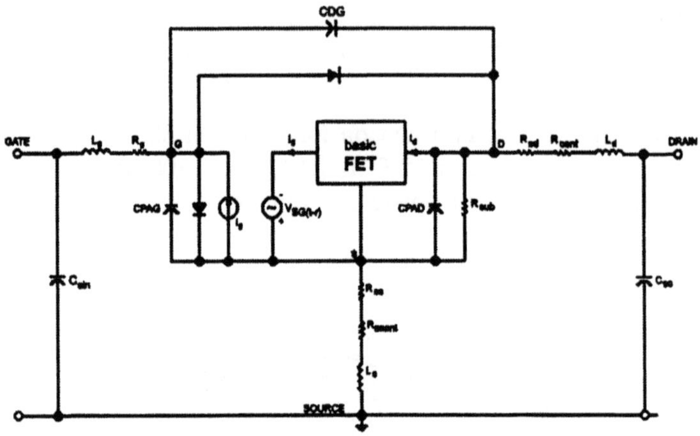

Fig. 10.9. Nonlinear, large-signal model for a GaAs MESFET [97, pages 105–106] (courtesy of the IEEE).

by the following 2-port field and transport equations, which are given in the original notation [97]:

$$I_g = \mathrm{GVSG}\frac{dV_{\mathrm{SG}}}{dt} + \mathrm{GVDS}\frac{dV_{\mathrm{DS}}}{dt},$$

$$I_D = I_{\mathrm{con}} + \mathrm{DVSG}\frac{dV_{\mathrm{SG}}}{dt} + \mathrm{DVDS}\frac{dV_{\mathrm{DS}}}{dt}.$$

The coefficients and the drain-source conduction current I_{con} are functions of the port voltages V_{SG} and V_{DS}. The nonlinear FET is still given as an admittance matrix but has nonlinear coefficients. As described by Madjar [97], the surrounding circuit models "the passive parasitic elements of the device, including the diodes that represent the source-gate and drain-gate junctions."

Digital communications can require highly linear but low-power amplifiers. For example, Orthogonal Frequency Division Multiplexing (OFDM) is a modulation format that is relatively immune to multipath but adversely affected by the amplifier's gain and phase nonlinearities. A nonlinear amplifier will compress the amplitude and rotate the phase of the OFDM signal. These distortions degrade the OFDM detectors and system performance suffers.

The amplifier literature contains a variety of schemes addressing this problem. One technique searches for matching circuits that not only maximize gain, minimize noise, and guarantee stability (AMP-1, AMP-2, AMP-3 of Chapter 2), the input and output VSWR (AMP-4 of Section 10.8) and also [52], [84]:

AMP-5 minimize power gain distortion,
AMP-6 minimize phase distortion.

Because these distortions are functions of the matching circuits, the H^∞ techniques should apply to the nonlinear amplifier problem also. Thus, best possible trade-offs of gain, noise, and distortion available from any stable matching should be computable from the H^∞ theory.

The challenge for the mathematician is to bring the nonlinear amplifier into the H^∞ theory. The first problem is to understand what sort of device the nonlinear amplifier is. In the nonlinear FET equations, the coefficient GVSG models the effect of dV_{SG}/dt on the gate displacement current as

$$\text{GVSG}(V_{\text{SG}}) := \begin{cases} \text{GVSG}_- & V_{\text{SG}} \leq -V_F \\ C_0\sqrt{1+V_{\text{SG}}/\phi} & V_{\text{SG}} \in [-V_F, V_{\text{cut}}] \\ \text{GVSG}_+ & V_{\text{SG}} \geq V_{\text{cut}} \end{cases} .$$

Ignoring the cutoffs, the effect is approximated as a Volterra series for the admittance operator

$$I_g = \{a_0 + a_1 V_{\text{SG}} + a_1 V_{\text{SG}}^2 \ldots\}\frac{dV_{\text{SG}}}{dt} + \ldots$$

in the time domain. The Laplace transform turns this into a nontrivial collection of convolution operators. These convolution operators mix frequencies so the straight forward multiplication of the linear admittance is lost. Nevertheless, we can still Cayley transform this admittance operator and call the result a scattering operator.

Question 33. In what sense does this nonlinear amplifier admit a scattering matrix?

What kind of properties does it retain? How can we compute it? We can raise the question of existence.

Question 34. Suppose a multiport is only causal and time-invariant. Does it still admit a scattering matrix?

From a practical point of view, it is tempting to force the nonlinear amplifier into the multiport theory of Chapter 9. However, it is the amplifier's linearity that let us compute the noise figure. Only a relatively small literature computes the noise figure for nonlinear amplifiers.

Question 35. What is the noise figure of a multiport with nonlinear amplifiers attached?

These basic questions show us that nonlinear amplifiers quickly get us to nontrivial research questions. Since the 1990s, these nonlinear models have been part of software design packages used by electrical engineers [122]. The challenge to the mathematics community is to bring the mathematical theory of these nonlinear models up to the level of computation now routinely available in software.

10.10 H^∞ Optimization

This section focuses on the H^∞ multiobjective problem using the Amplifier Matching Problem as an arbiter of good taste. The H^∞ multiobjective problem was laid out by Helton and Vityaev [75]. The multiobjective function $\gamma : H^\infty(\mathbf{C}_+, \mathbf{C}^M) \to \mathbf{R}_+^N$ has the form

$$\gamma(\mathbf{h}) := \begin{bmatrix} \gamma_1(\mathbf{h}) \\ \vdots \\ \gamma_N(\mathbf{h}) \end{bmatrix}.$$

The individual objective functions are

$$\gamma_m(\mathbf{h}) := \sup\{\Gamma_n(j\omega, \mathbf{h}(j\omega)) : \omega \in \mathbf{R}\}.$$

The performance functions $\Gamma_n : j\mathbf{R} \times \mathbf{C} \to \mathbf{R}_+$ are assumed to be continuous and bounded from below. The problem is to minimize

$$\min\{\gamma(\mathbf{h}) : \mathbf{h} \in H^\infty(\mathbf{C}_+, \mathbf{C}^M)\}.$$

From Section 6.5, the deep problem is to compute all Pareto minima [33]. From Section 7.4, the computation of the suboptimal solutions or sublevel sets subsumes the Pareto minima and is a valuable engineering approach. That is, given the design vector $\gamma_u \in \mathbf{R}^N$, parameterize the sublevel set

$$\begin{aligned} [\gamma \le \gamma_u] &:= \{\mathbf{h} \in H^\infty(\mathbf{C}_+, \mathbf{C}^M) : \gamma(\mathbf{h}) \le \gamma_u\} \\ &= [\gamma_1 \le \gamma_{u,1}] \cap [\gamma_2 \le \gamma_{u,2}] \cap \ldots \cap [\gamma_N \le \gamma_{u,N}]. \end{aligned}$$

There is also the problem of generalizing the finite-dimensional multiobjective optimizers. For example, the Goal Attainment Method generalizes as follows:

$$\min\{t \in \mathbf{R}\}$$

subject to $\mathbf{h} \in H^\infty(\mathbf{C}_+, \mathbf{C}^M)$ and

$$\gamma(\mathbf{h}) - t\mathbf{w} \le \gamma_u.$$

The associated research problems are the existence and characterization of the H^∞ minimizers and an implementation on a computer.

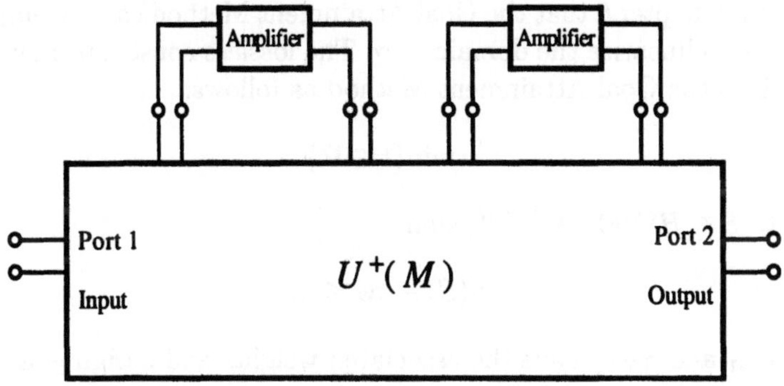

Fig. 10.10. Multiport matching over $U^+(M)$.

The Multiple Amplifier Matching Problem of Chapter 9 offers a generalization to matrix-valued H^∞ functions in the spirit of Helton and Merino [74]. Figure 10.10 illustrates the setup. There are N_A amplifiers with scattering matrices $S_{\text{amp}}(p)$ to be connected to a lossless multiport with scattering matrix $S \in U^+(M)$. The active load is

$$S_A(p) := S_{\text{amp}}(p) \oplus \ldots \oplus S_{\text{amp}}(p).$$

Partition the scattering matrix $S \in U^+(M)$ of the multiport as

$$S(p) = \begin{bmatrix} S_{11}(p) & S_{12}(p) \\ S_{21}(p) & S_{22}(p) \end{bmatrix},$$

where Ports 1 and 2 index the first two indices of $S(p)$. The scattering matrix of the resulting 2-port is

$$\mathcal{F}(S, S_A) = S_{11} + S_{12}S_A(I - S_{22}S_A)^{-1}S_{21}.$$

From this 2-port, we can define the gain and stability functions. Without going into the full derivation, the noise figure also is computable (using the noise parameters of the individual amplifiers from Chapter 9). For an immediate generalization, consider the multiobjective function $\gamma : U^+(M) \to \mathbf{R}_+^N$

$$\gamma(S) := \begin{bmatrix} \gamma_1(S) \\ \vdots \\ \gamma_N(S) \end{bmatrix}.$$

The domain $U^+(N)$ is a nonlinear subset of $H^\infty(\mathbf{C}_+, \mathbf{C}^{N\times N})$. From Chapter 6, the Goal Attainment Method encompasses this lossless constraint as follows. Add the objective functions

$$\gamma_{N+1}(S) := \|I - S^H S\|_\infty, \quad \gamma_{N+1}(S) := -\|I - S^H S\|_\infty.$$

Recall from Chapter 6 that the Goal Attainment Method can encompass some constraints to linearize the domain of γ. The lossless constraint may be incorporated into the Goal Attainment Method as follows:

$$\min\{t \in \mathbf{R}\}$$

subject to $S \in H^\infty(\mathbf{C}_+, \mathbf{C}^{M\times M})$ and

$$\gamma(S) - t\mathbf{w} \le \gamma_u,$$

where γ_{N+1} and γ_{N+2} have the associated weights and design goals:

$$\begin{aligned} w_{N+1} &= 0, \quad & \gamma_{u,N+1} &= 1, \\ w_{N+2} &= 0, \quad & \gamma_{u,N+2} &= -1. \end{aligned}$$

10.11 Epilogue

The rapid developments on a plethora of fronts in electrical engineering have shortened the half-life of most electrical engineers to five years or less. In light of this frenetic pace, the following observation by Helton provides a long-term perspective [66]:

> ... *while technologies come and go some of the mathematics underlying them has an eerie permanence.*

The permanent feature is Nehari's Theorem. This theorem is the computational workhorse of this monograph. One of the joys of applied mathematics is to see how such a clean theorem cracks messy, real-world problems. Nehari's Theorem computes the norm of a Hankel operator $\mathcal{H}_\phi$ as the distance between its symbol $\phi \in L^\infty$ and the Hardy subspace H^∞:

$$\|\mathcal{H}_\phi\| = \inf\{\|\phi - h\|_\infty : h \in H^\infty\}.$$

By linking H^∞ to electrical circuits, Helton brought this eigenvalue computation to bear on a host of problems in electrical engineering and control theory. These problems, in turn, led Helton to deep and original extensions of operator theory, geometry, convex analysis, and optimization theory.

For amplifier matching, the basic matching circuit of Figure 2.1 provided the bulk of the analysis. This matching problem led to the H^∞ Multidisk Method, the H^∞ Multiobjective Approach, and a State-Space Method. But a single amplifier in the specialized matching circuit topology of $U^+(2) \times U^+(2)$ cannot perform as well the general matching circuit of $U^+(4)$. Likewise, the single amplifier matched over $U^+(4)$ does not perform as well as N_{amp} amplifiers matched over $U^+(2N_{\mathrm{amp}} + 2)$. The State-Space Methods do generalize and demand that H^∞ theory catch up. Moreover, there are many problems regarding the existence of the scattering matrix, both for a particular class, its representations, and its realization as a practical electrical circuit.

The upshot is that the Amplifier Matching Problem directs the applied mathematician, the electrical engineer, or the emerging H^∞ engineer, to fascinating mathematics with substantial real-world applications. It is this interplay between mathematics and applications that simultaneously deepens the mathematics and provides new approaches to real-world problems.

A

The Axioms of Electric Circuits

This appendix reviews circuit axioms for the applied mathematician. Our goal is to make explicit the assumptions that underly the existence of the scattering matrix stated in Chapter 1, repeated here for convenience:

> EXISTENCE OF THE SCATTERING MATRIX [6], [8], [73]: *Any linear, causal, passive, time-invariant, solvable N-port $\mathcal{N}$ always admits a* scattering matrix $S \in \overline{B}H^\infty(\mathbb{C}_+, \mathbb{C}^{N\times N})$ *so that*
>
> $$\mathcal{N} = \left\{ \begin{bmatrix} \mathbf{a} \\ S\mathbf{a} \end{bmatrix} : \mathbf{a} \in L^2(j\mathbf{R}, \mathbb{C}^N) \right\}.$$

Our point of view takes the circuit axioms as invariants of subspaces in a Krein space. It is rather surprising that such "soft" functional analysis can lead to the "hard" identification of a circuit as an element of H^∞.

Starting in the late 1950s and continuing through the 1960s, electrical engineers had a running debate on the axioms of network theory. The debate centered on listing the minimal properties of an N-port $\mathcal{N}$ to obtain the standard formalisms of traditional circuit theory. In hindsight, this debate was less about circuit theory and more about the geometry of Krein spaces—it turns out that $\mathcal{N}$ is well described a positive subspace of a Krein space $\mathcal{K}$. Table A.1 provides a handy correspondence between the circuit properties of $\mathcal{N}$ and their counterparts in operator theory. We march through Table A.1

Table A.1. Translating circuit properties to operator theory.

Circuits	Operators
Linearity	$\mathcal{N}$ is a linear subspace
Passivity	$\mathcal{N}$ is a positive subspace
Solvability	$\mathcal{N}$ is a maximal positive subspace
Time invariance	$\mathcal{N}$ is an invariant subspace for a shift operator
Causality	$\mathcal{N}$ is an invariant subspace for a resolution of the identity

examining this correspondence. For example, a consequence of linearity and passivity is that $\mathcal{N}$ is characterized as the graph of a scattering operator S. The *existence* of S is a natural consequence of the geometry of a Krein space and its angle operators.

A.1 Krein Spaces and Angle Operators

This section is lifted from the preliminary pages of Ball and Helton [12]. The reader is referred to this paper for a beautiful treatment of the Beurling-Lax Theory in indefinite inner-product spaces or *Krein spaces*. For our purposes, we only need the definition of a Krein space and an *angle operator*.

Definition A.1.1 [12] *A complex vector space $\mathcal{K}$ equipped with a Hermitian bilinear form $[\ ,\ ]$ is called a Krein space if and only if there exist Hilbert spaces $\mathcal{K}_+$ and $\mathcal{K}_-$ such that*

(K-1) $\mathcal{K}$ can be written as the direct sum $\mathcal{K} = \mathcal{K}_+ \oplus \mathcal{K}_-$.
(K-2) The bilinear form factors as

$$[\mathbf{x}, \mathbf{y}] = \langle \mathbf{x}_+, \mathbf{y}_+ \rangle_+ - \langle \mathbf{x}_-, \mathbf{y}_- \rangle_-,$$

where $\mathbf{x} = \mathbf{x}_+ + \mathbf{x}_-$ and $\mathbf{y} = \mathbf{y}_+ + \mathbf{y}_-$ are the decompositions of $\mathbf{x}$ and $\mathbf{y}$ under the direct sum.

Given a Krein space $\mathcal{K}$, it is also a Hilbert space with inner product

$$\langle \mathbf{x}, \mathbf{y} \rangle = \langle \mathbf{x}_+, \mathbf{y}_+ \rangle_+ + \langle \mathbf{x}_-, \mathbf{y}_- \rangle_-.$$

A subspace $\mathcal{P}$ of a Krein space $\mathcal{K}$ is called a *positive subspace* provided

$$0 \leq [\mathbf{k}, \mathbf{k}], \qquad (\mathbf{k} \in \mathcal{P}).$$

Positive subspaces are modeled by their angle operators.

Lemma A.1.1 [12] *Let $\mathcal{K} = \mathcal{K}_+ \oplus \mathcal{K}_-$ be a Krein space. Let $\mathcal{P}$ be a linear subspace, not necessarily closed. Then the following are equivalent:*

(a) $\mathcal{P}$ is a positive subspace.
(b) There is a linear subspace $D \subseteq \mathcal{K}_+$ and a contraction operator $T : D \subseteq \mathcal{K}_+ \to \mathcal{K}_-$ such that

$$\mathcal{P} = \mathcal{G}(T|D) = \left\{ \begin{bmatrix} \mathbf{k}_+ \\ T\mathbf{k}_+ \end{bmatrix} : \mathbf{k}_+ \in D \subseteq \mathcal{K}_+ \right\}.$$

Proof:
(a⇒b) With respect to the decomposition $\mathcal{K} = \mathcal{K}_+ \oplus \mathcal{K}_-$, each element of $\mathcal{P}$ has the form

$$\mathbf{k} = \begin{bmatrix} \mathbf{k}_+ \\ \mathbf{k}_- \end{bmatrix} \in \mathcal{P}.$$

Let $P_\pm$ denote the orthogonal projections onto $\mathcal{K}_\pm$. Set $D = P_+(\mathcal{P})$ and define $T : D \to K_-$ by $T\mathbf{k}_+ = \mathbf{k}_-$. To verify that T is well defined, suppose

$$\begin{bmatrix} \mathbf{k}_+ \\ \mathbf{k}_- \end{bmatrix}, \begin{bmatrix} \mathbf{k}'_+ \\ \mathbf{k}'_- \end{bmatrix} \in \mathcal{P}.$$

Because $\mathcal{P}$ is a linear subspace, it follows that the negative vector

$$\begin{bmatrix} 0 \\ \mathbf{k}_- - \mathbf{k}'_- \end{bmatrix} \in \mathcal{P}.$$

Because $\mathcal{P}$ is positive, the Krein-space decomposition forces $\|\mathbf{k}_- - \mathbf{k}'_-\|_{\mathcal{K}_-} \leq 0$ or $\mathbf{k}_- = \mathbf{k}'_-$. Thus, T is well defined. Because $\mathcal{P}$ is a linear subspace, it follows that T is also linear. Because $\mathcal{P}$ is positive, T is a contraction. It remains to see how $\mathcal{G}(T|D)$ occupies $\mathcal{P}$. If

$$\mathbf{k} = \begin{bmatrix} \mathbf{k}_+ \\ \mathbf{k}_- \end{bmatrix} \in \mathcal{P} \setminus \mathcal{G}(T|D)$$

then $\mathbf{k}_+$ belongs to D by construction. This forces $\mathbf{k}_- = T(\mathbf{k}_+)$ so that $\mathbf{k}$ must belong to $\mathcal{G}(T|D)$. Thus, $\mathcal{P} = \mathcal{G}(T|D)$ even when $\mathcal{P}$ is not closed.
(a⇐b) Follows because T is a contraction. ///

Theorem A.1.1 *[12] Let $\mathcal{K} = \mathcal{K}_+ \oplus \mathcal{K}_-$ be a Krein space. Let $\mathcal{P}$ be a linear subspace. Then the following are equivalent:*

(a) $\mathcal{P}$ is a maximal positive subspace.
(b) There is a contraction operator $T : \mathcal{K}_+ \to \mathcal{K}_-$ such that

$$\mathcal{P} = \mathcal{G}(T) = \left\{ \begin{bmatrix} \mathbf{k}_+ \\ T\mathbf{k}_+ \end{bmatrix} : \mathbf{k}_+ \in \mathcal{K}_+ \right\}.$$

Proof:
(a⇒b) By Lemma A.1.1, there holds

$$\mathcal{P} = \mathcal{G}(T|D) = \left\{ \begin{bmatrix} \mathbf{k}_+ \\ T\mathbf{k}_+ \end{bmatrix} : \mathbf{k}_+ \in D \subseteq \mathcal{K}_+ \right\}.$$

We claim $D = \mathcal{K}_+$. Suppose first that D is not closed. By continuity, T admits an extension $\overline{T}$ onto $\overline{D}$. Then

$$\mathcal{P} = \mathcal{G}(T|D) \subset \mathcal{G}(\overline{T}|\overline{D}).$$

Continuity implies that $\mathcal{G}(\overline{T}|\overline{D})$ is positive. Thus, $\mathcal{P}$ is not maximal positive. This contradiction forces D to be closed. To show D is all of $\mathcal{K}_+$, suppose $\mathbf{k}_+ \in \mathcal{K}_+ \ominus D$. Then

$$\begin{bmatrix} \mathbf{k}_+ \\ 0 \end{bmatrix} \perp \mathcal{P}.$$

Consequently,

$$\begin{bmatrix} \mathbf{k}_+ \\ 0 \end{bmatrix} \oplus \mathcal{P}$$

is a positive subspace that strictly contains $\mathcal{P}$. This contradicts $\mathcal{P}$'s maximality. Thus, D must be all of $\mathcal{K}_+$.
(a⇐b) Suppose T is a contraction and $\mathcal{P} = \mathcal{G}(T)$. Then $\mathcal{P}$ is also positive because T is a contraction:

$$\left[\!\left[\begin{bmatrix} \mathbf{k}_+ \\ T\mathbf{k}_+ \end{bmatrix}, \begin{bmatrix} \mathbf{k}_+ \\ T\mathbf{k}_+ \end{bmatrix} \right]\!\right] = \langle \mathbf{k}_+, \mathbf{k}_+ \rangle_+ - \langle T\mathbf{k}_+, T\mathbf{k}_+ \rangle_- \geq 0.$$

Suppose $\mathcal{P}$ is not maximal. Then there must exist a nonzero vector $\mathbf{k} \in \mathcal{K}$ such that $\mathbf{k} + \mathcal{P}$ is a positive subspace. Because linear combinations of this space are positive, the following sum is positive:

$$0 \leq \begin{bmatrix} \mathbf{k}_+ \\ \mathbf{k}_- \end{bmatrix} - \begin{bmatrix} \mathbf{k}_+ \\ T\mathbf{k}_+ \end{bmatrix} = \begin{bmatrix} 0 \\ \mathbf{k}_- - T\mathbf{k}_+ \end{bmatrix}.$$

This negative vector can be positive only if $\mathbf{k}_- = T\mathbf{k}_+$. But this forces $\mathbf{k}$ to belong to the graph of T in contradiction to its selection. Thus, $\mathcal{P}$ must be maximal positive. ///

A.2 N-Ports $\Longleftrightarrow$ Angle Operators

The basic object under discussion is an N-port. This is simply a box with N pairs of wires sticking out of it. The use of the word "port" means that each pair of wires obey a "conservation of current": the current flowing in one wire of the pair equals the current flowing out of the other wire. Figure A.1 illustrates the setup for a 2-port.

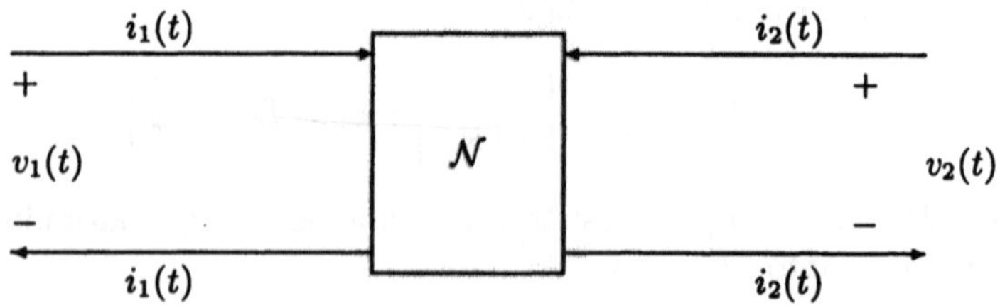

Fig. A.1. A 2-port.

The N-port is characterized as the collection of voltage vectors $\mathbf{v}(t)$ and current vectors $\mathbf{i}(t)$ that can appear on its ports:

$$\mathbf{v}(t) = \begin{bmatrix} v_1(t) \\ v_2(t) \\ \vdots \\ v_N(t) \end{bmatrix}, \quad \mathbf{i}(t) = \begin{bmatrix} i_1(t) \\ i_2(t) \\ \vdots \\ i_N(t) \end{bmatrix}.$$

The following definition of an N-port has been adapted from [92, page 3] and [4] to incorporate this notion.

Definition A.2.1 *Let $\mathcal{V}$ and $\mathcal{I}$ both denote the Hilbert space $L^2(\mathbf{R}, \mathbf{C}^N)$. An N-port network $\mathcal{N}$ is a subset of $\mathcal{V} \times \mathcal{I}$.*

Remarks: (*i*) The voltage and current are required to exist over all time. Kuh and Rohrer [92] are more general by allowing the voltage and current vectors to exist only over an observation interval. This finer structure is not important for our exposition. (*ii*) We have also made the decision to have the voltage and current to be square integrable to admit a Hilbert-space formalism. Both Wohlers [128] and Anderson and Newcomb [4] examine networks using generalized functions.

There are a number of "formalisms" associated with N-ports. Each formalism designates input and output variables. For example, the impedance formalism takes current as the input and voltage as output:

$$\mathbf{v}(t) = z(\mathbf{i}, t).$$

The admittance formalism take voltage as input and current as output:

$$\mathbf{i}(t) = y(\mathbf{v}, t).$$

The formalisms are designed to force the N-port to determine an input-to-output operator. The following examples show that these operators exist in some formalisms but not others.

Example A.2.1 (Resistor) *Figure A.2 shows a 1-port with a fixed resistor r. The impedance formalism takes the current as input and voltage as output so that $v(t) = ri(t)$ or*

$$\mathcal{R} = \left\{ \begin{bmatrix} ri \\ i \end{bmatrix} : i \in \mathcal{H} \right\}.$$

Example A.2.2 (Short Circuit) *The short circuit of Figure A.3 is a special case of the 1-port resistor with $r = 0$:*

$$\mathcal{S} = \left\{ \begin{bmatrix} 0 \\ i \end{bmatrix} : i \in \mathcal{H} \right\}.$$

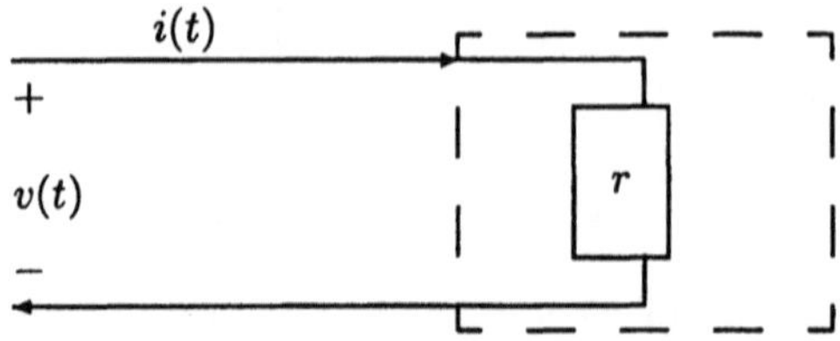

Fig. A.2. The 1-port resistor network: $v(t) = ri(t)$.

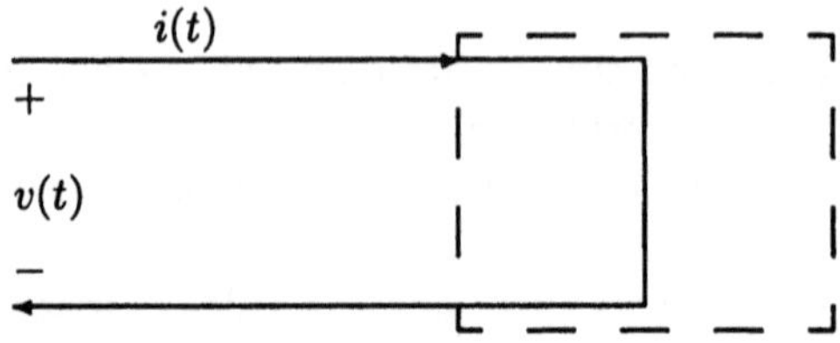

Fig. A.3. The 1-port short circuit: $v(t) = 0$.

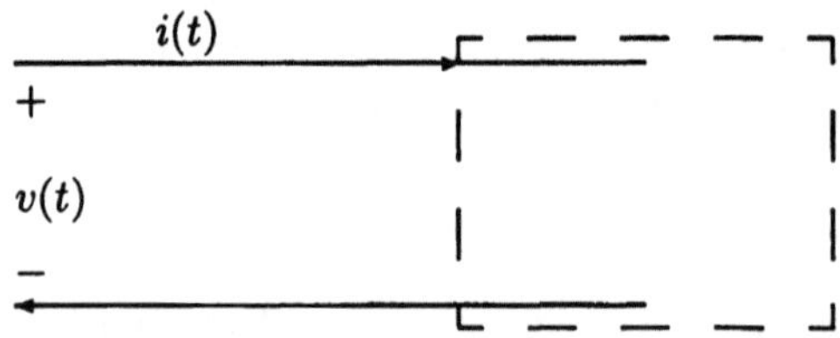

Fig. A.4. The 1-port open circuit: $i(t) = 0$.

Example A.2.3 (Open Circuit) *The open circuit of Figure A.4 is a special case of the 1-port admittance with zero admittance:*

$$\mathcal{O} = \left\{ \begin{bmatrix} v \\ 0 \end{bmatrix} : v \in \mathcal{H} \right\}.$$

One common characteristic of the preceding examples is that $\mathcal{N}$ is a linear subspace of $\mathcal{V} \times \mathcal{I}$.

Definition A.2.2 *An N-port $\mathcal{N}$ is called a* linear *provided $\mathcal{N}$ is a linear subspace of $\mathcal{V} \times \mathcal{I}$.*

Another common characteristic of the preceding examples is that $\mathcal{N}$ is a "graph." In the impedance formalism, the resistor $\mathcal{R}$ is the graph of the resistor. The short circuit $\mathcal{S}$ is the graph of the zero map in the impedance formalism. However, $\mathcal{S}$ does **not** admit an admittance operator because the single input 0 would have to map to every $i \in \mathcal{I}$. Likewise, the open circuit $\mathcal{O}$

is the graph of the zero admittance operator but does not admit an impedance operator.

Given a linear subspace $\mathcal{N}$, when is it the graph of an operator? The answer is to find a subspace $\mathcal{A}$ of equal dimension that contains no vectors orthogonal to $\mathcal{N}$. Consider now $\mathcal{N}$ with respect to such a decomposition $\mathcal{A}\times\mathcal{A}^{\perp}$. For each point $[\mathbf{a}\ \mathbf{b}]^T \in \mathcal{N}$, define the map $S : \mathcal{A} \to \mathcal{A}^{\perp}$ by $S\mathbf{a} = \mathbf{b}$. The selection of $\mathcal{A}$ forces S to be well defined. Indeed, S cannot map 0 to nonzero vectors because $[0\ \mathbf{b}]^T \in \mathcal{N}$ implies $\mathbf{b} = 0$. But there is always a trivial way to associate a linear subspace with an operator—simply take $\mathcal{A}$ to be $\mathcal{N}$ and the operator to be the zero map. The notion of a passive N-port introduces a nontrivial decomposition of $\mathcal{K}$.

Definition A.2.3 *$\mathcal{N}$ is passive provided $\Re\langle \mathbf{v}, \mathbf{i}\rangle \geq 0$ for each $\begin{bmatrix}\mathbf{v}\\ \mathbf{i}\end{bmatrix} \in \mathcal{N}$.*

Remarks: (*i*) Recall that current $\mathbf{i}(t) = dq(t)/dt$ has units of charge per time (Q/T) while voltage $\mathbf{v}(t)$ has units of energy per charge (W/Q). Thus, $\mathbf{v}^H(t)\mathbf{i}(t)$ has units of energy per time (W/T). Integration over all time is the total complex energy delivered to the network. (*ii*) Youla, Castriota, and Carlin [131], Kuh and Rohrer [92] have a stricter definition of a passive network. Wohlers [128] contains a fine exposition comparing both definitions and favors the more general.

It was Helton's insight to see that passivity is a nonpositive bilinear form in the scattering formalism. From Chapter 1, recall the scattering formalism:

- Incident signal: $\mathbf{a} = \frac{1}{2}(\mathbf{v} + \mathbf{i})$.
- Reflected signal: $\mathbf{b} = \frac{1}{2}(\mathbf{v} - \mathbf{i})$.

The parallelogram law permits us to write the power consumed by $\mathcal{N}$ as the bilinear form

$$\Re\langle \mathbf{v}, \mathbf{i}\rangle = [\mathbf{a}, \mathbf{b}] = \|\mathbf{a}\|_+^2 - \|\mathbf{b}\|_-^2.$$

This bilinear form turns $\mathcal{V}\times\mathcal{I}$ into the Krein space $\mathcal{K} = \mathcal{A}\oplus\mathcal{B}$ under the map:

$$\begin{bmatrix}\mathcal{A}\\ \mathcal{B}\end{bmatrix} = \frac{1}{2}\begin{bmatrix}1 & 1\\ 1 & -1\end{bmatrix}\begin{bmatrix}\mathcal{V}\\ \mathcal{I}\end{bmatrix},$$

where $\mathcal{A} = \mathcal{K}_+$ denotes the maximal positive space and $\mathcal{B} = \mathcal{K}_-$ denotes the maximal negative space. Lemma A.1.1 immediately applies to link an N-port to its scattering operator—with restricted domain.

Theorem A.2.1 *Let $\mathcal{N}$ be a linear and passive network. Then there is a linear subspace $\mathcal{D}$ of $\mathcal{A}$ and a linear contraction $S : \mathcal{D} \to \mathcal{B}$ such that*

$$\mathcal{N} = \left\{\begin{bmatrix}\mathbf{a}\\ S\mathbf{a}\end{bmatrix} : \mathbf{a} \in \mathcal{D} \subseteq \mathcal{A}\right\}.$$

We are suppressing the "change of basis" matrix and writing $\mathcal{N}$ in the scattering formalism. The next step is to remove the restriction on the domain of the scattering matrix. The electrical engineers use the following notion of solvability adapted from Kuh and Rohrer [92] and Youla, Castriota, and Carlin [131].

Definition A.2.4 *$\mathcal{N}$ is S-solvable if and only if for each $\mathbf{a} \in \mathcal{A}$ there exists exactly one*

$$\begin{bmatrix} \mathbf{v} \\ \mathbf{i} \end{bmatrix} \in \mathcal{N}$$

such that

$$\mathbf{a} = \frac{1}{2}(\mathbf{v} + \mathbf{i}).$$

$\mathcal{N}$ is completely S-solvable if and only if $\mathcal{N}$ is S-solvable and for each sequence $\{\mathbf{a}_k\} \in \mathcal{A}$ converging to $\mathbf{a}$ it follows that there exists exactly one

$$\begin{bmatrix} \mathbf{v} \\ \mathbf{i} \end{bmatrix} \in \mathcal{N}$$

such that

$$\mathbf{a} = \frac{1}{2}(\mathbf{v} + \mathbf{i}).$$

S-solvability is rigged to make the scattering formalism "work." Completely S-solvable is rigged to make $\mathcal{N}$ closed. Both notions are subsumed in the definition of a maximal positive subspace.

Definition A.2.5 *$\mathcal{N}$ is called maximal passive if and only if $\mathcal{N}$ is passive and contained in no larger passive subspace.*

A maximal passive subspace is a closed subspace that projects onto $\mathcal{A}$ in the sense of being completely S-solvable. Consequently, maximal passive handles the closure problems in a natural way and permits the immediate application of Theorem A.1.1 to give the N-port as the graph of the scattering matrix.

Theorem A.2.2 *$\mathcal{N}$ is linear and maximal passive if and only if there is a linear contraction $S : \mathcal{A} \to \mathcal{B}$ such that*

$$\mathcal{N} = \left\{ \begin{bmatrix} \mathbf{a} \\ S\mathbf{a} \end{bmatrix} : \mathbf{a} \in \mathcal{A} \right\}.$$

Remarks: (*i*) While mathematicians cheerfully add the voltage and current vectors $\mathbf{v}(t)$ and $\mathbf{i}(t)$, their units do not permit such a summation. Consequently, a normalization of the units is required. This normalization corresponds to the physical setup of augmenting the N-port with unit resistors [92, page 29–30], [131]. Kuh and Rohrer [92, Chapter 6] and Balabanian and Bickart [8] both have extensive developments of this augmentation and its implications for network interconnection. (*ii*) Wohlers [128, Section 2.2] also discusses the augmented network. (*iii*) A recent reference on the scattering matrix and the augmented network is [125].

A.3 Time Invariance $\Longleftrightarrow$ Convolution

The preceding sections show that a maximal passive N-port corresponds to a contraction operator S in the scattering formalism. The next structure to impose is *time invariance*—the notion that a network's behavior is not tied to a fixed point in time. Equivalently, the network is invariant with respect to time shifts. The time-shift operators on $L^2(\mathbf{R})$ are

$$T_\tau h(t) = h(t-\tau).$$

Each T_τ is unitary. The entire collection is a one-parameter unitary group $\{T_\tau\}$. The following subspace definition of time invariance is in the spirit of Kuh and Rohrer [92] and immediately leads to a commutation condition for S.

Definition A.3.1 *$\mathcal{N}$ is time invariant if and only if $\mathcal{N}$ is invariant under all the time shifts: $T_\tau \mathcal{N} \subseteq \mathcal{N}$ for all $\tau \in \mathbf{R}$.*

Lemma A.3.1 *Let $\mathcal{N}$ be a linear and maximal passive N-port with scattering operator S. Then $\mathcal{N}$ is time invariant if and only if S commutes with $\{T_\tau\}$.*

Proof: $\mathcal{N}$ is time invariant if and only if for all $\tau \in \mathbf{R}$ there holds

$$T_\tau \begin{bmatrix} \mathbf{a} \\ \mathbf{b} \end{bmatrix} \in \mathcal{N}.$$

Use Theorem A.2.2 to establish that $\mathcal{N}$ is time invariant if and only if for all $\tau \in \mathbf{R}$ there holds

$$\begin{bmatrix} T_\tau \mathbf{a} \\ T_\tau S\mathbf{a} \end{bmatrix} = \begin{bmatrix} \mathbf{a}' \\ S\mathbf{a}' \end{bmatrix}$$

that is equivalent to $T_\tau S\mathbf{a} = ST_\tau \mathbf{a}$ for all $\mathbf{a} \in \mathcal{A}$. ///

Any time operators commute, a natural question to ask is if the diagonalization of one operator carries over to the other? We use the Fourier transform in the electrical engineer's formalism:

$$\widehat{h}(j\omega) = \int_{-\infty}^{\infty} e^{-i\omega t} h(t)dt.$$

This transform is also known as the two-sided Laplace that extends into the right half plane $\mathbf{C}_+$:

$$\widehat{h}(p) = \int_{-\infty}^{\infty} e^{-pt} h(t)dt \quad (p \in \mathbf{C}_+).$$

The Fourier transform maps the time-domain functions of $L^2(\mathbf{R})$ into the frequency domain $L^2(j\mathbf{R})$:

$$\widehat{} : L^2(\mathbf{R}) \to L^2(j\mathbf{R}).$$

Because the Fourier transform diagonalizes the time shifts

$$\widehat{T_\tau h}(p) = e^{-p\tau}\widehat{h}(p),$$

the Fourier transform also diagonalizes S. This is the content of Bochner's L^2 Theorem.

Theorem A.3.1 (Bochner) [18] *Let S be a bounded linear operator on $L^2(\mathbf{R})$ Then the following are equivalent*

(a) S is time-invariant.
(b) S is equivalent to multiplication by an $s \in L^\infty(j\mathbf{R})$ under the Fourier transform.

Diagonalizing the time-invariant scattering operator of Theorem A.2.2 gives the representation of the N-port as a multiplication operator.

Theorem A.3.2 *$\mathcal{N}$ is a linear, maximal passive, time invariant N-port if and only if there is $s \in L^\infty(j\mathbf{R}, \mathbf{C}^{N\times N})$ with $\|s\|_\infty \leq 1$ such that*

$$\widehat{\mathcal{N}} = \left\{ \begin{bmatrix} \widehat{\mathbf{a}} \\ s\widehat{\mathbf{a}} \end{bmatrix} : \mathbf{a} \in \mathcal{A} \right\}.$$

Remarks: (*i*) Credit for applying Bochner's L^2 Theorem to network theory belongs to Youla, Castriota, and Carlin [131]. (*ii*) Wohlers [128] uses the distributional approach that starts by identifying a linear continuous system as a mapping $K : D_K \to \mathcal{D}'(\mathbf{R})$. The Kernel Theorem of Schwartz then permits the representation

$$K(\mathbf{a})(t) = \int k(t,\tau)\mathbf{a}(\tau)d\tau$$

for $\mathbf{a} \in C_0^\infty(\mathbf{R})$ and $k \in \mathcal{D}'(\mathbf{R}^2)$ [128, Theorem 1.1]. Time invariance turns this two-dimensional kernel into a convolution [128, Theorem 1.2]. The Fourier theory for distributions then turns this convolution into a multiplication operator. (*iii*) Kuh and Rohrer [92] step from the subspace formalism to a formal integral representation.

A.4 Causality $\Longleftrightarrow$ Analyticity

The last property we impose on the network $\mathcal{N}$ is causality—the notion that the present value of the network's output does not depend on future values of the input. One approach to causality is Saeks's *resolution space*.

Definition A.4.1 [111] *A resolution space $(\mathcal{H}, \{E^t\})$ is a Hilbert space $\mathcal{H}$ and a resolution of the identity $\{E^t\}$ on $\mathcal{H}$. That is, $\{E^t\}$ is a right-continuous family of orthogonal projections with $t \in \mathbf{R}$ such that*

(RS-1) *If* $s \le t$ *then* $E^s \le E^t$.
(RS-2) $\text{s}-\lim_{t\to-\infty} E^t = 0$.
(RS-3) $\text{s}-\lim_{t\to+\infty} E^t = I$.

A causal operator preserves some of the resolution-space structure.

Definition A.4.2 [111] *Let* $(\mathcal{H}, \{E^t\})$ *be a resolution space and let* S *be an operator on* $\mathcal{H}$. *Then* S *is causal if and only if for all* x, $y \in \mathcal{H}$ *and all* $t \in \mathbf{R}$ $E^t x = E^t y$ *implies* $E^t S x = E^t S y$.

Remarks: (*i*) Wohlers [128, Theorem 2.3] obtains the representation of a linear, time-invariant causal system in the context of distribution theory. (*ii*) Kuh and Rohrer [92] and Youla, Castriota, and Carlin [131] use a more restrictive version of passivity to obtain causality as a consequence of linearity and passivity. Wohlers [128] compares both approaches and opts for the more general notion of passivity. (*iii*) Observe that causality is defined by operators rather than subspaces.

A causal operator admits a lower-triangular form or an equivalent commutation relation with the projection operators.

Theorem A.4.1 [111] *S be an operator on a resolution space* $(\mathcal{H}, \{E^t\})$. *Let* $\mathcal{N}_t$ *denote the null space of* E^t. *The following are equivalent:*

(a) S *is causal.*
(b) $E^t x = 0$ *implies* $E^t S x = 0$ *for all* $t \in \mathbf{R}$.
(c) $S(\mathcal{N}_t) \subseteq \mathcal{N}_t$ *for all* $t \in \mathbf{R}$.
(d) $E^t S = E^t S E^t$ *for all* $t \in \mathbf{R}$.

With respect to the decomposition $\mathcal{H} = \mathcal{N}_t^{\perp} \oplus \mathcal{N}_t$, a causal linear operator S is lower triangular:

$$S = \begin{bmatrix} S_{11} & 0 \\ S_{21} & S_{22} \end{bmatrix}$$

for all $t \in \mathbf{R}$. In the time domain, causality takes the following form.

Definition A.4.3 *Define the resolution of the identity* $\{P_\tau\}$ *on* $L^2(\mathbf{R}, \mathbf{C}^N)$ *by*

$$P_\tau \mathbf{a}(t) = 1_{(-\infty,\tau]}(t) \times \mathbf{a}(t).$$

Let S *be a bounded linear operator on* $L^2(\mathbf{R}, \mathbf{C}^N)$. *Then* S *is causal if and only if* $P_\tau \mathbf{a} = 0$ *implies* $P_\tau S a = 0$ *for all* $\tau \in \mathbf{R}$ *and all* $\mathbf{a} \in L^2(\mathbf{R}, \mathbf{C}^N)$.

For example, a convolution operator on $L^2(\mathbf{R})$

$$C_h a(t) = h * a(t) = \int_{-\infty}^{\infty} h(t-\tau) a(\tau) d\tau$$

is causal if and only if $h(t) = 0$ for $t < 0$. This is the content of the Paley-Weiner Theorems [110, Theorem 19.2].

Theorem A.4.2 *Let C_h be a bounded convolution operator on $L^2(\mathbf{R})$. The following are equivalent:*

(a) C_h is causal.
(b) $\widehat{h} \in H^\infty(\mathbf{C}_+)$.

A.5 Existence

With some work, the causality of Theorem A.4.2 couples with Theorem A.3.2 to give the representation of the N-port as the graph of its H^∞ scattering matrix.

Theorem A.5.1 (Existence of the Scattering Matrix) [48], [128], [92], [131], [66], [68] *$\mathcal{N}$ is linear, maximal passive, time-invariant, causal N-port if and only if there is $s \in H^\infty(\mathbf{C}_+, \mathbf{C}^{N\times N})$ with $\|s\|_\infty \leq 1$ such that*

$$\widehat{\mathcal{N}} = \left\{ \begin{bmatrix} \widehat{\mathbf{a}} \\ s\widehat{\mathbf{a}} \end{bmatrix} : \mathbf{a} \in \mathcal{A} \right\}.$$

For brevity, the main text always worked in the frequency domain so the Fourier transform "hat" could be suppressed. Likewise, the "change-of-basis" to the scattering formalism is also suppressed. Finally, the descriptor "maximal passive" substitutes for the more well-known "solvable."

This classic axiomatic approach *appears* to address the correspondence between a network $\mathcal{N}$ and its scattering operator. However, it leaves one "floating in air" with "maximal passive" as a rigging to make the scattering formalism work. Consider applying this approach on a more practical level. Suppose one were handed the loop equations for a lumped circuit? Could one readily see "maximal passivity" in the loop equations? More lethal examples to test this axiomatic approach are the *nonlinear partial differential equations* associated with microwave devices. Can one glance at such a system and assert the existence of a credible domain for the scattering operator? Does "maximal positive" or "solvable" pop out of the system? Consequently, the foundations of network theory are not as sturdy as they appear.

B

Taylor's Expansion and the Descent Lemma

For brevity, Chapter 6 omitted the technical discussion of the Descent Lemma and Taylor's expansion. For the reader's convenience, this appendix proves both the lemma and the expansion.

B.1 Taylor's Expansion

The real numbers are denoted by $\mathbf{R}$. Real N-dimensional space is denoted by $\mathbf{R}^N$. The nonnegative real numbers are denoted by $\mathbf{R}_+$. The closed positive cone of $\mathbf{R}^N$ is $\mathbf{R}^N_+$. The complex numbers are denoted by $\mathbf{C}$. Complex N-dimensional space is denoted by $\mathbf{C}^N$. Throughout this appendix, Z denotes a compact subset of $\mathbf{C}$. The open unit disk

$$\mathbf{D} := \{z \in \mathbf{C} : |z| < 1\}$$

has the unit circle $\mathbf{T}$ as its boundary:

$$\mathbf{T} := \{z \in \mathbf{C} : |z| = 1\}.$$

If E is a Banach space with norm $\|\circ\|_E$, $C(Z,E)$ denotes the set of continuous E-valued functions on Z with norm

$$\|h\|_\infty := \sup\{\|h(z)\|_E : z \in Z\}.$$

Consequently,

- $C(Z,\mathbf{R})$ denotes the set of continuous real-valued functions on Z.
- $C(Z,\mathbf{C})$ denotes the set of continuous complex-valued functions on Z.
- $C(Z,\mathbf{C}^M)$ denotes the continuous $\mathbf{C}^M$-valued functions on Z.

If $\Gamma \in C(Z \times \mathbf{C},\mathbf{R})$ is continuous, it lifts to the mapping $\widetilde{\Gamma} : C(Z,\mathbf{C}) \to C(Z,\mathbf{R})$ given by

$$\widetilde{\Gamma}(h;z) = \Gamma(z,h(z))$$

and induces the *objective* function $\gamma : C(Z, \mathbf{C}) \to \mathbf{R}$

$$\gamma(h) := \sup\{\Gamma(z, h(z)) : z \in Z\}.$$

The associated *critical set* of $h \in C(Z, \mathbf{C})$ is

$$\mathrm{crit}[\gamma(h)] := \{z \in Z : \gamma(h) = \Gamma(z, h(z))\}.$$

If $\mathcal{H}$ is a subspace of $C(Z, \mathbf{C})$, a *nonzero* $\Delta h \in \mathcal{H}$ is called a *direction of nonincrease* [26] or *direction of descent* for γ provided for all $t > 0$ sufficiently small

$$\gamma(h + t\Delta h) \le \gamma(h).$$

The function $h \in \mathcal{H}$ is called a *local minimum* for γ provided for all $\Delta h \in \mathcal{H}$ sufficiently small there holds

$$\gamma(h + \Delta h) \ge \gamma(h).$$

The derivative on $\mathbf{C}$ can be written as [110]

$$\partial = \frac{\partial}{\partial z} = \frac{1}{2}\left\{\frac{\partial}{\partial x} - i\frac{\partial}{\partial y}\right\}.$$

If $\Gamma : \mathbf{C} \to \mathbf{R}$ is C^2, Taylor's expansion is

$$\begin{aligned}\Gamma(h + \Delta h) &= \Gamma(h) + \frac{\partial \Gamma}{\partial x}(h)\Delta u + \frac{\partial \Gamma}{\partial y}(h)\Delta v + \mathcal{O}[|\Delta h|^2] \\ &= \Gamma(h) + 2\Re[\partial\Gamma(h)\Delta h] + \mathcal{O}[|\Delta h|^2],\end{aligned}$$

where $\Delta h = \Delta u + i\Delta v \in \mathbf{C}$. The Omega Lemma lifts this expansion to the corresponding expansion for $\widetilde{\Gamma}$ operating on $C(Z, \mathbf{C})$.

Lemma B.1.1 (Omega) [1] *Let E and F be Banach spaces. Let $U \subseteq E$ be open. Assume $g : U \subseteq E \to F$ is a C^r map $(r > 0)$ with first variation $Dg : E \to F$. Let M be a compact topological space. Then the map $\widetilde{g} : C(M, U) \to C(M, F)$ defined by*

$$\widetilde{g}(h; m) := g(h(m))$$

is also C^r. The derivative of $\widetilde{g}$ at $h \in C(M, U)$ is denoted $D\widetilde{g}(h)$ and is the linear map $D\widetilde{g}(h) : C(M, E) \to C(M, F)$

$$D\widetilde{g}(h)[\Delta h; m] := Dg(h(m))[\Delta h(m)].$$

The only adjustment that we need to make is to account for the fact that Γ maps from $Z \times \mathbf{C}$. Let

$$\partial_1 \Gamma(z_1, z_2) = \frac{\partial \Gamma}{\partial z_1}(z_1, z_2), \qquad \partial_2 \Gamma(z_1, z_2) = \frac{\partial \Gamma}{\partial z_2}(z_1, z_2).$$

Lemma B.1.2 (Γ) *Let $Z \subset \mathbf{C}$ be compact. Let $U \subseteq \mathbf{C}$ be an open subset containing Z. Let $\Gamma : U \times \mathbf{C} \to \mathbf{R}$ be C^r ($r > 0$) with first variation*

$$D\Gamma(z_1, z_2) = [\partial_1 \Gamma(z_1, z_2) \ \ \partial_2 \Gamma(z_1, z_2)].$$

Then the map $\widetilde{\Gamma} : C(Z, \mathbf{C}) \to C(Z, \mathbf{R})$ defined by

$$\widetilde{\Gamma}(h; z) := \Gamma(z, h(z))$$

is also C^r. The derivative of $\widetilde{\Gamma}$ at $h \in C(Z, \mathbf{C})$ is the linear map $D\widetilde{\Gamma}(h)$: $C(Z, \mathbf{C}) \to C(Z, \mathbf{R})$

$$D\widetilde{\Gamma}(h)[\Delta h; z] := 2\Re[\partial_2 \Gamma(z, h(z)) \Delta h(z)].$$

The Taylor expansion exists on $C(Z, \mathbf{C})$ as

$$\Gamma(z, h(z) + \Delta h(z)) = \Gamma(z, h(z)) + 2\Re[\partial_2 \Gamma(z, h(z)) \Delta h(z)] + \mathcal{O}[\|\Delta h\|_\infty^2],$$

where $\mathcal{O}[\|\Delta h\|_\infty^2]$ does not depend on $z \in Z$.

Proof: Let $\mathbf{h} \in C(Z, \mathbf{C}^2)$ be written as

$$\mathbf{h}(z) = \begin{bmatrix} h_1(z) \\ h_2(z) \end{bmatrix} = \begin{bmatrix} u_1(z) + iv_1(z) \\ u_2(z) + iv_2(z) \end{bmatrix}$$

and with the corresponding notation for $\Delta \mathbf{h}(z)$. The Omega Lemma gives that $\widetilde{\Gamma} : C(Z, \mathbf{C}^2) \to C(Z, \mathbf{R})$ defined by $\widetilde{\Gamma}(\mathbf{h}; z) := \Gamma(h_1(z), h_2(z))$ is C^r with derivative

$$\begin{aligned} D\widetilde{\Gamma}(\mathbf{h})[\Delta \mathbf{h}; z] &= D\Gamma(\mathbf{h}(z)) \Delta \mathbf{h}(z) \\ &= \begin{bmatrix} \frac{\partial \Gamma}{\partial x_1}(\mathbf{h}(z)) & \frac{\partial \Gamma}{\partial y_1}(\mathbf{h}(z)) & \frac{\partial \Gamma}{\partial x_2}(\mathbf{h}(z)) & \frac{\partial \Gamma}{\partial y_2}(\mathbf{h}(z)) \end{bmatrix} \begin{bmatrix} \Delta u_1(z) \\ \Delta v_1(z) \\ \Delta u_2(z) \\ \Delta v_2(z) \end{bmatrix} \\ &= 2\Re[\partial_1 \Gamma(\mathbf{h}(z)) \Delta h_1(z)] + 2\Re[\partial_2 \Gamma(\mathbf{h}(z)) \Delta h_2(z)]. \end{aligned}$$

Restrict $\widetilde{\Gamma}$ to the affine space

$$\mathcal{M} = \{\mathrm{id}\} \times C(Z, \mathbf{C}) = \left\{ \begin{bmatrix} z \\ h(z) \end{bmatrix} : h \in C(Z, \mathbf{C}) \right\}.$$

$\mathcal{M}$ has tangent space

$$T\mathcal{M} = \{0\} \times C(Z, \mathbf{C}) = \left\{ \begin{bmatrix} 0 \\ \Delta h(z) \end{bmatrix} : \Delta h \in C(Z, \mathbf{C}) \right\}.$$

$\widetilde{\Gamma}$ restricted to $\mathcal{M}$ has derivative

$$D(\widetilde{\Gamma}|\mathcal{M})(\mathbf{h})[\Delta\mathbf{h}; z] = D\Gamma(\mathbf{h}) \begin{bmatrix} 0 \\ \Delta h(z) \end{bmatrix} = 2\Re[\partial_2 \Gamma(z, h(z))\Delta h(z)].$$

Taylor's expansion follows from this first variation. ///

That is, for each $h \in C(Z)$ there is a $K > 0$ and an $\epsilon > 0$ such that for all $\Delta h \in C(Z)$ with $\|\Delta h\|_\infty < \epsilon$ there holds

$$\|\Gamma(h + \Delta h) - \Gamma(h) - 2\Re[\partial\Gamma(h)\Delta h]\|_\infty \leq K\|\Delta h\|_\infty^2.$$

B.2 The Kolmogorov Criterion

The Kolmogorov Criterion provides a slick characterization for convex functionals [20, pages 6–11]. For our nonconvex functionals, the first variation "almost" convexifies the problem. Nonetheless, the nonlinearity splits the necessary and sufficient conditions of the Kolmogorov Criterion into two lemmas. The necessary condition is the easy part of the Kolmogorov Criterion [20, pages 6–11]. Although the result holds for arbitrary sets via tangent and contingent cones, we state it only for subspaces and restate here for the reader's convenience.

Lemma B.2.1 *(Descent) Let $Z \subset \mathbf{C}$ be compact. Let $\mathcal{H}$ be a closed linear subspace of $C(Z, \mathbf{C})$. Let U be an open subset containing Z. Let $\Gamma : U \times \mathbf{C} \to \mathbf{R}$ be C^2. Define $\gamma : \mathcal{H} \to \mathbf{R}$ by*

$$\gamma(h) := \sup\{\Gamma(z, h(z)) : z \in Z\}.$$

Assume $\mathcal{H}$ is boundedly compact. If $h \in \mathcal{H}$ is not a local minimum, there is a nonzero $\Delta h \in \mathcal{H}$ such that

$$0 \geq \Re[\partial_2\Gamma(z, h(z))\Delta h(z)] \quad (z \in \text{crit}[\gamma(h)]).$$

Proof: If $h \in \mathcal{H}$ not a local minimum then there is a nonzero sequence $\{\Delta h_n\} \subset \mathcal{H}$ converging to zero, such that $\gamma(h + \Delta h_n) \leq \gamma(h)$. Set $t_n := \|\Delta h_n\|_\infty > 0$ and $u_n := t_n^{-1}\Delta h_n$. Compactness of $\mathcal{H}$ implies the bounded sequence $\{u_n\}$ contains a convergent subsequence. By relabeling, $u_n \to \Delta h \in \mathcal{H}$. Because u_n has unit norm, $\Delta h \neq 0$. For all $z \in$ crit, Lemma B.1.2 provides the expansion:

$$\begin{aligned} \gamma(h + \Delta h_n) &\geq \Gamma(z, h(z) + \Delta h_n(z)) \\ &= \Gamma(z, h(z)) + 2\Re\left[\partial_2\Gamma(z, h(z))\Delta h_n(z)\right] + \mathcal{O}[t_n^2] \\ &= \gamma(h) + 2\Re\left[\partial_2\Gamma(z, h(z))\Delta h_n(z)\right] + \mathcal{O}[t_n^2]. \end{aligned}$$

Subtract $\gamma(h)$ from both sides, divide by $t_n > 0$ to get

$$0 \geq \Re\left[\partial_2\Gamma(z, h(z))u_n(z)\right] + \mathcal{O}[t_n].$$

Letting $n \to \infty$ gives the result. ///

The Descent Lemma (Lemma 6.6.1) has a clean proof that reveals why boundedly compact supplies a "direction of descent." It also supplies various points-of-departure for more sophisticated results. For example, we obtain a minimization test provided we handle the "=" in the "≥" with care.

Lemma B.2.2 (Min Test) *Let $Z \subset \mathbf{C}$ be compact. Let $\mathcal{H}$ be a closed linear subspace of $C(Z, \mathbf{C})$. Let U be an open subset containing Z. Let $\Gamma : U \times \mathbf{C} \to \mathbf{R}$ be C^2. Define $\gamma : \mathcal{H} \to \mathbf{R}$ by*

$$\gamma(h) := \sup\{\Gamma(z, h(z)) : z \in Z\}.$$

Let $h \in \mathcal{H}$. If there exists a $\Delta h \in \mathcal{H}$ such

$$0 > \Re[\partial_2 \Gamma(z, h(z)) \Delta h(z)] \quad (z \in \text{crit}[\gamma(h)]),$$

$h \in \mathcal{H}$ cannot be a local minimum for γ.

Proof: Compactness of Z and continuity give the existence of a $\delta > 0$ such that $\Re[\partial_2 \Gamma(z, h)\Delta h] \leq -\delta < 0$ on crit. Continuity then gives an open neighborhood U of crit such that for all $z \in U$ there holds:

$$\Re[\partial_2 \Gamma(z, h)\Delta h(z)]] \leq -\delta/2 < 0.$$

Then for $z \in U$ and for $t > 0$ sufficiently small there holds

$$\begin{aligned} \Gamma(z, h(z) + t\Delta h(z)) &= \Gamma(z, h(z)) + t2\Re[\partial_2 \Gamma(z, h(z))\Delta h(z)] + \mathcal{O}[t^2] \\ &\leq \gamma(h) - \delta t + \mathcal{O}[t^2] \\ &< \gamma(h). \end{aligned}$$

The first equality is obtained by taking $t > 0$ so small that $t\Delta h \in B(0, \epsilon)$ applying Lemma B.1.2. The first inequality follows from the δ bound on U. The last inequality follows by taking $t > 0$ small enough so that the first-order term dominates the second-order term. For $z \in Z \setminus U$, continuity forces $\Gamma(z, h(z)) < \gamma(h)$. Continuity of Γ and compactness of $Z \setminus U$ imply

$$\Gamma(z, h(z) + t\Delta h(z)) < \gamma(h)$$

for $t > 0$ sufficiently small. Thus, $\gamma(h + t\Delta h) < \gamma(h)$ for all $t > 0$ sufficiently small. Then h cannot be a local minimum of γ. ///

The Minimum Test (Lemma B.2.2) tells us that $h \in \mathcal{H}$ cannot be a local minimum if we can find a $\Delta h \in \mathcal{H}$ that "interpolates" the first variation $\partial_2 \Gamma(h)$. That is, $\Re[\partial_2 \Gamma(z, h(z))\Delta h(z)]$ must be both positive and negative on crit for any $\Delta h \in \mathcal{H}$. Put another way, a local minimum will force $\partial_2 \Gamma(z, h(z))\Delta h(z)$ to wind around zero. Thus, even at this abstract level, the winding numbers appear in the characterization of minima.

The Descent Lemma (Lemma 6.6.1) uses a "$\leq$". The Minimum Test (Lemma B.2.2) needs a "$<$". The necessary and sufficient conditions fail on the "$=$". Consequently, the bulk of one's efforts are devoted to bridging this gap.

One basic idea is to exploit the interpolating properties of the subspaces. For example, the polynomials are the classic interpolating space. Therefore, it is instructive to consider the minimization problem for the real polynomials $\mathcal{P}^N$ of degreee N:

$$\mathcal{P}^N := \{p_0 + p_1 x + p_2 x^2 \ldots + p_N x^N : p_n \in \mathbf{R}\}.$$

Using both the Descent Lemma and the Minimum Test, the following result on nonlinear polynomial optimization readily follows.

Corollary B.2.1 *Let U be an open subset containing $[0,1]$. Let $\Gamma : U \times \mathbf{R} \to \mathbf{R}$ be C^2. Define the mapping $\gamma : C([0,1], \mathbf{R}) \to \mathbf{R}$ by*

$$\gamma(h) := \sup\{\Gamma(x, h(x)) : x \in [0,1]\}.$$

Assume $h \in \mathcal{P}^N$ and $\partial_2 \Gamma(x, h(x)) \neq 0$ for $x \in \mathrm{crit}[\gamma(h)]$. Then the following are equivalent:

(a) $h \in \mathcal{P}^N$ is a local minimum.
(b) $\partial_2 \Gamma(x, h(x))$ admits alternating sequence of length $N+2$ on $\mathrm{crit}[\gamma(h)]$. That is, there are at least $N+2$ points $x_n \in \mathrm{crit}[\gamma(h)]$

$$0 \leq x_1 < \ldots x_2 \ldots < x_{N+2} \leq 1$$

such that

$$\mathrm{sign}(\partial_2 \Gamma(x_n, h(x_n))) = -\mathrm{sign}(\partial_2 \Gamma(x_{n+1}, h(x_{n+1}))).$$

The generalization of the corollary to *multiobjective optimization* on the real-valued polynomials is open to the best of my knowledge. Likewise, the multiobjective version of the Descent Lemma and the Minimum Test also await development.

References

1. Abraham, Ralph; Jerrold E. Marsden; Tudor Ratiu [1983] *Manifolds, Tensor Analysis and Applications*, Addison-Wesley, MA.
2. Allen, J. C. and D. F. Schwartz [2001] Optimal Impedance Matching by Lossless 2-Ports of Specified Degree Independent of Circuit Topology, in *Applied Computational Electromagnetics*, 17th Annual Conference Proceedings, pages 588-593.
3. Allen, J. C. and Dennis Healy [2003] Nehari's Theorem and Electric Circuits, in *Modern Signal Processing*, edited by Daniel Rockmore and Dennis Healy, Springer-Verlag, Berlin, New York.
4. Anderson, Brian D. O. and R. W. Newcomb [1976] Linear Passive Networks: Functional Theory, *Proceedings of the IEEE*, Volume 64, Number 1, pages 2–88.
5. Arov, D. Z. [1972] Darlington's Method for Dissipative Systems, *Doklady* Volume 16, Number 11, pages 954–956.
6. Baher, H. [1984] *Synthesis of Electrical Networks*, John Wiley, New York.
7. Bakonyi, Mihály, M. and T. Constantinescu [1992] *Schur's Algorithm and Several Applications*, Longman Scientific and Technical, Harlow, UK.
8. Balabanian, Norman and Theodore A. Bickart [1981] *Linear Network Theory*, Matrix Publishers, Beaverton, OR.
9. Ball, Joseph A. and J. William Helton [1979] Interpolation with Outer Functions and Gain Equalization in Amplifiers, in *International Symposium on Mathematical Theory of Networks and Systems*, Delft University of Technology, The Netherlands.
10. Ball, Joseph A. and J. William Helton [1981] Subinvariants for Analytic Mappings on Matrix Balls, *Analysis*, Volume 1, pages 217–226.
11. Ball, Joseph A. and J. William Helton [1982] Lie Groups Over the Field of Rational Functions, Signed Spectral Factorization, Signed Interpolations, and Amplifier Design, *Journal of Operator Theory*, Volume 8, pages 19–64.
12. Ball, Joseph A. and J. William Helton [1983] A Beurling-Lax Theorem for the Lie Group $U(m,n)$ Which Contains Most Classical Interpolation Theory, *Journal of Operator Theory*, Volume 9, Number 1, pages 107–142.
13. Ball, Joseph A. and J. William Helton [1984] Beurling-Lax Representations Using Classical Lie Groups With Many Applications II: $GL(n,\mathbb{C})$ and Wiener-Hopf Factorization, *Journal of Integral Equations and Operator Theory*, Volume 7, Number 3, pages 291–309.

14. Ball, Joseph A. and J. William Helton [1986] Beurling-Lax Representations Using Classical Lie Groups With Many Applications III: Groups Preserving Two Bilinear Forms, *American Journal of Mathematics*, Volume 108, Number 1, pages 95–174.
15. Ball, Joseph A., Ciprian Foias, J. William Helton, and Allen Tannenbaum [1989] A Poincaré-Dulac Approach to a Nonlinear Beurling-Lax-Halmos Theorem, *Journal of Mathematical Analysis and Applications*, Volume 139, pages 496–514.
16. Banyamin, Ben Y. and Michael Berwick [2000] Analysis of the Performance of Four-Cascaded Single-Stage Distributed Amplifiers, *IEEE Transactions on Microwave Theory and Techniques*, Volume 48, Number 12, pages 2657–2663.
17. Belevitch V. and Y. Genin [1971] Cascade Decomposition of Lossless 2-Ports, *Philips Research Reports*, Volume 26, pages 326–340.
18. Bochner, S. and K. Chandrasekharan [1949] *Fourier Transforms*, Princeton University Press, Princeton, NJ.
19. Boyd, Stephen and Lieven Vandenberghe [1999] *Convex Optimization*, preprint from Stanford University.
20. Braess, Dietrich [1986] *Nonlinear Approximation Theory*, Springer-Verlag, Berlin, New York.
21. Branch, Mary Ann and Andrew Grace [1996], *Optimization Toolbox*, The MathWorks Inc. Natick, MA.
22. Brayton, Robert K., Gary D. Hachtel, and Alberto L. Sangiovanni-Vincentelli [1981] A Survey of Optimization Techniques for Integrated Circuit Design, *Proceedings of the IEEE*, Volume 69, pages 1334–1362.
23. Bruccoleri, Federico, Erik A. M. Klumperink, and Bram Nauta [2001] Generating *All* Two-MOS-Transistor Amplifiers Leads to New Wide-Band LNAs, *IEEE Journal of Solid-State Circuits*, Volume 36, Number 7, pages 1032–1040.
24. Carleson, Lennart and Sigvard Jacobs [1972] Best Uniform Approximation by Analytic Functions, *Arkiv foer Matematik*, Volume 10, pages 219–229.
25. Carson, Ralph S. [1982] *High-Frequency Amplifiers*, John Wiley, New York.
26. Censor, Yair [1977] Pareto Optimality in Multiobjective Problems, *Applied Mathematics and Optimization*, Volume 4, pages 41–59.
27. Chen, G. Y. and B. D. Craven [1994] Existence and Continuity of Solutions for Vector Optimization, *Journal of Optimization Theory and Applications*, Volume 81, Number 3, pages 459–468.
28. Chen, Wai-Kai [1976] *Theory and Design of Broadband Matching Networks*, Pergamon Press, New York.
29. Cheney, E. W. [1982] *Approximation Theory*, Chelsea, New York.
30. Choi, Man-Duen [1975] Positive Semidefinite Biquadratic Forms, *Linear Algebra and Its Applications*, Volume 12, pages 95–100.
31. Cioffi, Kenneth R. [1989] Broad-Band Distributed Amplifier Impedance-Matching Techniques, *IEEE Transactions on Microwave Theory and Techniques*, Volume 37, Number 12, pages 1870–1876.
32. Clarke, Frank W. [1983] *Optimization and Nonsmooth Anaylsis*, John Wiley and Sons, New York.
33. Das, Indraneel and J. E. Dennis [1998] Normal-Boundary Intersecting: A New Method for Generating the Pareto Surface in Nonlinear Multicriteria Optimization Problems, *SIAM Journal on Optimization*, Volume 8, Number 3, pages 631–657.

34. Decker, Eva [1994] On the Boundary Behavior of Singular Inner Functions, *Michigan Mathematics Journal*, Volume 41, Number 3, pages 547–562.
35. Delsarte, Ph., Y. Genin, and Y. Kamp [1981] On the Role of the Nevanlinna-Pick Problem in Circuit and System Theory, *Circuit Theory and Applications*, Volume 9, pages 177–187.
36. Deng, Sien [1997] On Approximate Solutions in Convex Vector Optimization, *SIAM Journal on Control and Optimization*, Volume 35, Number 6, pages 2128–2136.
37. Dobrowolski, Janusz A. and Wojciech Ostrowski [1996] *Computer-Aided Analysis, Modeling, and Design of Microwave Networks*, Artech House, Boston, MA.
38. Douglas, R. G. [1972] *Banach Algebra Techniques in Operator Theory*, Academic Press, New York.
39. Douglas, R. G. and J. W. Helton [1973a] The Precise Theoretical Limits of Causal Darlington Synthesis, *IEEE Transactions on Circuit Theory*, Volume CT-20, Number 3, page 327.
40. Douglas, R. G. and J. W. Helton [1973b] Inner Dilations of Analytic Matrix Functions and Darlington Synthesis, *Acta Scientiarum Mathematicarum*, Szeged, Hungary, Volume 43, pages 61–67.
41. Dunford, Nelson and Jacob T. Schwartz [1967] Linear Operators, Part I, Interscience Publisher, New York.
42. Duren, Peter L. [1970] *Theory of H^p Spaces*, Academic Press, New York.
43. Dym, H. and H. P. McKean [1972] *Fourier Series and Integrals*, Academic Press, New York.
44. Dym, Harry, J. William Helton, and Orlando Merino [1999] Algorithms for Solving Multidisk Problems in H^∞ Optimization, *Proceedings of the 38th Conference on Decision and Control*, pages 3156–3161.
45. Efimov, A. V. and V. P. Potapov [1973] *J*-Expanding Matrix Functions and Their Role in the Analytical Theory of Electrical Circuits, *Russian Mathematical Surveys*, Usp. Volume 28.
46. Engberg, Jakob and T. Larsen [1995] *Noise Theory of Linear and Nonlinear Circuits*, Johns Wiley, New York.
47. Fialkow, Aaron D. [1979] Inductance, Capacitance Networks Terminated in Resistance, *IEEE Transactions on Circuits and Systems*, Volume CAS-26, Number 8, pages 603–641.
48. Fourès Y. and I. E. Segal [1954] Causality and Analyticity, *Transactions of the American Mathematical Society*, Volume 78, pages 386–405.
49. Fuhrmann, Paul A. [1981] *Linear Systems and Operators in Hilbert Space*, McGraw-Hill, New York.
50. Fujisawa, T. [1955] Realisability Theorem for Mid-Series or Mid-Shunt Low-Pass Ladders Without Mutual Induction, *IRE Transactions on Circuit Theory*, Volume CT-2, pages 320–325.
51. Garnett, J. B. [1981] *Bounded Analytic Functions*, Academic Press, New York.
52. Giannini, Franco, Giorgio Leuzzi, Ernesto Limiti, Jaroslaw S. Mroz, and Lucio Scucchia [1998] Nonlinear Mixed Analysis/Optimization Algorithm for Microwave Power Amplifier Design, *IEEE Transactions on Microwave Theory and Techniques*, Volume 43, Number 3, pages 552–558.
53. Gonzalez, Guillermo [1997] *Microwave Transistor Amplifiers*, 2nd Edition, Prentice-Hall, Upper Saddle River, NJ.

54. Grosch, Theodore O. and Lynn A. Carpenter [1993] Two-Port to Three-Port Noise-Wave Transformations for CAD Applications, *IEEE Transactions on Microwave Theory and Techniques*, Volume 41, Number 9, pages 1543–1548.
55. Guillemin, Ernst A. [1957] *Synthesis of Passive Networks*, John Wiley and Sons, London.
56. Hasler, Martin and Jacques Neirynck [1986] *Electric Filters*, Artech House, Dedham, MA.
57. Hayt, William H. and Jack E. Kemmerly [1971] *Engineering Circuit Analysis*, 2nd edition, McGraw-Hill, New York.
58. Helson, Henry [1964] *Lectures on Invariant Subspaces*, Academic Press, New York.
59. Helson, Henry [1983] *Harmonic Analysis*, Addison-Wesley, Reading, MA.
60. Helton, J. W. [1972] The Characteristic Functions of Operator Theory and Electrical Network Realization, *Indiana University Mathematics Journal*, Volume 22, Number 5, pages 403–414.
61. Helton, J. W. and A. H. Zemanian [1972] The Cascade Loading of Passive Hilbert Ports, *SIAM Journal on Applied Mathematics*, Volume 23, Number 3, pages 292–306.
62. Helton, J. W. [1974] Discrete Time Systems, Operator Models, and Scattering Theory, *Journal of Functional Analysis*, Volume 16, Number 1, pages 15–37.
63. Helton, J. W. [1976] Systems with Infinite-Dimensional State Space: The Hilbert Space Approach, *Proceedings of the IEEE*, Volume 64, Number 1, pages 145–160.
64. Helton, J. W. [1980] The Distance of a Function to H^∞ in the Poincaré Metric: Electrical Power Transfer, *Journal of Functional Analysis*, Volume 38, pages 273–314.
65. Helton, J. W. [1981] Broadbanding: Gain Equalization Directly from Data, *IEEE Transactions on Circuits and Systems*, Volume CAS-28 Number 12, pages 1125–1137.
66. Helton, J. W. [1982] Non-Euclidean Functional Analysis and Electronics, *Bulletin of the American Mathematical Society*, Volume 7, Number 1, pages 1–64.
67. Helton, J. W. [1983] A Systematic Theory of Worst-Case Optimization in the Frequency Domain: High-Frequency Amplifiers, *IEEE ISCAS*, Newport Beach, CA.
68. Helton, J. W. [1985] Worst Case Analysis in the Frequency Domain: The $H^\infty(\mathbb{C}_+)$ Approach to Control, *IEEE Transactions on Automatic Control*, Volume AC-30, Number 12, pages 1154–1170.
69. Helton, J. W. and Roger E. Howe [1986] A Bang-Bang Theorem for Optimization Over Spaces of Analytic Functions, *Journal of Approximation Theory*, Volume 47, pages 101–121.
70. Helton, J. W. and Donald E. Marshall [1990] Frequency-Domain Design and Analytic Selections, *Indiana University Mathematics Journal*, Volume 39, Number 1, pages 157–183.
71. Helton, J. W., C. Foias, B. Frances, H. Kwakernaak, and J. B. Person [1991] *H^∞ Control Theory*, Lecture Notes in Mathematics, Volume 1496, pages 1–222.
72. Helton, J. W., O. Merino, T. E. Walker [1993] Algorithms for Optimizing over Analytic Functions, *Indiana University Mathematics Journal*, Volume 42, Number 3, pages 839–874.

73. Helton, J. W. [1987] *Operator Theory, Analytic Functions, Matrices, and Electrical Engineering*, Regional Conference Series in Mathematics, Number 68, American Mathematical Society, Providence, RI.
74. Helton, J. W. and O. Merino [1998] *Classical Control Using $H^\infty(\mathbb{C}_+)$ Methods*, Society for Industrial and Applied Mathematics, Philadelphia.
75. Helton, J. W. and A. E. Vityaev [1997] Analytic Functions Optimizing Competing Constraints, *SIAM Journal on Mathematical Analysis*, Volume 30, Number 3, pages 749–767.
76. Helton, J. W. and Marshall A. Whittlesey [2000] Global Uniqueness Tests for H^∞ Optima, *Proceedings of the 39th IEEE Conference on Decision and Control*, Sydney, Australia, pages 1043–1048.
77. Henrici, Peter [1979] *Fast Fourier Methods in Computational Complex Analysis, SIAM Review*, Volume 21, Number 4, pages 481–527.
78. Hintzman, William [1975a] Best Uniform Approximations via Annihilating Measures, *Bulletin of the American Mathematical Society*, Volume 76, pages 1062–1066.
79. Hintzman, William [1975b] On the Existence of Best Analytic Approximations, *Journal of Approximation Theory*, Volume 14, pages 20–22.
80. Ho, S. L., Shiyou Yang, Guangzheng Ni, and H. C. Wong [2002] A Tabu Method to Find the Pareto Solutions to Multiobjective Optimal Design Problems in Electromagnetics, *IEEE Transactions on Magnetics*, Volume 38, Number 2, pages 1013–1016.
81. Hoffman, K. [1962] *Banach Spaces of Analytic Functions*, Prentice-Hall, Englewood Cliffs, NJ.
82. Hewlett-Packard [1972] *S-Parameter Design*, Application Note 154.
83. Hewlett-Packard [1983] *Fundamentals of RF and Microwave Noise Figure Measurements*, Application Note 57-1.
84. Iwai, Taisuke, Shiro Ohara, Hiroshi Yamada, Yasuhiro Yamaguchi, Kenji Imanishi, and Kazukiyo Joshin [1998] High Efficiency and High Linearity InGaP/GaAs HBT Power Amplifiers: Matching Techniques of Source and Load Impedance to Improve Phase Distortion and Linearity, *IEEE Transactions on Electron Devices*, Volume 45, Number 6, pages 1196–1200.
85. Jonckheere, Edmond A. and J. W. Helton [1985] Power Spectrum Reduction by Optimal Hankel Norm Approximation of the Phase of the Outer Spectral Factor, *IEEE Transactions on Automatic Control*, Volume AC-30, Number 12, pages 1192–1201.
86. Jung, Wen-Lin and Jingshown Wu [1990] Stable Broad-Band Microwave Amplifier Design, *IEEE Transactions on Microwave Theory and Techniques*, Volume 38, Number 8, pages 1079–1085.
87. Katsnelson, V. E. and B. Kirstein [1997] On the Theory of Matrix-Valued Functions Belonging to the Smirnov Class, *Topics in Interpolation Theory*, edited by H. Dym, Birkhäuser Verlag, Basel.
88. Kerherve, Eric, Pierre Jarry, and Pierre-Marie Martin [1998] Design of Broad-Band Matching Network With Lossy Junctions Using the Real-Frequency Technique, *IEEE Transactions on Microwave Theory and Techniques*, Volume 46, Number 3, pages 242–249.
89. Kline, Ronald [1993] Harold Black and the Negative-Feedback Amplifier, *IEEE Control Systems Magazine*, pages 82–85.

90. Ko, Beom Kyu and Kwyro Lee [1997] A New Simultaneous Noise and Input Power Matching Technique for Monolithic LNA's Using Cascode Feedback, *IEEE Transactions on Microwave Theory and Techniques*, Volume 45, Number 9, pages 1627–1630.
91. Koosis, Paul [1980] *Introduction to H_p Spaces*, Cambridge University Press, Cambridge, UK.
92. Kuh, Ernest S. and R. A. Rohrer [1967] *Theory of Linear Active Networks*, Holden-Day, San Francisco.
93. Leach, W. Marshall [1994] Fundamentals of Low-Noise Analog Circuit Design, *Proceedings of the IEEE*, Volume 82, Number 10, pages 1515–1538.
94. Leung, Ka Nang and Philip K. T. Mok [2001] Analysis and Multistage Amplifier-Frequency Compensation, *IEEE Transactions on Circuits and Systems–I*, Volume 48, Number 9, pages 1041–1056.
95. Light, W. A. and E. W. Cheney [1985] *Approximation Theory in Tensor Product Spaces*, Lecture Notes in Mathematics, Volume 1169, Springer-Verlag, New York.
96. Luenberger, David G. [1969] *Optimization by Vector Space Methods*, John-Wiley, New York.
97. Madjar, Asher [1988] A Fully Analytical AC Large-Signal Model of the GaAs MESFET for Nonlinear Network Analysis and Design, *IEEE Transactions on Microwave Theory and Techniques*, Volume 36, Number 1, pages 61–67.
98. NEC Data Sheet [1999] *Hetero-Junction Field-Effect Transistor NE321000*, Document Number P1420EJ2V0DSOO, NEC Corporation.
99. Newcomb, Robert W. [1966] *Linear Multiport Synthesis*, McGraw-Hill, New York.
100. Niclas, Karl B. and Ramon R. Pereira [1989] On the Design and Performance of a 6–18 GHz Three-Tier Matrix Amplifier, *IEEE Transactions on Microwave Theory and Techniques*, Volume 36, Number 1, pages 11–20.
101. Osipenkov, Vyacheslav and Sergei G. Vesnin [1994] Microwave Filters of Parallel-Cascade Structure, *IEEE Transactions on Microwave Theory and Techniques*, Volume 42, Number 7, pages 1360–1367.
102. Paoloni, Claudio [2000] HEMT-HBT Matrix Amplifier, *IEEE Transactions on Microwave Theory and Techniques*, Volume 48, Number 8, pages 1308–1312.
103. Papoulis, Athanasios [1984] *Probability, Random Variables, and Stochastic Processes*, McGraw-Hill, New York.
104. Partington, J. R. [1988] *An Introduction to Hankel Operators*, Cambridge University Press, Cambridge, UK.
105. Peller, Vladimir V. and Sergei R. Treil [1999] Approximation by Analytic Matrix Functions. The Four Block Problem, *preprint*.
106. Pozar, David M. [1998] *Microwave Engineering*, 2nd edition, John Wiley, New York.
107. Rizzoli, V. and A. Lipparini [1985] Computer-Aided Noise Analysis of Linear Multiport Networks of Arbitrary Topology, *IEEE Transactions on Microwave Theory and Techniques*, Volume MTT-33, Number 12, pages 1507–1512.
108. Rosenblum, Marvin and James Rovnyak [1985] *Hardy Classes and Operator Theory*, Oxford University Press, New York.
109. Rudin, Walter [1973] *Functional Analysis*, McGraw-Hill, New York.
110. Rudin, Walter [1974] *Real and Complex Analysis*, McGraw-Hill, New York.
111. Saeks, R. [1970] Causality in Hilbert Space, *SIAM Review*, Volume 12, Number 3, pages 357–383.

112. Saleh, Adel A. [1981] Frequency-Independent and Frequency-Dependent Models of TWT Amplifiers, *IEEE Transactions on Communications*, Volume COM-29, pages 1715–1720.
113. Schwartz, David F. [1984] Optimization Over Families of Bounded Analytic Functions, Ph.D. Thesis, University of California, San Diego.
114. Sertbaş, Ahmet, Ahmet Aksen, and B. Siddik Yarman [1998] Construction of Some Classes of Two-Variable Lossless Ladder Networks with Simple Lumped Elements and Uniform Transmission Lines, *IEEE Asia-Pacific Conference on Circuits and Systems* (APCCAS), pages 295–298.
115. Stoer, J. and R. Bulirsch [1980] *Introduction to Numerical Analysis*, Springer-Verlag, New York.
116. Sz.-Nagy, Béla and Ciprian Foiaş [1970] *Harmonic Analysis of Operators on Hilbert Space*, North-Holland, Amsterdam.
117. Tanzi, Nebil [1994] Design of Broad-Band, Low-Noise Microwave Transistor Amplifiers with Input and Output VSWR Constraints using CAD Tools, *Proceedings of the 37th Midwest Symposium on Circuits and Systems*, IEEE Circuits and Systems Society, pages 1215–1219.
118. Teeter, Douglas A., Jack R. East, Richard K Mains, and George I. Haddad [1993] Large-Signal Numerical and Analytical HBT Models, *IEEE Transactions on Electronic Devices*, Volume 40, Number 5, pages 837–845.
119. Temes, Gabor C. and Jack W. LaParta [1977] *Circuit Synthesis and Design*, McGraw-Hill, New York.
120. Uchida, Hiromitsu, Sumire Takatsu, Kazuhiko Nakahara, Takayuki Katoh, Yasushi Itoh, Ryoichi Imai, Minoru Yamamoto, and Naoto Kadowaki [1999] *Ka*-Band Multistage MMIC Low-Noise Amplifier Using Source Inductors with Different Values for Each Stage, *IEEE Microwave and Guided Wave Letters*, Volume 9, Number 2, pages 71–72.
121. Vendelin, George D. [1982] *Design of Amplifiers and Oscillators by the S-Parameter Method*, John Wiley, New York.
122. Vendelin, George D. [1990] Evaluating Nonlinear Models of Microwave GaAs-FETs, *IEEE Spectrum*, September, pages 48–50.
123. Vongpanitlerd, Sumeth [1970] Reciprocal Lossless Synthesis Via State-Variable Techniques, *IEEE Transactions on Circuit Theory*, Volume CT-17 Number 4, pages 630–632.
124. Vongpanitlerd, Sumeth and B. D. O. Anderson [1970] Scattering Matrix Synthesis Via Reactance Extraction, *IEEE Transactions on Circuit Theory*, CT-17 (4), pages 511–517.
125. Wan, Jin-Liang and Wai-Kai Chen [1993] On Consistency of the Complex Normalized Scattering Matrices of Multi-port Networks, *Journal of the Franklin Institute*, Volume 330, Number 3, pages 441–451.
126. Wedge, Scott W. and David B. Rutledge [1992] Wave Techniques for Noise Modeling and Measurement, *IEEE Transactions on Microwave Theory and Techniques*, Volume 40, Number 11, pages 2004–2012.
127. Widom, Harold [1965] Toeplitz Matrices, in *Studies in Real and Complex Analysis*, Volume 3, pages 179–209, Mathematical Association of America.
128. Wohlers, M. Ronald [1969] *Lumped and Distributed Passive Networks*, Academic Press, New York.
129. Yagle, Andrew [1991] Fast Algorithms for Nevanlinna-Pick Interpolation, *IEEE Transactions on Signal Processing*, Volume 39, Number 10, pages 2363–2365.

130. Yengst, William C. [1964] *Procedures of Modern Network Synthesis*, Macmillan, New York.
131. Youla, Dante C., L. J. Castriota, and H. J. Carlin [1959] Bounded Real Scattering Matrices and the Foundations of Linear Passive Network Theory, *IRE Transactions on Circuit Theory*, pages 102–124.
132. Youla, Dante C. [1961] On the Factorization of Rational Matrices, *IRE Transactions on Information Theory*, Volume IT-7, pages 172–189.
133. Youla, Dante C. and M. Saito [1967] Interpolation by Positive-Real Functions, *Journal of the Franklin Institute*, Volume 284, pages 77–108.
134. Youla, Dante C. [1971] A Tutorial Exposition of Some Key Network-Theoretic Ideas Underlying Classical Insertion-Loss Filter Design, *Proceedings of the IEEE*, Volume 59, Number 5, pages 760–799.
135. Youla, Dante C. and Nerses N. Kazanjian [1978] Bauer-Type Factorization of Positive Matrices and the Theory of Matrix Polynomials Orthogonal on the Unit Circle, *IEEE Transactions on Circuits and Systems*, Volume CAS-25, Number 2, pages 57–69.
136. Youla, Dante C. [1980] A Maximum Modulus Theorem for Spectral Radius and Absolutely Stable Amplifiers, *IEEE Transactions on Circuits and Systems*, Volume 25, pages 541–560.
137. Youla, Dante C., H. J. Carlin, and B. S. Yarman [1984] Double Broadband Matching and the Problem of Reciprocal Reactance $2n$-Port Cascade Decomposition, *International Journal of Circuit Theory and Applications*, Volume 12, pages 269–281.
138. Youla, Dante C., F. Winter, and S. U. Pillai [1997] A New Study of the Problem of Compatible Impedances, *International Journal of Circuit Theory and Applications*, Volume 25, pages 57–69.
139. Young, N. [1988] *An introduction to Hilbert space*, Cambridge University Press, Cambridge, UK.
140. Young, Paul [1998] *Electronic Communications Techniques*, 4th edition, Prentice Hall, Upper Saddle River, NJ.
141. Zeheb, E. and A. Lempel [1966] Interpolation in the Network Sense, *IEEE Transactions on Circuit Theory*, pages 118–119.
142. Zeidler, Eberhard [1985] *Nonlinear Functional Analysis and Its Applications III*, Springer-Verlag, New York.
143. Zhu, Lizhong, Boxiu Wu, and Chuyu Sheng [1998] Real Frequency Technique Applied to the Synthesis of Lumped Broad-Band Matching Networks with Arbitrary Nonuniform Losses for MMIC's, *IEEE Transactions on Microwave Theory and Techniques*, Volume 36, Number 12, pages 1614–1620.
144. Zhu, Kehe [1990] *Operator Theory in Function Spaces*, Marcel Dekker, New York.

Index

MIX
Papier aus verantwortungsvollen Quellen
Paper from responsible sources
FSC® C105338

If you have any concerns about our products,
you can contact us on
ProductSafety@springernature.com

In case Publisher is established outside the EU,
the EU authorized representative is:
Springer Nature Customer Service Center GmbH
Europaplatz 3, 69115 Heidelberg, Germany

Printed by Libri Plureos GmbH
in Hamburg, Germany